Kurzweil-Henstock Integral
in Riesz spaces

Antonio Boccuto
Beloslav Riečan
Marta Vrábelová

2009

FOREWORD

In 1956 Jaroslav Kurzweil was involved in special phenomena occurring in ordinary differential equations with fast oscillating entries which have not been justified by theories and approaches known at those times. For explaining the observed results he constructed a *tool*, which reminded in some aspects strongly the way how the Perron integral using minor and major functions was constructed. The story ended by a success and the tool became an independent, self-contained *object*, the *generalized Perron integral*.

Since the generalized Perron integral occurred to be very interesting and, due to the need of research, it was described via integral sums of Riemann type, Jaroslav Kurzweil described it in detail in his very first paper [163] on the topic and used it in a series of subsequent writings on ordinary differential equations. Only a restricted number of people knew at this times about the existence of a newly defined integral. There is no reason to be surprised by this fact, looking at the title of [163] nobody can expect deep interest of mathematicians involved in integration theory in this paper.

In the same time Ralph Henstock worked on variational approaches to integrals, no existing connection to Kurzweil in those days. For the first time the possible relation is mentioned cautiously in Henstock's booklet [153].

It was discovered early in the sixties that in the case of real valued functions both approaches (that of Henstock and of Kurzweil) are equivalent and, of course, the definition of the very general non-absolutely convergent integral based on Riemann-type integral sums came to the foreground.

The distinctive individual life of an integration theory, the Kurzweil-Henstock integral, started in the second half of seventies. With all of its advantages and drawbacks coming to general awareness.

One of the interesting points of integration theories is the problem when functions with values in general spaces have to be integrated. This problem is of interests especially in the case of infinite dimensional spaces equipped with some topology, the models are e. g. the Bochner, Dunford and Pettis integrals of Banach space-valued functions based on the Lebesgue approach.

The book of A. Boccuto, B. Riečan and M. Vrábelová is oriented in this direction. The functions which are integrated have values in Riesz spaces in general. The combination of techniques used in Riesz spaces with the more or less algebraic approach which is in charge for integrals based on Riemann type integral sums makes the presentation interesting and inspirative for further research. Besides their own research the authors present also short insights into applications and less known theories. All this makes the value of the present book, which should reach the reader in an unorthodox form. I am sure we are facing an inspirative text, with many new information and maybe also a nice reference text in various fields of analysis.

Štefan Schwabik, Prague

PREFACE

Kurzweil and Henstock's idea to construct a new type of integral turned out to be both surprisingly successful and extremely useful, not only from the didactic but also from the scientific point of view. It has very promising applications, for example in differential equations and surface integrals. Riesz spaces, on the other hand, offer a very important tool in modern mathematics and have many practical applications, for example in economics. Recall that the eminent mathematician and Nobel Prize L. V. Kantorovich was the founder of the theory of Riesz spaces. This monograph is concerned with both the theory of the Kurzweil-Henstock integral and the basic facts on Riesz spaces.

Another important application of this theory was discovered recently. In 2002 D. Kahneman received the Nobel Prize in Economics. While he was looking for a theoretical basis on his economical theory, he found an appropriate mathematical model: the so-called Šipoš integral, one of the topics we present in this monograph.

It is well-known that integration theory with values in ordered spaces cannot be reduced to the analogous theory for locally convex spaces. This fact justifies the main goal of this book: to investigate and develop a measure and integration theory of the Kurzweil-Henstock type for functions with values in ordered spaces.

The first chapter offers to the reader a self-contained treatment of the real-valued theory of the Kurzweil-Henstock integral. Namely, the extremely simple definition enables us to use a very concise and effective theory.

The following chapter on Riesz spaces should also be accessible to a large class of readers. We not only mention slightly more general structures such as lattice ordered groups, but also some basic facts about MV-algebras: these are important for multivalued logic. The general theory of the Kurzweil-Henstock integral in Riesz spaces is presented in the third chapter. As discovered by J. D. M. Wright and D. Fremlin, there exists a sufficient and necessary condition for the possibility to extend Riesz space-valued Daniell integrals from the set of simple functions to the set of integrable functions, or a Riesz space-valued measure from an algebra to the generated σ-algebra. This condition, which is called weak σ-distributivity, holds in any probability MV-algebra.

Chapters 4 - 6 contain new and, in our opinion, important results on convergence theorems and multiple integrals. These chapters also contain a systematic exposition of the Kurzweil-Henstock integral theory for functions defined on abstract topological spaces. Recall that most papers on the Kurzweil-Henstock integral use as a domain only Euclidean spaces.

Some more special topics are treated in chapters 7 - 10 and 13, namely improper integrals, SL-integrals, the Pettis and Choquet approach, and integration in metric semigroups. The Choquet integral (or its Šipoš symmetric variant) is of particular importance in non-additive measures.

In chapters 11 and 12 we are concerned with some applications to probability theory. In particular, it is important to observe that a probability theory on the so-called intuitionistic fuzzy sets can be constructed by considering them as embedded in an appropriate MV-algebra.

As we mentioned before, the aim of this monograph is two-fold. First, it can be understood as an introductory textbook to the Kurzweil-Henstock integral as well as to some algebraic structures (Riesz spaces, l-groups, MV-algebras) which are important from the viewpoint of applications to integration and probability theory. Second, it offers some possibilities of further developments including important open problems in this attractive area, with a glimpse of the diversity of directions in which the current research is moving.

The first author wants to dedicate this book to the loving memory of his mother Teresa who passed away on August 2nd, 2004, while Antonio was in Slovakia for cooperating with Prof. Riečan. She always helped him, not only in overcoming many difficulties in his personal life, but she also encouraged him to leave Italy in order to broaden his fields of interest and to enrich his personal experiences. That is why he decided to participate in the Winter School on Measure Theory in Liptovský Ján in 1993, which marked the beginning of a fruitful cooperation with his Slovak colleagues and friendship with marvellous people from Slovakia. The present book is the outcome of this wonderful cooperation and friendship which, hopefully, will continue for still many years to come.

The third author wants to thank Antonio Boccuto and Beloslav Riečan for the nice teamwork and friendship and Prof. Riečan for the scientific upbringing.

We would like to thank Prof. Štefan Schwabik for writing the foreword and Bentham Science Publishers, particularly Manager Bushra Siddiqui, for their support and efforts.

iv

KURZWEIL-HENSTOCK INTEGRAL
IN RIESZ SPACES

Antonio Boccuto, Beloslav Riečan, Marta Vrábelová

Abstract: This monograph is concerned with both the theory of the Kurzweil-Henstock integral and the basic facts on Riesz spaces. Moreover even the so-called Šipoš integral, which has several applications in economy, is illustrated. The aim of this book is two-fold. First, it can be understood as an introductory textbook to the Kurzweil-Henstock integral as well as to some algebraic structures which are important from the viewpoint of applications to integration and probability theory. Second, it discusses some possibilities of further developments including recent results and open problems.

CONTRIBUTORS

Antonio Boccuto, Doc., Ph.D.
Assistant Professor, Dipartimento di Matematica e Informatica, via Vanvitelli, 1 I-06123 Perugia, Italy Email: boccuto@yahoo.it

Beloslav Riečan, Prof. RNDr., DrSc.

Professor, Matej Bel University, Tajovského 40, SK-97401 Banská Bystrica, Slovakia,
and Mathematical Institute, Slovak Academy of Science, Štefánikova 49, SK-81473 Bratislava, Slovakia Email: riecan@fpv.umb.sk

Marta Vrábelová, Doc. RNDr., CSc.

Professor, Constantine the Philosopher University, Tr. A. Hlinku 1, SK-94974 Nitra, Slovakia Email: mvrabelova@ukf.sk

Address correspondence to
Antonio Boccuto, Dipartimento di Matematica e Informatica, via Vanvitelli, 1 I-06123 Perugia, Italy, fax +39 075 5855024 Email: boccuto@yahoo.it

1 Elementary Introduction to Kurzweil-Henstock Integral

Abstract

In this chapter we introduce the theory of the Kurzweil-Henstock integral for real-valued functions, defined on a bounded interval of the real line.
The main properties are illustrated, the Fundamental Theorem of Calculus and some convergence theorems are proved; moreover some examples and exercises are given.

1.1 Introduction

In this section we introduce the Kurzweil-Henstock integral (or the gauge integral or the generalized Riemann integral) for real-valued functions f defined on a bounded interval $[a,b] \subset \mathrm{R}$, only in Subsection 1.9 the interval $[a,b]$ is taken unbounded.

The integral theory, which we are speaking about, was, independently, introduced by Ralph Henstock (1955) and Jaroslav Kurzweil (1957). This integral is simpler than the Lebesgue integral and as strong as the Denjoy-Perron integral. The definition is similar to the definition of the Riemann integral.

This chapter does not fetch new results. Its aim is that the reader became familiar with the Kurzweil-Henstock integral theory. Our intention is especially to point out some differences between the Kurzweil-Henstock integral and the Riemann integral. The proofs and the solutions are given in details and the section is readable for a reader with an elementary knowledge of the calculus. For a deeper study of this theory there are a few interesting books. We mention, for example, the books written by Robert McLeod [188], Lee Peng Yee and Rudolf Výborný [170], and Charles Swartz [254].

A *division* of a compact interval $[a,b]$ is a finite, ordered sequence of points

$$a = a_0 < a_1 < ... < a_n = b.$$

By choosing a number $t_i \in [a_{i-1}, a_i]$ for $i = 1, 2, ..., n$ a partition P of $[a,b]$ is defined,

$$P = \{([a_{i-1}, a_i], t_i) : i = 1, 2, ..., n\}.$$

If $\delta > 0$, then a partition P with

$$t_i - \delta < a_{i-1} \leq t_i \leq a_i < t_i + \delta$$

for $i = 1, 2, ..., n$ is called a δ-*fine partition* of $[a,b]$.

Lets suppose that f is a bounded real-valued function defined on $[a,b]$. A Riemann sum for the function f and the partition $P = \{([a_{i-1}, a_i], t_i) : i = 1, 2, ..., n\}$ of the interval $[a,b]$ is the number

$$\sum_P f = \sum_{i=1}^{n} f(t_i)(a_i - a_{i-1}).$$

Generally, a partition P of $[a,b]$ is a set of the type

$$P = \{(E_i, t_i) : i = 1, 2, ..., n\},$$

where E_i are non-overlapping, closed intervals, $t_i \in E_i$ for $i = 1, 2, ..., n$ and $\bigcup_{i=1}^{n} E_i = [a,b]$. The *Riemann sum* is the quantity

$$\sum_P f = \sum_{i=1}^{n} f(t_i) | E_i |,$$

where $| E_i |$ is the length of the interval E_i, $i = 1, ..., n$.

Antonio Boccuto / Beloslav Riečan / Marta Vrábelová

Definition 1.1 A function $f:[a,b] \to \mathbb{R}$ is *Riemann integrable* if there is a real number I such that for every $\varepsilon > 0$ there is $\delta > 0$ such that

$$\left| \sum_P f - I \right| < \varepsilon$$

for every δ-fine partition P.

In this case, the number I is called the *Riemann integral* of f on $[a,b]$ and its standard notation is $\int_a^b f(x)dx$.

The definition fails for an unbounded function. If f is a Riemann integrable function then f is bounded.

Suppose $f(x) \geq 0$ for $x \in [a,b]$. Then the Riemann sum is an approximation to the area of the region under the graph of f. To have a good approximation, the width of the interval $[a_{i-1}, a_i]$ must be small whenever the graph of f is steep, but it can be wider where the graph of f is more horizontal.

It is a good idea to take the points $t_1, t_2, \ldots, t_n$ from $[a,b]$ and with respect to the behavior of f to choose numbers $\delta(t_i) > 0$ and intervals $[a_{i-1}, a_i]$, $i = 1, 2, \ldots, n$, such that $f(t_i)(a_i - a_{i-1})$ is a good approximation to the area of the strip under the graph between the lines $x = a_{i-1}$ and $x = a_i$ for $i = 1, 2, \ldots, n$.

In other words, the function $\delta:[a,b] \to \mathbb{R}^+$ is taken instead of the number $\delta > 0$. Then a δ-*fine partition* P of the interval $[a,b]$ is the set

$$P = \{([a_{i-1}, a_i], t_i) : t_i \in [a_{i-1}, a_i] \subset (t_i - \delta(t_i), t_i + \delta(t_i)), \; i = 1, 2, \ldots, n\}.$$

A construction of the function δ will be showed in the following example.

Example 1.2 Let $f(\frac{1}{m}) = m$ for every $m \in \mathbb{N}$ and $f(x) = 0$ for $x \in [0,1] \setminus \{\frac{1}{m} : m \in \mathbb{N}\}$. We will define a function $\delta:[0,1] \to \mathbb{R}^+$ such that

$$\left| \sum_P f - \int_0^1 f(x)dx \right| < \varepsilon$$

for every δ-fine partition $P = \{([a_{i-1}, a_i], t_i), \; i = 1, 2, \ldots, n\}$ of $[0,1]$.

Since the area of the region under the graph of f is zero, $\int_0^1 f(x)dx = 0$ and we need a function δ such that $\sum_P f < \varepsilon$ for any δ-fine partition P of $[0,1]$. The function δ, in general, cannot be a small constant. Indeed,

$$\sum_{i=1}^n f(t_i)(a_i - a_{i-1}) < 2\delta \sum_{i=1}^n f(t_i)$$

and $\sum_{i=1}^n f(t_i)$ is not bounded on $[0,1]$. The value $f(t_i)$ is nonzero in the case $t_i = 1/m$ for some $m \in \mathbb{N}$ only if $1/m$ is equal to at most two t_i, namely t_{i-1} and t_i for some i. Hence, it is appropriate to take

$$\delta\left(\frac{1}{m}\right) = \frac{\varepsilon}{m \, 2^{m+2}}$$

for $m \in \mathbb{N}$ and $\delta(x)$ can be any positive number for $x \in [0,1]$, $\{1/m : m \in \mathbb{N}\}$, say $\delta(x) = 1$. Put $M = \{m \in \mathbb{N} : t_i = 1/m, i = 1, 2, ..., n\}$. Then

$$\sum_P f = \sum_{i=1}^n f(t_i)(a_i - a_{i-1}) \le \sum_{m \in M} 2m \frac{\varepsilon}{m2^{m+1}} < \sum_{m=1}^{\infty} \frac{\varepsilon}{2^m} = \varepsilon.$$

The δ-fine partition P that we have used in the previous example exists, for example $P = \{([0,1], \sqrt{2}/2)\}$.

The compatibility theorem (or Cousin's lemma) guarantees the existence of a δ-fine partition for every function $\delta : [a,b] \to \mathbb{R}^+$.

Theorem 1.3 *Let $\delta : [a,b] \to \mathbb{R}^+$ and $a \le c < d \le b$. Then there exists a δ-fine partition of $[c,d]$.*

Proof: By contradiction, suppose that there is no δ-fine partition of $[c,d]$. Divide the interval $[c,d]$ into the intervals

$$\left[c, \frac{c+d}{2}\right], \quad \left[\frac{c+d}{2}, d\right].$$

Then one of them has no δ-fine partition. Denote that interval by $[c_1, d_1]$.

The continuation by this manner creates a sequence of nested intervals

$$[c_n, d_n], \ d_n - c_n = (d-c)/2^n.$$

Then there is one point $e \in \bigcap_{n=1}^{\infty} [c_n, d_n]$. Since $\delta(e) > 0$, there is $n_0 \in \mathbb{N}$ such that $d_n - c_n < \delta(e)$ for every $n > n_0$. Hence,

$$P = \{([c_n, d_n], e)\} \ (n > n_0)$$

is a δ-fine partition of $[c_n, d_n]$, that is a contradiction.

Exercise 1.4 Construct a δ-fine partition P of $[0,1]$ when

$$\delta\left(\frac{m}{m+1}\right) = \frac{\varepsilon}{2^{m+2}}, \quad \delta(x) = \min\left\{x - \frac{m-1}{m}, \frac{m}{m+1} - x\right\}$$

for $x \in \left(\frac{m-1}{m}, \frac{m}{m+1}\right) (m = 0, 1, 2, ...)$ and $\delta(1) = \varepsilon/2$.

Hint. Take n for which $n/(n+1) > 1 - \varepsilon/2$.

All points $m/(m+1) = t_{2m+1}$ $(m = 0, 1, ..., n)$ and 1 must be tags. Take bounds of intervals such that

$$a_1 < \frac{\varepsilon}{4}, \quad \frac{m}{m+1} - \frac{\varepsilon}{2^{m+2}} < a_{2m} < \frac{m}{m+1} < a_{2m+1} < \frac{m}{m+1} + \frac{\varepsilon}{2^{m+2}} \quad (m = 1, 2, ..., n).$$

Furthermore, put $t_{2m} = ((m-1)/m + m/(m+1))/2$ $(m = 1, 2, ..., n)$.

1.2 Definition of Kurzweil-Henstock Integral

Definition 1.5 A function $f : [a,b] \to \mathbb{R}$ is *Kurzweil–Henstock integrable*, if there is a real number I such that for every $\varepsilon > 0$ there is a function $\delta : [a,b] \to \mathbb{R}^+$ such that

$$\left| \sum_P f - I \right| < \varepsilon$$

for every δ-fine partition P of $[a,b]$.

We will denote the *Kurzweil–Henstock integral* ((KH)-*integral*) I of f on $[a,b]$ by the symbol

$$(KH) \int_a^b f(x)dx.$$

(KH) in the notation of (KH)-integral will be omitted in the proofs and the examples of this section.

Example 1.6 Let $f(0) = 0$ and $f(x) = \frac{1}{\sqrt{x}}$ for $x \in (0,1]$. Then f is unbounded on $[0,1]$ and f is not Riemann integrable. We will prove that for every $\varepsilon > 0$ there exists $\delta : [0,1] \to \mathrm{R}^+$ such that

$$\left| \sum_P f - 2 \right| < \varepsilon$$

for every δ-fine partition of $[0,1]$. The function δ is defined in [188], p. 13 by the formula

$$\delta(0) = \frac{\varepsilon^2}{16}, \quad \delta(x) = \frac{\varepsilon x \sqrt{x}}{4} \quad \text{for } x \in (0,1].$$

The δ-fine partition P of $[0,1]$ has $t_1 = 0$. Otherwise,

$$t_1 \in (0, a_1] \subset \left(t_1 - \frac{\varepsilon t_1 \sqrt{t_1}}{4}, t_1 + \frac{\varepsilon t_1 \sqrt{t_1}}{4} \right)$$

and hence

$$t_1 \le a_1 = a_1 - 0 < \frac{2\varepsilon t_1 \sqrt{t_1}}{4} < t_1$$

for $0 < \varepsilon < 1$. Now

$$\left| \sum_P f - 2 \right|$$

$$= \left| f(0)(a_1 - a_0) - (2\sqrt{a_1} - 2\sqrt{a_0}) + \sum_{i=2}^{n} \frac{1}{\sqrt{t_i}}(a_i - a_{i-1}) - \sum_{i=2}^{n}(2\sqrt{a_i} - 2\sqrt{a_{i-1}}) \right|$$

$$\le 2\sqrt{a_1} + \sum_{i=2}^{n} \left| \frac{1}{\sqrt{t_i}}(a_i - a_{i-1}) - 2(\sqrt{a_i} - \sqrt{a_{i-1}}) \right|.$$

Furthermore, $2\sqrt{a_1} < 2\sqrt{\frac{\varepsilon^2}{16}} = \frac{\varepsilon}{2}$ and

$$\sum_{i=2}^{n} \left| \frac{1}{\sqrt{t_i}}(a_i - a_{i-1}) - 2(\sqrt{a_i} - \sqrt{a_{i-1}}) \right|$$

$$= \sum_{i=2}^{n} \left| \frac{(a_i - a_{i-1})(\sqrt{a_i} + \sqrt{a_{i-1}}) - 2\sqrt{t_i}(a_i - a_{i-1})}{\sqrt{t_i}(\sqrt{a_i} + \sqrt{a_{i-1}})} \right| = \sum_{i=2}^{n} \left| \frac{(a_i - a_{i-1})(\sqrt{a_i} + \sqrt{a_{i-1}} - 2\sqrt{t_i})}{\sqrt{t_i}(\sqrt{a_i} + \sqrt{a_{i-1}})} \right|$$

$$\le \sum_{i=2}^{n} \left| \frac{(a_i - a_{i-1})(\sqrt{a_i} + \sqrt{a_{i-1}} - 2\sqrt{t_i})}{\sqrt{t_i}\sqrt{t_i}} \right| = \sum_{i=2}^{n} \left| \frac{(a_i - a_{i-1})(\sqrt{a_i} - \sqrt{t_i} + \sqrt{a_{i-1}} - \sqrt{t_i})}{\sqrt{t_i}\sqrt{t_i}} \right|$$

$$= \sum_{i=2}^{n} \frac{a_i - a_{i-1}}{\sqrt{t_i}\sqrt{t_i}} \left| \sqrt{a_i} - \sqrt{t_i} + \sqrt{a_{i-1}} - \sqrt{t_i} \right|$$

$$= \sum_{i=2}^{n} \frac{a_i - a_{i-1}}{\sqrt{t_i}\sqrt{t_i}} \left| \frac{a_i - t_i}{\sqrt{a_i} + \sqrt{t_i}} + \frac{a_{i-1} - t_i}{\sqrt{a_{i-1}} + \sqrt{t_i}} \right| \le \sum_{i=2}^{n} \frac{a_i - a_{i-1}}{\sqrt{t_i}\sqrt{t_i}} \left(\frac{|a_i - t_i|}{\sqrt{t_i}} + \frac{|a_{i-1} - t_i|}{\sqrt{t_i}} \right)$$

$$\le \sum_{i=2}^{n} \frac{a_i - a_{i-1}}{\sqrt{t_i}\sqrt{t_i}} \left(\frac{a_i - t_i}{\sqrt{t_i}} + \frac{t_i - a_{i-1}}{\sqrt{t_i}} \right) = \sum_{i=2}^{n} \frac{(a_i - a_{i-1})^2}{t_i \sqrt{t_i}} < \sum_{i=2}^{n} \frac{a_i - a_{i-1}}{t_i \sqrt{t_i}} \frac{2\varepsilon t_i \sqrt{t_i}}{4} = \frac{\varepsilon}{2}(1 - a_1) < \frac{\varepsilon}{2}.$$

Hence,

$$\left| \sum_P f - 2 \right| < \varepsilon.$$

Theorem 1.7 Let $f : [a,b] \to \mathrm{R}$ be a (KH)-integrable function. Then the (KH)-integral of f is defined uniquely.

Proof: Let I_1 and I_2 be two integrals of f on $[a,b]$. Let $\varepsilon > 0$. Then there are $\delta_1, \delta_2 : [a,b] \to \mathbb{R}^+$ such that

$$\left| \sum_{P_1} f - I_1 \right| < \frac{\varepsilon}{2}$$

for every δ_1-fine partition P_1 of $[a,b]$ and

$$\left| \sum_{P_2} f - I_2 \right| < \frac{\varepsilon}{2}$$

for every δ_2-fine partition P_2 of $[a,b]$. Put

$$\delta(x) = \min\{\delta_1(x), \delta_2(x)\} \text{ for } x \in [a,b].$$

Let P be a δ-fine partition of $[a,b]$. Then P is δ_1-fine and δ_2-fine, too. Now

$$|I_1 - I_2| \le \left| I_1 - \sum_P f + \sum_P f - I_2 \right| \le \left| \sum_P f - I_1 \right| + \left| \sum_P f - I_2 \right| < \varepsilon$$

for any arbitrary positive number ε. Then

$$|I_1 - I_2| = 0 \text{ and hence } I_1 = I_2.$$

Example 1.8 Let C be a countable infinite subset of $[a,b]$. Let $f(x) = K$ $(K \in \mathbb{R})$ for $x \in [a,b] \setminus C, f(x) \ne K$ for $x \in C$. We will prove that

$$(KH)\int_a^b f(x)dx = K(b-a)$$

and hence we will look for a function $\delta : [a,b] \to \mathbb{R}^+$ such that

$$\left| \sum_P f - K(b-a) \right| < \varepsilon$$

for every δ-fine partition P of $[a,b]$. Since

$$\left| \sum_P f - K(b-a) \right|$$

$$= \left| \sum_{i=1}^n f(t_i)(a_i - a_{i-1}) - K(b-a) \right|$$

$$= \left| \sum_{i=1}^n (f(t_i) - K)(a_i - a_{i-1}) \right| = \left| \sum_{t_i \in C} (f(t_i) - K)(a_i - a_{i-1}) \right|,$$

the value $\delta(x)$ for $x \in [a,b] \setminus C$ can be arbitrary, say $\delta(x) = 1$ in this case and

$$\delta(c_n) = \frac{\varepsilon}{2^{n+2} |f(c_n) - K|} \text{ for } c_n \in C, n \in \mathbb{N}.$$

Now

$$\left| \sum_{t \in C} (f(t_i) - K)(a_i - a_{i-1}) \right| \le \sum_{t_i \in C} |f(t_i) - K|(a_i - a_{i-1})$$

$$< \sum_{t_i = c_n} |f(t_i) - K| \frac{2\varepsilon}{2^{n+2}|f(t_i) - K|} < 2\sum_{n=1}^\infty \frac{\varepsilon}{2^{n+1}} = \varepsilon.$$

Exercise 1.9 Prove that every Riemann integrable function f on $[a,b]$ is (KH)-integrable on $[a,b]$.

1.3 Basic Theorems

Theorem 1.10   *If f is (KH)-integrable on $[a,b]$ and $c \in \mathbb{R}$, then cf is (KH)-integrable on $[a,b]$ and*

$$(KH)\int_a^b cf(x)dx = c(KH)\int_a^b f(x)dx.$$

Proof: The assertion of the theorem holds in the case $c = 0$. If $c \neq 0$ then for every $\varepsilon > 0$ there is $\delta : [a,b] \to \mathbb{R}^+$ such that

$$\left|\sum_P f - \int_a^b f(x)dx\right| < \frac{\varepsilon}{|c|}$$

whenever P is δ-fine, which implies

$$\left|\sum_P cf - c\int_a^b f(x)dx\right| = |c|\left|\sum_P f - \int_a^b f(x)dx\right| < \varepsilon.$$

Theorem 1.11   *If f,g are (KH)-integrable on $[a,b]$, then $f+g$ is (KH)-integrable on $[a,b]$ and*

$$(KH)\int_a^b (f(x)+g(x))dx = (KH)\int_a^b f(x)dx + (KH)\int_a^b g(x)dx.$$

Proof: Let $\varepsilon > 0$. Then there are functions δ_1 and δ_2 such that

$$\left|\sum_P (f+g) - \int_a^b f(x)dx - \int_a^b g(x)dx\right|$$
$$\leq \left|\sum_P f - \int_a^b f(x)dx\right| + \left|\sum_P g - \int_a^b g(x)dx\right|$$
$$< \frac{\varepsilon}{2} + \frac{\varepsilon}{2} = \varepsilon$$

whenever P is δ-fine, where $\delta = \min\{\delta_1, \delta_2\}$.

Theorem 1.12   *If f,g are (KH)-integrable on $[a,b]$ and $f(x) \leq g(x)$ for $x \in [a,b]$, then*
$$(KH)\int_a^b f(x)dx \leq (KH)\int_a^b g(x)dx.$$

Proof: Let $\delta_1, \delta_2 : [a,b] \to \mathbb{R}^+$ be such that

$$\left|\sum_{P_1} f - \int_a^b f(x)dx\right| < \varepsilon, \quad \left|\sum_{P_2} g - \int_a^b g(x)dx\right| < \varepsilon$$

for every δ_1-fine partition P_1 and every δ_2-fine partition P_2.

Put $\delta = \min\{\delta_1, \delta_2\}$, and let P be a δ-fine partition. Then

$$\int_a^b f(x)dx - \varepsilon < \sum_P f \leq \sum_P g < \int_a^b g(x)dx + \varepsilon$$

for every $\varepsilon > 0$ and hence $\int_a^b f(x)dx \leq \int_a^b g(x)dx$.

Corollary 1.13   *If f and $|f|$ are (KH)-integrable on $[a,b]$ then*

$$\left|(KH)\int_a^b f(x)dx\right| \le (KH)\int_a^b |f(x)|\,dx.$$

Proof: It is sufficient to use the inequalities $-|f| \le f \le |f|$ and Theorem 1.12.

Theorem 1.14 *If f is (KH)-integrable on $[a,c]$ and f is (KH)-integrable on $[c,b]$, then f is (KH)-integrable on $[a,b]$ and*

$$(KH)\int_a^b f(x)dx = (KH)\int_a^c f(x)dx + (KH)\int_c^b f(x)dx.$$

Proof: Let $\varepsilon > 0$. There are functions $\delta_1 : [a,c] \to \mathrm{R}^+, \delta_2 : [c,b] \to \mathrm{R}^+$ such that

$$\left|\sum_{P_1} f - \int_a^c f(x)dx\right| < \frac{\varepsilon}{2}, \quad \left|\sum_{P_2} f - \int_c^b f(x)dx\right| < \frac{\varepsilon}{2}$$

for every δ_1-fine partition P_1 of $[a,c]$ and every δ_2-fine partition P_2 of $[c,b]$.

We will choose a function $\delta : [a,b] \to \mathrm{R}^+$ such that point c is needed as a tag (that is $c = t_i$ for some i) in every δ-fine partition P of $[a,b]$ and we will split P into P_1, P_2, where P_1 is a δ_1-fine partition of $[a,c]$, P_2 is a δ_2-fine partition of $[c,b]$ with $\sum_P f = \sum_{P_1} f + \sum_{P_2} f$. Put

$$\delta(x) = \begin{cases} \min\left\{\delta_1(x), \dfrac{1}{2}(c-x)\right\} & \text{for } x \in [a,c), \\[2mm] \min\left\{\delta_2(x), \dfrac{1}{2}(x-c)\right\} & \text{for } x \in (c,b], \\[2mm] \min\left\{\delta_1(x)\delta_2(x)\right\} & \text{for } x = c. \end{cases}$$

If $P = \{([a_{i-1}, a_i], t_i) : i = 1, 2, \ldots, n\}$ is a δ-fine partition of $[a,b]$, then there is k, $1 \le k \le n$, such that $t_k = c = t_{k+1}$ and $a_k = c$ or $a_{k-1} < c < a_k, c = t_k$. In the first case,

$$\sum_P f = \sum_{i=1}^{k-1} f(t_i)(a_i - a_{i-1}) + f(c)(c - a_{k-1}) + f(c)(a_{k+1} - c)$$

$$+ \sum_{i=k+2}^{n} f(t_i)(a_i - a_{i-1}) = \sum_{P_1} f + \sum_{P_2} f$$

and

$$\left|\sum_P f - \int_a^c f(x)dx - \int_c^b f(x)dx\right|$$

$$\le \left|\sum_{P_1} f - \int_a^c f(x)dx\right| + \left|\sum_{P_2} f - \int_c^b f(x)dx\right|$$

$$< \frac{\varepsilon}{2} + \frac{\varepsilon}{2} < \varepsilon,$$

where $P_1 = \{([a,a_1], t_1), \ldots, ([a_{k-1}, c], c)\}$ is a δ_1-fine partition of $[a,c]$ and $P_2 = \{[c, a_{k+1}], t_{k+1}), \ldots, ([a_{n-1}, b], t_n)\}$ is a δ_2-fine partition of $[c,b]$.
In the second case, $f(c)(a_k - a_{k-1}) = f(c)(c - a_{k-1}) + f(c)(a_k - c)$ and

$$\sum_P f = \sum_{P_1} f + \sum_{P_2} f,$$

where $P_1 = \{([a,a_1], t_1), \ldots, ([a_{k-1}, c], c)\}$, $P_2 = \{([c, a_k], c), \ldots, ([c_{n-1}, b], t_n)\}$.

Proposition 1.15   *(Bolzano-Cauchy condition) The function f is (KH)-integrable on $[a,b]$ if and only if for every $\varepsilon > 0$ there is a function $\delta : [a,b] \rightarrow \mathbb{R}^+$ such that*

$$\left| \sum_P f - \sum_Q \right| < \varepsilon$$

whenever P, Q are δ-fine partitions of $[a,b]$.

Proof: Let I be the (KH)-integral of f on $[a,b]$. Then for every $\varepsilon > 0$ there is δ such that

$$\left| \sum_P f - I \right| < \frac{\varepsilon}{2}$$

for every δ-fine partition P of $[a,b]$. If P, Q are δ-fine, then

$$\left| \sum_P f - \sum_Q f \right| \leq \left| \sum_P f - I \right| + \left| \sum_Q f - I \right| < \varepsilon.$$

Prove the sufficiency. For $\varepsilon = 2/n (n \in N)$ there is δ_n such that

$$\left| \sum_P f - \sum_Q f \right| < \frac{2}{n}$$

whenever P, Q are δ_n-fine partitions of $[a,b]$. We can assume $\delta_{n+1} \leq \delta_n$, $n \in \mathbb{N}$. The set $S_n = \{ \sum_P f : P$ is a δ_n-fine partition of $[a,b] \}$ is bounded and $S_{n+1} \subset S_n$, $n \in \mathbb{N}$. Put

$$u_n = \inf S_n, \quad v_n = \sup S_n.$$

Then $[u_{n+1}, v_{n+1}] \subset [u_n, v_n], v_n - u_n < 2n^{-1} (n \in \mathbb{N})$ and there is one number I such that $\{I\} = \bigcap_{n=1}^{\infty} [u_n, v_n]$. Take $n \in \mathbb{N}$ such that $2n^{-1} < \varepsilon$ and a δ_n-fine partition P of $[a,b]$. Then $\sum_P f \in [u_n, v_n]$ and

$$\left| \sum_P f - I \right| < v_n - u_n < \frac{2}{n} < \varepsilon.$$

Theorem 1.16   *If f is (KH)-integrable on $[a,b]$, then f is (KH)-integrable on every $[\alpha, \beta] \subset [a,b]$.*

Proof: If f is (KH)-integrable on every $[a,c] \subset [a,b]$ and every $[c,b] \subset [a,b]$, then f is integrable on every $[\alpha, \beta] \subset [a,b]$.

   We will prove that (KH)-integrable function f on $[a,b]$ is (KH)-integrable on $[a,c]$, when $a < c < b$. The integrability of $[c,b]$ can be proved similarly. The Bolzano-Cauchy condition (Proposition 1.15) is a useful tool in this proof: there is $\delta : [a,b] \rightarrow \mathbb{R}^+$ such that

$$\left| \sum_P f - \sum_Q f \right| < \varepsilon$$

whenever P, Q are δ-fine partitions of $[a,b]$. Let P_1, Q_1 be two arbitrary δ-fine partitions of $[a,c]$ and P_2 be a δ-fine partition of $[c,b]$, then $P = P_1 \cup P_2$, $Q = Q_1 \cup P_2$ are δ-fine partitions of $[a,b]$ and

$$\left| \sum_{P_1} f - \sum_{Q_1} f \right| = \left| \sum_{P} f - \sum_{Q} f \right| < \varepsilon.$$

1.4 Continuity of the Integral

We have defined the (KH)-integral of the function f on the interval $[a,b]$, where $a < b$. Define

$$(KH)\int_a^a f(x)dx = 0, \quad (KH)\int_a^b f(x)dx = -(KH)\int_b^a f(x)dx \text{ for } a > b. \text{ Now, the function}$$

$$F(x) = (KH)\int_c^x f(t)dt$$

is meaningful for a fixed $c \in [a,b]$ and any $x \in [a,b]$.

Theorem 1.17 *(Continuity of the integral) Let f be a (KH)-integrable function on $[a,b]$. Then the function*

$$F(x) = (KH)\int_a^x f(t)dt$$

is continuous on $[a,b]$.

Proof: We have to prove $F(c) = \lim_{x \to c} F(x)$. We will prove the continuity of F at c from the left, that is, we will show that for every $\varepsilon > 0$ there is $\delta > 0$ such that

$$\left| \int_a^x f(t)dt - \int_a^c f(t)dt \right| < \varepsilon$$

whenever $x \le c, c - x < \delta$.

First, take a number γ such that

$$\left| f(c)(c-x) \right| < \frac{\varepsilon}{3}$$

for any $x \in (c - \gamma, c]$.

The function f is integrable on $[a,c]$ by Theorem 1.16, hence there is $\delta_1 : [a,c] \to \mathbf{R}^+$ such that

$$\left| \sum_{P_1} f - F(c) \right| < \frac{\varepsilon}{3}$$

for every δ_1-fine partition P_1 of $[a,c]$.

Put $\delta = \delta(c) = \min\{\delta_1(c), \gamma\}$ and take $x \in (c - \delta, c]$. From the integrability of f on $[a,x]$ there exists $\delta_2 : [a,x] \to \mathbf{R}^+$ such that

$$\left| \sum_{P_2} f - F(x) \right| < \frac{\varepsilon}{3}$$

for every δ_2-fine partition P_2 of $[a,x]$.

Let Q be an arbitrary δ-fine partition of $[a,x]$, where

$$\delta(t) = \begin{cases} \min\{\delta_1(t), \delta_2(t)\} & \text{for } t \in [a,x], \\ \delta_1(t) & \text{for } t \in (x,c), \\ \delta & \text{for } t = c. \end{cases}$$

Then $P = Q \cup \{([x,c],c)\}$ is a δ-fine partition of $[a,c]$. Now

$$\left| F(x) - F(c) \right| = \left| F(x) - \sum_Q f - f(c)(c-x) + \sum_P f - F(c) \right|$$

$$\leq |F(x) - \sum_Q f| + |f(c)(c-x)| + \left|\sum_P f - F(c)\right| < \varepsilon.$$

The continuity of F at c from the right can be proved by a similar manner (Exercise 1.18).

Exercise 1.18   Prove the theorem: If f is (KH)-integrable on $[a,b]$, then the function $F(x) = (KH)\int_a^x f(t)dt$ is continuous at c from the right for every $c \in [a,b]$.

Hint. See the proof of Theorem 1.17.

1.5 Fundamental Theorem of Calculus

If a function f is Riemann integrable on $[a,b]$ and F is a continuous function such that $F'(x) = f(x)$, for all $x \in (a,b)$, then

$$\int_a^b f(x)dx = F(b) - F(a).$$

The function F is called *antiderivative* (or *primitive*) of the function f. There are some generalizations of that theorem and the antiderivative in the Kurzweil–Henstock integral theory.

Definition 1.19   Let $[a,b]$ be an interval in R . Let $f:[a,b] \to$ R. A function $F:[a,b] \to$ R is a *primitive of f on $[a,b]$* if the following conditions hold:

  (i) F is continuous on $[a,b]$;

  (ii) $F'(x) = f(x)$ for $x \in [a,b] \setminus C$, where C is a countable subset of $[a,b]$.

Lemma 1.20   *If function $F:[a,b] \to$ R has a derivative at t, then for every $\varepsilon > 0$ there is $\delta > 0$ such that*

$$|F(v) - F(u) - F'(t)(v-u)| < \varepsilon|v-u|$$

for $t \in [u,v] \subset (t-\delta, t+\delta) \cap [a,b]$.

Proof: By the definition of the derivative we have

$$\lim_{x \to t} \frac{F(x) - F(t)}{x - t} = F'(t),$$

hence

$$|F(x) - F(t) - F'(t)(x-t)| < \varepsilon|x-t|,$$

whenever $|x-t| < \delta$ and $x \in [a,b]$, and by application of the triangle inequality we get

$$|F(v) - F(u) - F'(t)(v-u)|$$
$$= |F(v) - F(t) - F(u) + F(t) - F'(t)(v-t-u+t)|$$
$$\leq |F(v) - F(t) - F'(t)(v-t)| + |F(u) - F(t) - F'(t)(u-t)|$$
$$< \varepsilon(|v-t| + |u-t|) = \varepsilon|v-u|$$

for $a \leq u < t < v \leq b$, $t-u < \delta$, $v-t < \delta$.

Theorem 1. 21   *(Fundamental Theorem) If $f:[a,b] \to$ R has a primitive F on $[a,b]$, then f is (KH)-integrable and*

$$(KH)\int_a^b f(x)dx = F(b) - F(a).$$

Proof: Let P be a partition of $[a,b]$, $P = \{([a_{i-1}, a_i], t_i) : i = 1, 2, ..., n\}$. Then

$$\left| \sum_P f - (F(b) - F(a)) \right| = \left| \sum_{i=1}^n f(t_i)(a_i - a_{i-1}) - \sum_{i=1}^n (F(a_i) - F(a_{i-1})) \right|$$

$$\leq \sum_{i=1}^n \left| f(t_i)(a_i - a_{i-1}) - F(a_i) + F(a_{i-1}) \right|,$$

$a_0 = a, a_n = b$. Let $C = \{c_1, c_2, ...\}$ be a set such that $F'(t) = f(t)$ when $t \in [a,b] \setminus C$. Let $\varepsilon > 0$. Since F is continuous at c_n for every n, there is $\delta(c_n) > 0$ such that

$$|F(v) - F(u)| < \frac{\varepsilon}{2^{n+3}}$$

and such that

$$|f(c_n)(v - u)| < \frac{\varepsilon}{2^{n+3}}$$

whenever $c_n \in [u,v] \subset (c_n - \delta(c_n), c_n + \delta(c_n))$. It implies

$$|f(c_n)(v - u) - F(v) + F(u)| < \frac{\varepsilon}{2^{n+2}} \tag{1.1}$$

whenever $c_n \in [u,v] \subset (c_n - \delta(c_n), c_n + \delta(c_n))$.

By Lemma 1.20 for every $t \in [a,b] \setminus C$ and for $\varepsilon / (2(b-a)) > 0$ there exists $\delta(t) > 0$ such that

$$|f(t)(v - u) - F(v) + F(u)| < \frac{\varepsilon}{2(b-a)}(v - u) \tag{1.2}$$

when $t \in [u,v] \subset [a,b] \cap (t - \delta(t), t + \delta(t))$.

So we have defined the function $\delta : [a,b] \to \mathrm{R}^+$. Let P be a δ-fine partition of $[a,b]$. Let P_1 and P_2 be the subsets of P, defined as follows:

$$P_1 = \{([a_{i-1}, a_i], t_i) \in P : t_i \in C\}, \quad P_2 = \{([a_{i-1}, a_i], t_i) \in P : t_i \in [a,b] \setminus C\}.$$

If $t_i = c_n$ for some n, then by (1.1) we get

$$|f(t_i)(a_i - a_{i-1}) - F(a_i) + F(a_{i-1})| < \frac{\varepsilon}{2^{n+2}}.$$

Each c_n is the point of at most two intervals of the partition P. Consequently

$$\sum_{t_i \in C} |f(t_i)(a_i - a_{i-1}) - F(a_i) + F(a_{i-1})| < 2\sum_{n=1}^\infty \frac{\varepsilon}{2^{n+2}} = \frac{\varepsilon}{2}$$

and by (1.2)

$$\sum_{t_i \in [a,b] \setminus C} |f(t_i)(a_i - a_{i-1}) - F(a_i) + F(a_{i-1})| < \sum_{t_i \in [a,b] \setminus C} \frac{\varepsilon}{2(b-a)}(a_i - a_{i-1})$$

$$\leq \frac{\varepsilon}{2(b-a)}(b - a) = \frac{\varepsilon}{2}.$$

Hence

$$\left| \sum_P f - F(b) + F(a) \right| < \frac{\varepsilon}{2} + \frac{\varepsilon}{2} = \varepsilon.$$

We have proved

$$\int_a^b f(x)dx = F(b) - F(a).$$

Example 1.22 Let $f = F'$ be the derivative of the function

$$F(x) = x^2 \cos\left(\frac{\pi}{x^2}\right) \text{ for } x \in (0,1] \text{ and } F(0) = 0.$$

Then, by the Fundamental Theorem 1.21, f is (KH) - integrable on $[0,1]$ and

$$\int_0^1 f(x)dx = \int_0^1 F'(x)dx = F(1) - F(0) = -1.$$

Note that $|f|$ is not (KH) -integrable on $[0,1]$. Indeed,

$$\int_0^1 |f(x)|\,dx \geq \sum_{i=1}^n \int_{\frac{1}{\sqrt{i+1}}}^{\frac{1}{\sqrt{i}}} |F'(x)|\,dx \geq \sum_{i=1}^n \left| F\left(\tfrac{1}{\sqrt{i}}\right) - F\left(\tfrac{1}{\sqrt{i+1}}\right) \right|$$

$$= \sum_{i=1}^n \left| \frac{1}{i}\cos(i\pi) - \frac{1}{i+1}\cos((i+1)\pi) \right| = \sum_{i=1}^n \left(\frac{1}{i} + \frac{1}{i+1} \right)$$

for every n. Hence $\int_0^1 |f(x)|\,dx = +\infty$.

Corollary 1.23 (Integration by parts) Let F and G be the continuous functions on $[a,b]$, $F' = f$ and $G' = g$ on $[a,b] \backslash K$, $K \subset [a,b]$ be a finite set. Then

$$(KH)\int_a^b (Fg + Gf)(x)dx = F(b)G(b) - F(a)G(a). \qquad (1.3)$$

Moreover, if one of the integrals $(KH)\int_a^b F(x)g(x)dx$, $(KH)\int_a^b G(x)f(x)dx$ exists, then

$$(KH)\int_a^b F(x)g(x)dx = F(b)G(b) - F(a)G(a) - (KH)\int_a^b f(x)G(x)dx.$$

Proof: The function FG is continuous on $[a,b]$ and $(FG)' = fG + Fg$ on $[a,b] \backslash K$. The equation (1.3) follows from the Fundamental Theorem 1.21.

1.6 Improper Integrals

An integral $\int_a^b f(x)dx$ is improper in the Riemann sense when f is not Riemann integrable on $[a,b]$ but f is Riemann integrable on $[s,b]$ for all $s \in (a,b]$ (on $[a,t]$ for all $t \in [a,b)$) and

$$\lim_{s \to a} \int_s^b f(x)dx, \qquad \left(\lim_{t \to b} \int_a^t f(x)dx \right)$$

exists in R . The example 1.6 (and many other examples) shows that the improper Riemann integral can be viewed as an ordinary Kurzweil-Henstock integral. We will prove (Theorem 1.25) that there is no improper (KH) -integral.

Proposition 1.24 *Let f be (KH) -integrable on $[a,b]$. Let $\delta : [a,b] \to \mathrm{R}^+$ be such that*

$$\left| \sum_P f - (KH)\int_a^b f(x)dx \right| < \varepsilon$$

whenever P is any δ -fine partition of $[a,b]$. Then

$$\left| \sum_Q f - (KH)\int_c^d f(x)dx \right| < 2\varepsilon$$

when Q is any δ-fine partition of $[c,d] \subset [a,b]$.

Proof: By Theorem 1.16 the function f is (KH)-integrable on every bounded subinterval of $[a,b]$. Then there are

$$\delta_1 : [a,c] \to R^+, \qquad \delta_2 : [d,b] \to R^+$$

such that

$$\left| \sum_{P_1} f - \int_a^c f(x)dx \right| < \frac{\varepsilon}{2}, \qquad \left| \sum_{P_2} f - \int_d^b f(x)dx \right| < \frac{\varepsilon}{2}$$

when P_1 is any δ_1-fine partition of $[a,c]$ and P_2 is any δ_2-fine partition of $[d,b]$. Put

$$\gamma_1(x) = \min\{\delta(x), \delta_1(x)\} \text{ for } x \in [a,c],$$
$$\gamma_2(x) = \min\{\delta(x), \delta_2(x)\} \text{ for } x \in [d,b]$$

and take a γ_1-fine partition P_1 of $[a,c]$, a δ-fine partition Q of $[c,d]$ and a γ_2-fine partition P_2 of $[d,b]$. Then $P = P_1 \cup Q \cup P_2$ is a δ-fine partition of $[a,b]$ and by the additivity of the integral (Theorem 1.14) we have:

$$\left| \sum_Q f - \int_c^d f(x)dx \right| \le \left| \sum_P f - \int_a^b f(x)dx \right| + \left| \sum_{P_1} f - \int_a^c f(x)dx \right| + \left| \sum_{P_2} f - \int_d^b f(x)dx \right|$$

$$< \varepsilon + \frac{\varepsilon}{2} + \frac{\varepsilon}{2} = 2\varepsilon.$$

Theorem 1.25 *Let $f : [a,b] \to R$ be (KH)-integrable on all $[s,b] \subset (a,b]$. Then $(KH)\int_a^b f(x)dx$ exists if and only if $\lim_{s \to a}(KH)\int_s^b f(x)dx$ exists. Moreover,*

$$(KH)\int_a^b f(x)dx = \lim_{s \to a}(KH)\int_s^b f(x)dx.$$

Proof: If $I = \int_a^b f(x)dx$ exists, then $I = \lim_{s \to a} \int_s^b f(x)dx$ by the continuity of the integral (Theorem 1.17).

Suppose that $L = \lim_{s \to a} \int_s^b f(x)dx$ exists. We need to prove that for every $\varepsilon > 0$ there is $\delta : [a,b] \to R^+$ such that

$$| \sum_P f - L | < \varepsilon$$

whenever P is any δ-fine partition of $[a,b]$. According to the existence of the limit there exists $\gamma > 0$ such that

$$\left| \int_s^b f(x)dx - L \right| < \frac{\varepsilon}{2}, \quad | f(a)(s-a) | < \frac{\varepsilon}{2^2} \qquad (1.4)$$

for every $s \in (a, a+\gamma)$.

Let $(c_n)_{n=0}^{\infty}$ be a strictly decreasing sequence with $c_0 = b, \lim_n c_n = a$. Let $\varepsilon > 0$ be arbitrary. The function f is integrable on $[c_1, c_0]$ and $[c_n, c_{n-2}] (n = 2,3,...)$, hence there exist

$$\delta_1 : [c_1, c_0] \to R^+, \qquad \delta_n : [c_n, c_{n-2}] \to R^+$$

such that

$$\left| \sum_{P_1} f - \int_{c_1}^{c_0} f(x)dx \right| < \frac{\varepsilon}{2^4}, \qquad \left| \sum_{P_n} f - \int_{c_n}^{c_{n-2}} f(x)dx \right| < \frac{\varepsilon}{2^{n+3}}$$

for any δ_1-fine P_1 of $[c_1, c_0]$ and any δ_n-fine P_n of $[c_n, c_{n-2}]$ $(n \in N)$.

Construct the function $\delta : [a,b] \to R^+$ in the following way:

$$\delta(x) = \begin{cases} \min\{\delta_1(x), x-c\} & \text{for } x \in (c_1, c_0], \\ \min\{\delta_n(x), x-c_n, c_{n-2}-x\} & \text{for } x \in (c_n, c_{n-1}] \ (n=2,3,\ldots), \\ \gamma & \text{for } x = a. \end{cases}$$

Let P be a δ-fine partition of $[a,b]$, $P = \{([a_{i-1}, a_i], t_i) : i = 1,2,\ldots,m\}$. Then $t_1 = a$ and $a_1 < a + \gamma$. The partition

$$P' = \{([a_{i-1}, a_i], t_i) : i = 2,3,\ldots,m\}$$

is a δ-fine partition of $[a_1, b]$.

Let P_n be a subset of P' such that $t_i \in (c_n, c_{n-1}]$ $(n \in \mathbb{N})$. We get a finite number of nonempty sets P_n with $P_r \cap P_s = \varnothing$ $(r \neq s)$. Let

$$E_n = \bigcup_{t_i \in (c_n, c_{n-1}]} [a_{i-1}, a_i].$$

Then the E_n's are non-overlapping intervals, $E_1 \subset (c_1, c_0], E_n \subset (c_n, c_{n-2})$ $(n \in \mathbb{N})$ and P_n is a δ-fine partition of E_n. Applying Proposition 1.24 we get

$$\left| \sum_{P_n} f - \int_{E_n} f(x)dx \right| < 2\frac{\varepsilon}{2^{n+3}} = \frac{\varepsilon}{2^{n+2}}.$$

In the inequality

$$\left| \sum_P f - L \right| \le \left| \sum_P f - \int_{a_1}^b f(x)dx \right| + \left| \int_{a_1}^b f(x)dx - L \right|$$

we have estimated the first term on the right side, by additivity of the integral we estimate

$$\left| \sum_P f - \int_{a_1}^b f(x)dx \right| = \left| f(a)(a_1 - a) + \sum_{P'} f - \int_{a_1}^b f(x)dx \right|$$

$$\le |f(a)(a_1 - a)| + \sum_{n=1}^\infty \left| \sum_{P_n} f - \int_{E_n} f(x)dx \right| \qquad (1.5)$$

$$< \frac{\varepsilon}{2^2} + \sum_{n=1}^\infty \frac{\varepsilon}{2^{n+2}} = \frac{\varepsilon}{2}.$$

By (1.4) and (1.5) we get $\left| \sum_P f - L \right| < \varepsilon.$

Remark 1.26   If f is (KH)-integrable on $[a,t]$ for every $t \in (a,b)$, then f is (KH)-integrable on $[a,b]$ if and only if $\lim_{t \to b} (KH) \int_a^t f(x)dx$ exists. The integral is equal to this limit. The proof is similar to the proof of Theorem 1.25 (Exercise 1.27).

Exercise 1.27   Prove Remark 1.26.

1.7 Absolute Integrability

In the Riemann integral theory, integrability of the function $f : [a,b] \to R$ implies integrability of $|f|$. In the Kurzweil-Henstock integral theory $|f|$ need not be integrable even if f is integrable (see Example 1.22). We give a necessary and sufficient condition for the integrability of $|f|$. The following lemma plays an important role (not only) in this subsection.

Lemma 1.28 *(Henstock's lemma)　Let a function f be integrable on $[a,b]$. Let $\delta:[a,b]\to \mathbb{R}^+$ be a function such that $\left|\sum_P f-(KH)\int_a^b f(x)dx\right|<\varepsilon$ whenever P is any δ-fine partition of $[a,b]$. Let $P=\{([a_{i-1},a_i],t_i):i=1,2,...,n\}$　be　a　δ-fine　partition　of　$[a,b]$　and　let　$Q\subset P$, $Q=\{[a_{i_j-1},a_{i_j}],t_{i_j}):j=1,2,...,k,1\le k\le n\}$. Then*

$$\left|\sum_{j=1}^{k}\left(f(t_{i_j})(a_{i_j}-a_{i_j-1})-(KH)\int_{a_{i_j-1}}^{a_{i_j}}f(x)dx\right)\right|<\varepsilon$$

and

$$\sum_{j=1}^{k}\left|f(t_{i_j})(a_{i_j}-a_{i_j-1})-(KH)\int_{a_{i_j-1}}^{a_{i_j}}f(x)dx\right|<2\varepsilon.$$

Proof: There are $n-k$ intervals $[a_{r-1},a_r]\subset[a,b]$ with $([a_{r-1},a_r],t_r)\in P\setminus Q$. Denote by M the set of those r, $M=\{r:([a_{r-1},a_r],t_r)\in P\setminus Q\}$. Let $\varepsilon'>0$. By the integrability on subintervals (Theorem 1.16) there exists δ_r defined on $[a_{r-1},a_r]$, $\delta_r\le\delta$, such that

$$\left|\sum_{P_r}f-\int_{a_{r-1}}^{a_r}f(x)dx\right|<\frac{\varepsilon'}{n-k}$$

whenever P_r is a δ_r-fine partition of $[a_{r-1},a_r]$, $r\in M$. Put $P'=\bigcup_{r\in M}P_r\cup Q$. Then P' is a δ-fine partition of $[a,b]$ and $\sum_{P'}f=\sum_Q f+\sum_{r\in M}\sum_{P_r}f$. By the additivity of integral we get

$$\left|\sum_{j=1}^{k}\left(f(t_{i_j})(a_{i_j}-a_{i_j-1})-\int_{a_{i_j-1}}^{a_{i_j}}f(x)dx\right)\right|$$

$$\le\left|\sum_{P'}f-\int_a^b f(x)dx\right|+\sum_{r\in M}\left|\sum_{P_r}f-\int_{a_{r-1}}^{a_r}f(x)dx\right|$$

$$<\varepsilon+(n-k)\frac{\varepsilon'}{n-k}=\varepsilon+\varepsilon'.$$

Since $\varepsilon'>0$ is arbitrary, the proof of the first inequality is complete.

To prove the second inequality of the Henstock's lemma, split Q into Q_1,Q_2, where

$$Q_1=\left\{([a_{i_j-1},a_{i_j}],t_{i,j})\in Q:f(t_{i_j})(a_{i_j}-a_{i_j-1})-\int_{a_{i_j-1}}^{a_{i_j}}f(x)dx\ge 0\right\},$$

$$Q_2=\left\{([a_{i_j-1},a_{i_j}],t_{i,j})\in Q:f(t_{i_j})(a_{i_j}-a_{i_j-1})-\int_{a_{i_j-1}}^{a_{i_j}}f(x)dx< 0\right\}$$

and denote $M^+=\{i_j:([a_{i_j-1},a_{i_j}],t_{i_j})\in Q_1\}$, $M^-=\{i_j:([a_{i_j-1},a_{i_j}],t_{i_j})\in Q_2\}$. Since $Q_1\subset Q,Q_2\subset Q$, by the first inequality

$$\sum_{j=1}^{k}\left|f(t_{i_j})(a_{i_j}-a_{i_j-1})-\int_{a_{i_j-1}}^{a_{i_j}}f(x)dx\right|$$

$$=\sum_{i_j\in M^+}\left(f(t_{i_j})(a_{i_j}-a_{i_j-1})-\int_{a_{i_j-1}}^{a_{i_j}}f(x)dx\right)$$

$$-\sum_{i_j\in M^-}\left(f(t_{i_j})(a_{i_j}-a_{i_j-1})-\int_{a_{i_j-1}}^{a_{i_j}}f(x)dx\right)<2\varepsilon.$$

If a partition $Q = \{([b_{j-1}, b_j], z_j) : j = 1, 2, \ldots, m\}$ is a refinement of a partition $P = \{([a_{i-1}, a_i], t_i) : i = 1, 2, \ldots, n\}$, that is, each interval in Q is a subset of an interval of P (the interval $[a_{i-1}, a_i]$ is divided into a finite number of intervals), then

$$\sum_{i=1}^{n} \left| \int_{a_{i-1}}^{a_i} f(x)dx \right| \leq \sum_{j=1}^{m} \left| \int_{b_{j-1}}^{b_j} f(x)dx \right| \qquad (1.6)$$

by the additivity of integral and the triangle inequality.

Theorem 1.29 *(Integration of absolute values) Let f be an integrable function on $[a,b]$. Then $|f|$ is integrable on $[a,b]$ if and only if*

$$A = \sup_D \sum_{i=1}^{n} \left| (KH) \int_{a_{i-1}}^{a_i} f(x)dx \right| < +\infty,$$

where $D = \{a_0, a_1, \ldots, a_n\}$ is any division of $[a,b]$. Moreover $\int_a^b |f(x)| \, dx = A$.

Proof: Let f and $|f|$ be integrable on $[a,b]$. According to Corollary 1.13 and additivity of integral we have

$$\sum_{i=1}^{n} \left| \int_{a_{i-1}}^{a_i} f(x)dx \right| \leq \sum_{i=1}^{n} \int_{a_{i-1}}^{a_i} |f(x)| \, dx = \int_a^b |f(x)| \, dx.$$

Hence, $A < +\infty$.

Conversely, let $A < +\infty$. We have to prove $\int_a^b |f(x)| \, dx = A$. Let $\varepsilon > 0$ be arbitrary. Then

(i) there is a division $D = \{a_0, a_1, \ldots, a_n\}$ such that

$$A - \varepsilon < \sum_{i=1}^{n} \left| \int_{a_{i-1}}^{a_i} f(x)dx \right| \leq A,$$

(ii) there is a function $\delta : [a,b] \to \mathrm{R}^+$ such that $\left| \sum_P f - I \right| < \varepsilon$
  whenever P is a δ-fine partition of $[a,b]$.

Take $\delta' \leq \delta$ such that $(x - \delta'(x), x + \delta'(x))$ does not have any a_i when $x \neq a_i$, $i = 1, 2, \ldots, n$. Let P be a δ'-fine partition of $[a,b]$. Then every a_i $(i = 1, 2, \ldots, n)$ is a tag in P. Make a refinement P' of P by adding a_i $(i = 1, 2, \ldots, n)$ as the dividing points into P and if $([b_{k-1}, b_k], a_i) \in P$ then $([b_{k-1}, a_i], a_i) \in P'$ and $([a_i, b_k], a_i) \in P'$ $(i = 1, 2, \ldots, n)$. It has no effect to the Riemann sum $\sum_P f$ (see the proof of Theorem 1.14). By the inequality $\| |a| - |b| \| \leq |a - b|$ and Henstock's lemma we get

$$\sum_{(t_j, E_j) \in P'} \left\| \left| \int_{E_j} f(x)dx \right| - |f(t_j)| \| E_j \right\| \leq \sum_{(t_j, E_j) \in P'} \left\| \int_{E_j} f(x)dx - f(t_j) | E_j \right\| \leq 2\varepsilon. \quad (1.7)$$

Hence,

$$\left| \sum_P |f| - A \right| = \left| \sum_{P'} |f| - A \right| \leq \left| A - \sum_{(t_j, E_j) \in P'} \left| \int_{E_j} f(x)dx \right| \right|$$

$$+ \left| \sum_{(t_j, E_j) \in P'} \left| \int_{E_j} f(x)dx \right| - \sum_{P'} |f| \right|$$

$$< \varepsilon + 2\varepsilon$$

according to (1.6), (i) and (1.7).

Corollary 1.30 *(Comparison test for absolute integrability) Let* $f:[a,b] \to \mathbb{R}$*,* g *be a* (KH)*-integrable function on* $[a,b]$ *and* $|f| \le g$*. Then* f *is* (KH)*-integrable and*

$$\left| (KH) \int_a^b f(x)dx \right| \le (KH) \int_a^b |f(x)|\,dx \le (KH) \int_a^b g(x)dx.$$

Proof: Let $D = \{a = a_0, a_1, \ldots, a_n = b\}$ be any division of $[a,b]$. By Theorems 1.16 and 1.12

$$\left| \int_{a_{i-1}}^{a_i} f(x)dx \right| \le \int_{a_{i-1}}^{a_i} g(x)dx \ (i=1,2,\ldots,n).$$

Theorem 1.14 implies

$$\sum_{i=1}^n \left| \int_{a_{i-1}}^{a_i} f(x)dx \right| \le \int_a^b g(x)dx.$$

Hence, $A < +\infty$ and Theorem 1.29 guarantee (KH)-integrability of $|f|$. The desired inequalities follow from Theorem 1.12.

Remark 1.31 If function F is a primitive of a function f on $[a,b]$, then

$$A = \sup_D \sum_{i=1}^n \left| \int_{a_{i-1}}^{a_i} f(x)dx \right| = \sup_D \sum_{i=1}^n |F(a_i) - F(a_{i-1})|$$

by Theorem 1.21. The number A is called the *total variation of F* on $[a,b]$ and, if $A < +\infty$, then the function F is said to be *of bounded variation.*

The following example is an example of non-absolutely integrable function f and simultaneously of a function F which is not of bounded variation.

Example 1.32 The series $\sum_{n=1}^\infty a_n$ with members

$$1, -\frac{1}{2}, \frac{1}{2}, -\frac{1}{3}, \frac{1}{3}, -\frac{1}{4}, \frac{1}{4}, \ldots$$

is convergent. The partial sum of this series $s_{2n-1} = 1, s_{2n} = 1 - 1/(n+1)$ $(n \in \mathbb{N})$. The sum of the series $\sum_{n=1}^\infty a_n = \lim_n s_n = 1$. Put $s_0 = 0$. Define

a) $(c_n)_{n=0}^\infty = \left(\dfrac{n}{n+1} \right)_{n=0}^\infty$;

b) $f(x) = \dfrac{s_n - s_{n-1}}{c_n - c_{n-1}}$ for $x \in [c_{n-1}, c_n)$ $(n \in \mathbb{N})$ and $f(1) = 1$;

c) $F(c_n) = s_n$, $F(x) = \dfrac{s_n - s_{n-1}}{c_n - c_{n-1}}(x - c_n) + s_n$ for $x \in (c_{n-1}, c_n)$ $(n \in \mathbb{N})$, $F(1) = 1$ (The Figure 1.1 obtains the part of the graph of F).

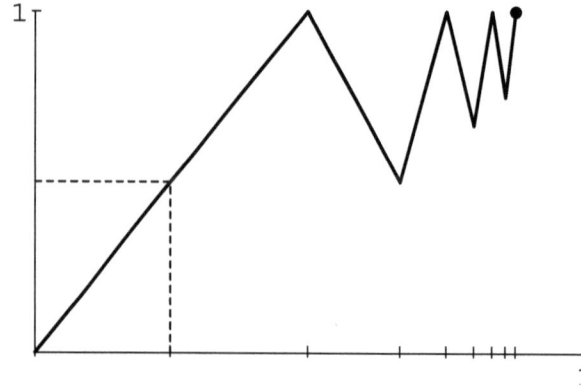

Fig. 1.1 The part of the graph of the function F

Then $F'(x) = f(x)$ for $x \neq c_n$ $(n \in \mathbb{N})$, that is F is a primitive of f on $[0,1]$, and by the Fundamental Theorem and continuity of integral (Theorems 1.21 and 1.17) we have

$$\int_0^1 f(x)dx = \lim_n \int_0^{c_n} f(x)dx$$

$$= \lim_n \sum_{k=1}^n \int_{c_{k-1}}^{c_k} f(x)dx = \lim_n \sum_{k=1}^n \left(F(c_k) - F(c_{k-1}) \right)$$

$$= \lim_n \sum_{k=1}^n (s_k - s_{k-1}) = \lim_n \sum_{k=1}^n a_k = 1.$$

The function f is not absolutely (KH)-integrable because F is not of bounded variation, indeed $D_n = \{c_0 = 0, c_1, \ldots, c_n, 1\}$ is a division of $[0,1]$ for every $n \in \mathbb{N}$ and hence

$$A \geq \lim_n \left(\sum_{i=1}^n \left| F(c_i) - F(c_{i-1}) \right| + \left| F(1) - F(c_n) \right| \right) = \sum_{n=1}^\infty | a_n | = +\infty.$$

1.8 Convergence Theorems

If $(f_n)_{n=1}^\infty$ is a sequence of Riemann integrable functions on $[a,b]$, which converges uniformly to a function f (that is, for every $\varepsilon > 0$ there is $n_0 \in \mathbb{N}$ such that $| f_n(x) - f(x) | < \varepsilon$ for all $x \in [a,b]$ and every $n > n_0$), then

$$\int_a^b f(x)dx = \lim_n \int_a^b f_n(x)dx.$$

In the Kurzweil-Henstock integral theory, besides the uniform convergence theorem, also the monotone convergence theorem and the dominated convergence theorem hold.

Theorem 1.33 *(Uniform convergence theorem)*   Let $(f_n)_{n=1}^\infty$ be a uniformly convergent sequence of (KH)-integrable functions on an interval $[a,b]$ to a function f. Then f is (KH)-integrable and

$$(KH)\int_a^b f(x)dx = \lim_n (KH)\int_a^b f_n(x)dx.$$

Proof: Choose $n_0 \in \mathbb{N}$ such that $|f_n(x) - f(x)| < \varepsilon$ for any $x \in [a,b]$ and any $n > n_0$. Since $|f_m(x) - f_n(x)| < 2\varepsilon$ for any $x \in [a,b]$ and any $m,n > n_0$, the sequence of the integrals is Cauchy, indeed

$$\left| \int_a^b f_m(x)dx - \int_a^b f_n(x)dx \right| \le \int_a^b |f_m(x) - f_n(x)|\,dx < 2\varepsilon(b-a).$$

Hence $L = \lim_n \int_a^b f_n(x)dx$ exists and there is $n_1 \ge n_0$ such that

$$\left| L - \int_a^b f_n(x)dx \right| < \varepsilon$$

for every $n > n_1$.

The sequence of Riemann sums converges uniformly, indeed

$$\left| \sum_{i=1}^k f_n(t_i)(a_i - a_{i-1}) - \sum_{i=1}^k f(t_i)(a_i - a_{i-1}) \right| \le \sum_{i=1}^k |f_n(t_i) - f(t_i)|(a_i - a_{i-1}) < \varepsilon(b-a)$$

for any $n > n_0$ and every partition $P = \{([a_{i-1},a_i],t_i) : i = 1,2,\ldots,k\}$. By the integrability of f_n there is $\delta_n : [a,b] \to \mathbb{R}^+$ such that

$$\left| \sum_P f_n - \int_a^b f_n(x)dx \right| < \varepsilon$$

when P is a δ_n-fine partition of $[a,b]$. For fixed $n > n_1$ put $\delta = \delta_n$. Take an arbitrary δ-fine P. Then

$$\left| \sum_P f - L \right| \le \left| \sum_P f - \sum_P f_n \right|$$
$$+ \left| \sum_P f_n - \int_a^b f_n(x)dx \right| + \left| \int_a^b f_n(x)dx - L \right|$$
$$< \varepsilon(b-a) + \varepsilon + \varepsilon. \qquad \square$$

A sufficient condition for interchanging the limit and the (KH)-integral is obtained in the following lemma:

Lemma 1.34 *Suppose that $(f_n)_{n=1}^\infty$ is a sequence of (KH)-integrable functions on $[a,b]$, $\lim_n f_n(t) = f(t)$ is finite for every $t \in [a,b]$ and for every $\varepsilon > 0$ there is $\delta : [a,b] \to \mathbb{R}^+$ such that*

$$\left| \sum_P f_n - (KH)\int_a^b f_n(x)dx \right| < \varepsilon$$

for every δ-fine partition P of $[a,b]$ and $n \in \mathbb{N}$. Then the function f is (KH)-integrable on $[a,b]$ and

$$(KH)\int_a^b f(x)dx = \lim_n (KH)\int_a^b f_n(x)dx.$$

Proof: Let $\varepsilon > 0$ be arbitrary and $P = \{([a_{i-1},a_i],t_i) : i = 1,2,\ldots,m\}$ be any partition of $[a,b]$. First we show that $\sum_P f_n$ converges to $\sum_P f$. For t_i in P there is $p(t_i) \in \mathbb{N}$ $(i = 1,2,\ldots,m)$ such that

$$|f_n(t_i) - f(t_i)| < \frac{\varepsilon}{b-a}$$

whenever $n \ge p(t_i)$. Put $n_P = \max\{p(t_1), p(t_2), \ldots, p(t_m)\}$, to obtain

$$\left| \sum_P f_n - \sum_P f \right| = \left| \sum_{i=1}^m f_n(t_i)(a_i - a_{i-1}) - \sum_{i=1}^m f(t_i)(a_i - a_{i-1}) \right|$$

$$\leq \sum_{i=1}^{m} \left| f_n(t_i) - f(t_i) \right| (a_i - a_{i-1}) < \frac{\varepsilon}{b-a}(b-a) = \varepsilon. \tag{1.8}$$

Furthermore, if P, Q are δ-fine partitions of $[a,b]$ and $n \geq \max\{n_P, n_Q\}$ is fixed, then

$$\left| \sum_P f - \sum_Q f \right| \leq \left| \sum_P f - \sum_P f_n \right|$$

$$+ \left| \sum_P f_n - \int_a^b f_n(x)dx \right| + \left| \int_a^b f_n(x)dx - \sum_Q f_n \right|$$

$$+ \left| \sum_Q f_n - \sum_Q f \right| < 4\varepsilon,$$

which implies that f is (KH)-integrable on $[a,b]$ by the Bolzano-Cauchy condition (Proposition 1.15). Since f is (KH)-integrable, there is $\delta_1 : [a,b] \to \mathrm{R}^+$ such that

$$\left| \sum_P f - \int_a^b f(x)dx \right| < \varepsilon \tag{1.9}$$

whenever P is δ_1-fine. Put $\delta' = \min\{\delta, \delta_1\}$. Then, by (1.9), (1.8) and the assumptions of the lemma, we have:

$$\left| \int_a^b f(x)dx - \int_a^b f_n(x)dx \right| \leq \left| \int_a^b f(x)dx - \sum_P f \right|$$

$$+ \left| \sum_P f - \sum_P f_n \right| + \left| \sum_P f_n - \int_a^b f_n(x)dx \right| < 3\varepsilon$$

for any δ'-fine partition P and $n \geq n_P$. Hence

$$\lim_n \int_a^b f_n(x)dx = \int_a^b f(x)dx.$$

Theorem 1.35 *(Monotone convergence theorem)   Let $(f_n)_{n=1}^{\infty}$ be a monotone sequence of (KH)-integrable functions. Let $\lim_n f_n = f$ be finite on $[a,b]$. Let the sequence $\left((KH)\int_a^b f_n(x)dx \right)_n$ be bounded. Then the function f is (KH)-integrable and*

$$(KH)\int_a^b f(x)dx = \lim_n (KH)\int_a^b f_n(x)dx.$$

Proof: Let $\varepsilon > 0$ be arbitrary. Since the sequence $\left(\int_a^b f_n(x)dx \right)_n$ is monotone and bounded, there exists $N \in \mathrm{N}$ such that

$$\left| \int_a^b f_N(x)dx - \int_a^b f_n(x)dx \right| < \frac{\varepsilon}{3} \tag{1.10}$$

whenever $n \geq N$.
From integrability of f_n there exists $\delta_n : [a,b] \to \mathrm{R}^+$ such that

$$\left| \sum_P f_n - \int_a^b f_n(x)dx \right| < \frac{\varepsilon}{3 \cdot 2^n} \tag{1.11}$$

for any δ-fine partition P of $[a,b]$.

Since $f_n(t)$ converges to $f(t)$ for every $t \in [a,b]$, there exists $p(t) \in \mathrm{N}, p(t) \geq N$ such that

$$\left| f(t) - f_k(t) \right| < \frac{\varepsilon}{6(b-a)} \qquad (1.12)$$

for every $k \geq p(t)$. Denote

$A_n = \{t \in [a,b] : p(t) = n,\ n \in \mathrm{N},\ n \geq N \text{ is as minimal as possible}\}$.

That is, for $t \in A_N$ the inequality (1.12) holds for $k \geq N$, in the case $t \in A_n, n > N$ the inequality (1.12) holds for $k \geq n$ but for $k = n-1$ the inequality does not hold. Thus the sets A_n disjoint and $\bigcup_{n=N}^{\infty} A_n = [a,b]$.

Put $\delta(t) = \min\{\delta_1(t), \delta_2(t), \ldots, \delta_n(t)\}$ for $t \in A_n$. We will prove that

$$\left| \sum_P f_n - \int_a^b f_n(x)dx \right| < \varepsilon$$

whenever P is a δ-fine partition and $n \in \mathrm{N}$. Take any δ-fine partition P of $[a,b]$. If $n \leq N$ then P is δ_n-fine and (1.11) holds. In the case $n > N$, split P into

$$P_j = \{([a_{i-1}, a_i], t_i) \in P : t_i \in A_j\}.$$

Then P_j is δ_1-fine, δ_2-fine, ..., δ_j-fine and hence, if $n \leq j$, then P_j is δ_n-fine; if $n > j$, then P_j is not δ_n-fine. Fix $n > N$. Denote

$$\tau = \{i : t_i \in A_j, j < n\}, \qquad \zeta = \{i : t_i \in A_j, j \geq n\}.$$

Estimate

$$\left| \sum_P f_n - \int_a^b f_n(x)dx \right| \leq \left| \sum_{i \in \tau} f_n(t_i)(a_i - a_{i-1}) - \sum_{i \in \tau} \int_{a_{i-1}}^{a_i} f_n(x)dx \right|$$

$$+ \left| \sum_{i \in \zeta} f_n(t_i)(a_i - a_{i-1}) - \sum_{i \in \zeta} \int_{a_{i-1}}^{a_i} f(x)dx \right|$$

$$\leq \left| \sum_{i \in \tau} f_n(t_i)(a_i - a_{i-1}) - \sum_{i \in \tau} \int_{a_{i-1}}^{a_i} f_n(x)dx \right| + \frac{\varepsilon}{3 \cdot 2^n}, \qquad (1.13)$$

because for $i \in \zeta$ we can apply Henstock's lemma. For $i \in \tau$ we can also apply Henstock's lemma to f_j and δ_j, hence

$$\left| \sum_{t_i \in A_j} f_j(t_i)(a_i - a_{i-1}) - \sum_{t_i \in A_j} \int_{a_{i-1}}^{a_i} f_j(x)dx \right| < \frac{\varepsilon}{3 \cdot 2^j}. \qquad (1.14)$$

Now

$$\left| \sum_{i \in \tau} f_n(t_i)(a_i - a_{i-1}) - \sum_{i \in \tau} \int_{a_{i-1}}^{a_i} f_n(x)dx \right|$$

$$\leq \sum_{j=N}^{n-1} \left| \sum_{t_i \in A_j} f_n(t_i)(a_i - a_{i-1}) - \sum_{t_i \in A_j} f_j(t_i)(a_i - a_{i-1}) \right| \qquad (1.15)$$

$$+ \sum_{j=N}^{n-1} \left| \sum_{t_i \in A_j} f_j(t_i)(a_i - a_{i-1}) - \sum_{t_i \in A_j} \int_{a_{i-1}}^{a_i} f_j(x)dx \right| \qquad (1.16)$$

$$+ \sum_{j=N}^{n-1} \left| \sum_{t_i \in A_j} \int_{a_{i-1}}^{a_i} f_j(x)dx - \sum_{t_i \in A_j} \int_{a_{i-1}}^{a_i} f_n(x)dx \right|. \qquad (1.17)$$

When $t_i \in A_j$ then $p(t_i) = j \geq N$ and by (1.12) for $n > j$ we have

$$\left| f_n(t_i) - f_j(t_i) \right| \le \left| f_n(t_i) - f(t_i) \right| + \left| f(t_i) - f_j(t_i) \right| < \frac{\varepsilon}{3(b-a)}.$$

It implies that the sum (1.15) is less than $\varepsilon/3$ and by (1.14), the sum (1.16) is less than $\sum_{j=N}^{n-1} \varepsilon/(3 \cdot 2^j)$. We need to estimate the sum (1.17), which is equal to

$$\sum_{j=N}^{n-1} \left| \sum_{t_i \in A_j} \int_{a_{i-1}}^{a_i} \left(f_j(x) - f_n(x) \right) dx \right|. \tag{1.18}$$

The sequence $(f_n)_n$ is monotone. If the sequence is increasing (decreasing), then all the integrals in (1.18) have the negative (positive) sign. When $(f_n)_n$ is increasing we get

$$\sum_{j=N}^{n-1} \left| \sum_{t_i \in A_j} \int_{a_{i-1}}^{a_i} \left(f_j(x) - f_n(x) \right) dx \right|$$

$$= \sum_{j=N}^{n-1} \sum_{t_i \in A_j} \int_{a_{i-1}}^{a_i} \left(f_n(x) - f_j(x) \right) dx$$

$$\le \sum_{j=N}^{n-1} \sum_{t_i \in A_j} \int_{a_{i-1}}^{a_i} \left(f_n(x) - f_N(x) \right) dx$$

$$\le \int_a^b \left(f_n(x) - f_N(x) \right) dx = \left| \int_a^b \left(f_N(x) - f_n(x) \right) dx \right| < \frac{\varepsilon}{3} \tag{1.19}$$

by (1.10). The same estimation we get also when $(f_n)_n$ is decreasing. Finally, from the estimates (1.13), (1.15), (1.16) and (1.19), we get

$$\left| \sum_P f_n - \int_a^b f_n(x) dx \right| < \frac{\varepsilon}{3} + \sum_{j=N}^{n-1} \frac{\varepsilon}{3 \cdot 2^j} + \frac{\varepsilon}{3} + \frac{\varepsilon}{3 \cdot 2^n} < \varepsilon.$$

Thanks to Lemma 1.34, the proof is complete.

Remark 1.36 If f_n $(n \in \mathbb{N})$ and h are (KH)-integrable on $[a,b]$, $|f_n - f_m| \le h$ for all n and m and $\lim_n f_n(x) = f(x)$ for every $x \in [a,b]$, then f is integrable on $[a,b]$ and

$$(KH) \int_a^b f(x) dx = \lim_n (KH) \int_a^b f_n(x) dx.$$

The proof of this dominated convergence theorem is different from the proof of Theorem 1.35 only in the estimation of (1.18) and choosing N (Exercise 1.37).

Exercise 1.37   Prove Remark 1.36.

Hint. The function

$$h_j = \lim_{k \to \infty} \bigvee_{j \le m \le n \le k} | f_n - f_m |$$

is (KH)-integrable by the monotone convergence theorem, $\lim_j h_j(x) = 0$ for every $x \in [a,b]$ and by the same theorem

$$\lim_j \int_a^b h_j(x) dx = 0.$$

Choose N for which $\int_a^b h_N(x) dx < \varepsilon/3$.

In Chapter 6 we will see that this technique, with suitable small modifications, can be applied to prove a generalized version of the dominated convergence theorem in the context of Riesz spaces.

1.9 Unbounded Intervals

The Riemann integral of a function f on $[a, +\infty]$ is defined by $\lim_{b \to +\infty} \int_a^b f(x) dx$. The Kurzweil-Henstock integral over an unbounded interval can be defined directly.

Let $\tilde{R}$ be the extended real line, $\tilde{R} = R \cup \{-\infty, +\infty\}$, ordered in the natural way ($-\infty < x < +\infty$ for all $x \in R$). The intervals in $\tilde{R}$ consist of all the intervals in R and the intervals of the form $[-\infty, a), (a, +\infty], [-\infty, a], [a, +\infty]$, $[-\infty, +\infty] = \tilde{R}$. The interval in $\tilde{R}$ is unbounded when at least one endpoint is $-\infty$ or $+\infty$.

Example 1.38 We will show that $(KH) \int_1^{+\infty} x^{-2} dx$ exists and we will find its value. From the Riemann integral theory we have

$$\int_1^{+\infty} x^{-2} dx = \lim_{b \to +\infty} \int_1^b x^{-2} dx = \lim_{b \to +\infty} \left[-\frac{1}{x} \right]_1^b = \lim_{b \to +\infty} \left(-\frac{1}{b} \right) + 1 = 1.$$

If $P = \{([a_{i-1}, a_i], t_i) : i = 1, 2, \ldots, n\}$ is a partition of $[1, b]$, then

$$\int_1^{+\infty} x^{-2} dx = \sum_{i=1}^n \int_{a_{i-1}}^{a_i} x^{-2} dx + \int_b^{+\infty} x^{-2} dx = \sum_{i=1}^n \left(-\frac{1}{a_i} + \frac{1}{a_{i-1}} \right) + \int_b^{+\infty} x^{-2} dx.$$

We need to choose $b \in R$ sufficiently large to get

$$\int_b^{+\infty} x^{-2} dx = \frac{1}{b} < \frac{\varepsilon}{2}$$

and to define the function $\delta : [1, b] \to R^+$ so that

$$\left| \sum_P f - \left(1 - \frac{1}{b} \right) \right| < \frac{\varepsilon}{2}$$

for every δ-fine partition P of $[1, b]$. Calculate

$$\left| \sum_{i=1}^n t_i^{-2} (a_i - a_{i-1}) - \sum_{i=1}^n \left(\frac{1}{a_{i-1}} - \frac{1}{a_i} \right) \right|$$

$$\leq \sum_{i=1}^n \left| \frac{1}{t_i^2} (a_i - a_{i-1}) - \frac{1}{a_{i-1}} + \frac{1}{a_i} \right|$$

$$= \sum_{i=1}^n \left| \frac{a_i - a_{i-1}}{t_i} \left(\frac{1}{t_i} - \frac{t_i}{a_{i-1}(a_i - a_{i-1})} + \frac{t_i}{a_i(a_i - a_{i-1})} \right) \right|$$

$$\leq \sum_{i=1}^n \frac{a_i - a_{i-1}}{t_i} \left| \frac{a_{i-1} a_i (a_i - a_{i-1}) - t_i^2 (a_i - a_{i-1})}{t_i a_{i-1} a_i (a_i - a_{i-1})} \right|$$

$$= \sum_{i=1}^n \frac{a_i - a_{i-1}}{t_i} \left| \frac{1}{t_i} - \frac{t_i}{a_{i-1} a_i} \right| = \sum_{i=1}^n \frac{a_i - a_{i-1}}{t_i} \left(\frac{t_i}{a_{i-1} a_i} - \frac{1}{t_i} \right)$$

$$\leq \sum_{i=1}^n \frac{a_i - a_{i-1}}{t_i} \left(\frac{1}{a_{i-1}} - \frac{1}{a_i} \right).$$

Put $\delta(x) = \varepsilon x / 4$. Then

$$\left| \sum_P f - \left(1 - \frac{1}{b}\right) \right| < \frac{\varepsilon}{2} \sum_{i=1}^{n} \left(\frac{1}{a_{i-1}} - \frac{1}{a_i} \right) = \frac{\varepsilon}{2} \left(1 - \frac{1}{b} \right) < \frac{\varepsilon}{2}$$

and

$$\left| \sum_P f - 1 \right| = \left| \sum_P f - 1 + \frac{1}{b} - \frac{1}{b} \right|$$

$$\leq \left| \sum_P f - \left(1 - \frac{1}{b}\right) \right| + \frac{1}{b} < \varepsilon.$$

By the previous example, the definition of the (KH)-integral on an unbounded interval can be formulated in the following way.

Definition 1.39 Let $[A, B] \subset \tilde{R}$. A function $f : [A, B] \to R$ is *(KH)-integrable on* $[A, B]$, if there is a number I such that for every $\varepsilon > 0$ there is a number $h > 0$ and there is a function $\delta : [A, B] \to R^+$ such that

$$| \sum_P f - I | < \varepsilon$$

for every δ-fine partition P of a bounded interval $[a_h, b_h]$ with $[A, B] \cap [-h, h] \subset [a_h, b_h] \subset [A, B]$.

Exercise 1.40   Prove that Definition 1.39 is correct (see also Chapter 7).

Exercise 1.41   Prove that Proposition 1.15 holds for an unbounded interval $[A, B]$.

Remark 1.42   If a function f is integrable on an unbounded interval $[a, +\infty]$, then f is integrable on every bounded subinterval of $[a, +\infty)$ (Exercise 1.43).

Exercise 1.43   Prove Remark 1.42.

Theorem 1.44   *The function f is (KH)-integrable on $[a, +\infty]$ if and only if f is (KH)-integrable on $[a, b]$ for every $b > a$, $b \in R$ and*

$$\lim_{b \to +\infty} (KH) \int_a^b f(x) dx$$

exists. Moreover,

$$(KH) \int_a^{+\infty} f(x) dx = \lim_{b \to +\infty} (KH) \int_a^b f(x) dx.$$

We will give the proof of this theorem in Chapter 7.

Exercise 1.45   Prove the theorem: A real-valued function f is (KH)-integrable on $[-\infty, b]$ if and only if f is (KH)-integrable on $[a, b]$ for every $a < b, a \in R$ and

$$\lim_{a \to -\infty} (KH) \int_a^b f(x) dx$$

exists. Moreover,

$$(KH) \int_{-\infty}^b f(x) dx = \lim_{a \to -\infty} (KH) \int_a^b f(x) dx.$$

Hint. The proof is similar to the proof of Theorem 1.44.

2 Elementary Theory of Riesz Spaces

Abstract

In this chapter we deal with the fundamental properties of lattice groups and Riesz spaces. We introduce the concepts of order and *(D)*-convergence, weak σ-distributivity and Egorov property and prove some related results. We deal also with order bounded and order continuous linear functionals in the setting of Riesz spaces. Finally we introduce the Maeda-Ogasawara-Vulikh representation theorem.

2.1 Lattice Ordered Groups

Throughout this chapter, given $a < b \in \tilde{\mathbb{R}}$, we will denote by $[a,b]$ the closed interval $\{x \in \tilde{\mathbb{R}} : a \leq x \leq b\}$ and by (a,b) the open interval $\{x \in \tilde{\mathbb{R}} : a < x < b\}$.

Definition 2.1 A *partially ordered set* R is a nonempty set endowed with a reflexive, transitive and antisymmetric relation, denoted by $\leq$.

Given a nonempty subset $A \subset R$ and an element $s \in R$, we say that s is the *supremum* of A if for every element $a \in A$ we have $a \leq s$, and moreover we get $s \leq c$ whenever $c \in R$ and $c \geq b$ for all $b \in A$. Analogously, given $j \in R$, we say that j is the *infimum* of A if for each $a \in A$ we have $a \geq j$, and for all $d \in R$, such that $d \leq b \ \forall b \in A$, we get $j \geq d$. In this case, we write $s = \sup A$ and $j = \inf A$ respectively.

If Λ is any nonempty set and $(x_\lambda)_{\lambda \in \Lambda}$ is a family of elements in R, we denote also by $\bigvee_{\lambda \in \Lambda} x_\lambda$ and $\bigwedge_{\lambda \in \Lambda} x_\lambda$, or $\sup_{\lambda \in \Lambda} x_\lambda$ and $\inf_{\lambda \in \Lambda} x_\lambda$, the quantities $\sup\{x_\lambda : \lambda \in \Lambda\}$ and $\inf\{x_\lambda : \lambda \in \Lambda\}$ respectively, provided that they exist in R.

A partially ordered set R is said to be a *lattice* if for every two elements a, $b \in R$ there exist in R the *supremum* $s := a \vee b$ $(= \sup\{a,b\})$ and the *infimum* $j := a \wedge b$ $(= \inf\{a,b\})$. In a partially ordered set R, we say that a nonempty subset $A \subset R$ is *bounded from above,* if there exists $x \in A$ such that $a \leq x$, $\forall a \in A$; *bounded from below,* if there exists $y \in A$ such that $a \geq y$, $\forall a \in A$; *bounded,* if it is bounded both from above and from below.

A lattice R is said to be *Dedekind complete* if every nonempty subset of R, bounded from above (with respect to the relation $\leq$), admits supremum in R, and every nonempty subset of R, bounded from below, admits infimum in R. A Dedekind complete lattice R is said to be *super Dedekind complete* if, for any nonempty set $A \subset R$, bounded from above, there exists a countable subset $A^* \subset A$, such that $\sup A = \sup A^*$, and for every nonempty set $B \subset R$, bounded from below, there exists a countable subset $B^* \subset B$, such that $\inf B = \inf B^*$.

An element a of a lattice R is said to be *positive* if $a \geq 0$. Two positive elements $a, b \in R$ are said to be *disjoint* or *orthogonal* if $a \wedge b = 0$. A nonempty set $A \subset R$ is called a *disjoint system* if every element $a \in A$ is positive and $a \wedge b = 0 \ \forall a, b \in A$. Given any two elements a, $b \in R$, we say that $a < b$ or $b > a$ if $a \leq b$ and $a \neq b$. A *unit* of R is an element a such that $a > 0$.

A lattice R is said to be *laterally complete* if every disjoint system $A \subset R$ has a supremum in R. We say that R is *universally complete* if it is both Dedekind and laterally complete.

Definition 2.2 Let R be a lattice. A nonempty set $C \subset R$ is said to be *directed upwards* [*downwards*] if for every pair $a, b \in C$ there exists $c \in C$ such that $c \geq a$, $c \geq b$ [$c \leq a$, $c \leq b$].

Definition 2.3 An Abelian partially ordered group $(R,+,\le)$ is called a *lattice ordered group* (or briefly an *l-group*) if it is a lattice and the following implication holds:

$$[a \le b] \Rightarrow [a+c \le b+c] \quad \forall a,b,c \in R. \qquad (2.1)$$

The following properties hold in any *l*-group R (see [14], pp. 292-295).

Proposition 2.4 (Distributive laws) For every $a,b,x,y \in R$, we have:

$$a+(x \vee y) = (a+x) \vee (a+y),$$

$$a+(x \wedge y) = (a+x) \wedge (a+y),$$

and, more generally, for each family $(x_\lambda)_{\lambda \in \Lambda}$ of elements of R,

$$a + \left(\bigvee_{\lambda \in \Lambda} x_\lambda \right) = \bigvee_{\lambda \in \Lambda} (a + x_\lambda),$$

$$a + \left(\bigwedge_{\lambda \in \Lambda} x_\lambda \right) = \bigwedge_{\lambda \in \Lambda} (a + x_\lambda),$$

in the sense that the left member exists in R if and only if the right member exists in R too, and in this case the two involved quantities coincide.

Proposition 2.5 In any *l*-group R, we have

$$a-(a \wedge b)+b = b \vee a \quad \forall a,b \in R.$$

Definition 2.6 For every element r of an *l*-group R, set $r^+ = r \vee 0$, $r^- = (-r) \vee 0$; r^+ and r^- are called the *positive* and *negative part* of r respectively. Moreover, set $|r| = r \vee (-r)$; $|r|$ is called the *absolute value* of R.

Proposition 2.7 For each element r of an *l*-group, we have:

$$r = r^+ - r^-; \qquad |r| = r^+ + r^-.$$

Moreover, $|r| \ge 0$ and $|r| = 0$ if and only if $r = 0$.

Definition 2.8 Given $a \in R$ and $n \in \mathbb{N}$, we denote by na the element $a+...+a$ (n times).

The following results hold in any *l*-group (see [14], p. 296):

i) $|na| = |n||a| \quad \forall n \in \mathbb{N}, \quad \forall a \in R$;

ii) $|(a \vee c)-(b \vee c)|+|(a \wedge c)-(b \wedge c)| = (a \vee b)-(a \wedge b) \quad \forall a,b,c \in \mathbb{R}$;

iii) $|a+b| \le |a|+|b| \quad \forall a,b \in R$.

Definition 2.9 An *l*-group R is said to be *Archimedean* if, for every choice of $a,b \in R$, with $na \le b \quad \forall n \in \mathbb{N}$, we have: $a \le 0$.

Proposition 2.10 Every Dedekind complete l-group R is Archimedean.

Proof: (See also [14], p. 291) Let $a, b \in R$, with $na \leq b \quad \forall n \in \mathbb{N}$. By Dedekind completeness of R, the element $c := \bigvee_{n \in \mathbb{N}} na$ does exist in R. Thus, we get:

$$c + a = \bigvee_{n \in \mathbb{N}} (n+1)a \leq \bigvee_{n \in \mathbb{N}} na = c.$$

Hence, we obtain: $a \leq 0$.

Remark 2.11 In this book, we will investigate mainly *Dedekind complete* l-groups. We note that the definition of l-group can be given without requiring commutativity *a priori*: this, substantially, is not a restriction for our purposes. Indeed, by virtue of the Iwasawa theorem (see [14], p. 317), every Dedekind complete partially ordered group $(R, +, \leq)$ which is a lattice and satisfies implication (2.1) is an l-group according to Definition 2.3.

From now on, let R be an l-group and $\Phi = \mathbb{N}^{\mathbb{N}}$ be the set of all mappings, defined on $\mathbb{N}$ and taking values in $\mathbb{N}$.

We now introduce two kinds of convergence in l-groups. First of all, we begin with the following preliminary definitions.

Definition 2.12 A bounded double sequence $(a_{i,j})_{i,j}$ in R is called a (D)-*sequence* or *regulator* if

$$a_{i,j} \geq a_{i,j+1} \quad \forall i, j \in \mathbb{N} \quad \text{and} \quad \bigwedge_{j=1}^{\infty} a_{i,j} = 0 \, \forall i \in \mathbb{N}.$$

Definition 2.13 A sequence $(p_n)_n$ is called an (O)-*sequence* if $p_n \geq p_{n+1} \, \forall n \in \mathbb{N}$ and $\bigwedge_{n=1}^{\infty} p_n = 0$. In this case, we write also $p_n \downarrow 0$.

Definition 2.14 Given a sequence $(r_n)_n$ in R, we say that $(r_n)_n$ (D)-*converges* to an element $r \in R$ if there exists a (D)-sequence $(a_{i,j})_{i,j}$ in R, satisfying the following condition:

$$\forall \varphi \in \Phi, \text{ there exists an integer } n_0 \text{ such that}$$

$$|r_n - r| \leq \bigvee_{i=1}^{\infty} a_{i,\varphi(i)},$$

for all $n \geq n_0$. In this case, we write $(D) \lim_n r_n = r$.

Definition 2.15 Given a sequence $(r_n)_n$ in R, we say that $(r_n)_n$ (O)-*converges* to an element $r \in R$ if there exists an (O)-sequence $(p_n)_n$ in R, such that

$$|r_n - r| \leq p_n \qquad \forall n \in \mathbb{N}.$$

In this case, we write $(O) \lim_n r_n = r$.

Definition 2.16 Let R be a Dedekind complete Riesz space, and $(a_n)_n$ be a sequence in R. We call *series associated with* $(a_n)_n$ the sequence $(S_n)_n$, defined by setting

$$\begin{cases} S_1 = a_1, \\ S_n = S_{n-1} + a_n, \quad n \geq 2, \end{cases}$$

and we denote this series by the symbol $\sum\limits_{n=1}^{\infty} a_n$. We say that the series $\sum\limits_{n=1}^{\infty} a_n$ converges to $L \in R$ if $L = (O)\lim_n S_n$.

We now recall the very famous Fremlin Lemma (see also [130,227]), whose proof will be given in next chapter in the case R is a Riesz space.

Theorem 2.17 *Let R be a Dedekind complete l-group, $(a_{i,j}^{(n)})_{i,j}$, $n \in \mathbb{N}$, be a sequence of regulators in R. Then to every $a \in R$, $a \geq 0$, there exists a regulator $(a_{i,j})_{i,j}$ such that*

$$a \wedge \left(\sum_{n=1}^{k} \left(\bigvee_{i=1}^{\infty} a_{i,\varphi(i+n)}^{(n)} \right) \right) \leq \bigvee_{i=1}^{\infty} a_{i,\varphi(i)}$$

for any $k \in \mathbb{N}$ and $\varphi \in \Phi$.

Definition 2.18 Let $u \in R$, $u \geq 0$. We say that u has the *Egorov property* if, for every (D)-sequence $(a_{i,j})_{i,j}$ bounded from above by u, there exist an (O)-sequence $(b_m)_m$ and a sequence $(\varphi_m)_m$ of elements of Φ, such that, $\forall m \in \mathbb{N}$,

$$\bigvee_{i=1}^{\infty} a_{i,\varphi_m(i)} \leq b_m.$$

We say that R has the *Egorov property* if each positive element $u \in R$ has the Egorov property.

Proposition 2.19 (See also [175], Theorem 67.7 (i) and (ii), p. 467) *An element $u \in R$, $u \geq 0$, has the Egorov property if and only if, for each (D)-sequence $(a_{i,j})_{i,j}$ bounded from above by u, there exist an (O)-sequence $(b_n)_n$ and an element $\varphi \in \Phi$ such that, for every $n \in \mathbb{N}$,*

$$a_{n,\varphi(n)} \leq b_n.$$

Proof: The proof of the "only if" part is straightforward. We report the proof of the "if" part, for the sake of clearness.

Let $(a_{i,j})_{i,j}$ be a (D)-sequence bounded from above by u. For every $i, j \in \mathbb{N}$, set

$$\beta_{i,j} = a_{1,j} \vee a_{2,j} \vee \ldots \vee a_{i,j}.$$

It is easy to check that $(\beta_{i,j})_{i,j}$ is a (D)-sequence bounded from above by u too. Thus, by hypothesis, there exist an (O)-sequence $(b_n)_n$ and an element $\varphi \in \Phi$, such that, for every $n \in \mathbb{N}$,

$$\beta_{n,\varphi(n)} \leq b_n.$$

Fix arbitrarily $n \in \mathbb{N}$. For $m \geq n$ we have

$$a_{n,\varphi(m)} \leq \beta_{n,\varphi(m)} \leq \beta_{m,\varphi(m)} \leq b_m,$$

and for $m \leq n$ we get:

$$a_{n,\varphi(n)} \leq \beta_{n,\varphi(n)} \leq b_n \leq b_m.$$

For every $m, n \in \mathbb{N}$, set $\varphi_m(n) = \varphi(\max\{m, n\})$. We obtain:

$$\bigvee_{n=1}^{\infty} a_{n, \varphi_m(n)} \leq b_m,$$

that is the assertion.

We observe that, if $(X, \mathcal{B}, \mu)$ is a measure space, with μ σ-additive and σ-finite, then the space $L^0 = L^0(X, \mathcal{B}, \mu)$ of all measurable real-valued functions, with identification up to the complement of sets of measure μ zero, has the Egorov property (see also [175], p. 459).

2.2 Weak σ-distributivity

Definition 2.20 An l-group R is said to be *weakly σ-distributive* if for every bounded double sequence $(b_{i,j})_{i,j}$ with $b_{i,j} \geq b_{i,j+1}$ $\forall i, j$ one has:

$$\bigvee_{i=1}^{\infty}\left(\bigwedge_{j=1}^{\infty} b_{i,j}\right) = \bigwedge_{\varphi \in \Phi}\left(\bigvee_{i=1}^{\infty} b_{i,\varphi(i)}\right) \qquad (2.2)$$

(see [278]). The following characterization holds (see also [160], Proposition 2.1, pp. 156-157).

Proposition 2.21 *A Dedekind complete l-group R is weakly σ-distributive if and only if for every (D)-sequence $(a_{i,j})_{i,j}$ in R one has:*

$$\bigwedge_{\varphi \in \Phi}\left(\bigvee_{i=1}^{\infty} a_{i,\varphi(i)}\right) = 0. \qquad (2.3)$$

Proof: We first prove the "only if" part. Let $(a_{i,j})_{i,j}$ be a (D)-sequence in R. Then we have:

$$\bigwedge_{j=1}^{\infty} a_{i,j} = 0 \qquad \forall i \in \mathbb{N},$$

and hence

$$\bigvee_{i=1}^{\infty}\left(\bigwedge_{j=1}^{\infty} a_{i,j}\right) = 0.$$

Thus, by (2.2), we have:

$$\bigwedge_{\varphi \in \Phi}\left(\bigvee_{i=1}^{\infty} a_{i,\varphi(i)}\right) = 0, \qquad (2.4)$$

that is (2.3).

We now prove the "if" part. Fix arbitrarily a bounded double sequence $(b_{i,j})_{i,j}$, and put

$$l = \bigvee_{i=1}^{\infty}\left(\bigwedge_{j=1}^{\infty} b_{i,j}\right), v = \bigwedge_{\varphi \in \Phi}\left(\bigvee_{i=1}^{\infty} b_{i,\varphi(i)}\right);$$

moreover, for every $i, j \in \mathbb{N}$, set

$$c_{i,j} = (b_{i,j} \vee l) \wedge v.$$

Since R is infinitely distributive, we get

$$l = \bigvee_{i=1}^{\infty} \left(\bigwedge_{j=1}^{\infty} c_{i,j} \right), v = \bigwedge_{\varphi \in \Phi} \left(\bigvee_{i=1}^{\infty} c_{i,\varphi(i)} \right).$$

It is easy to check that the double sequence $(c_{i,j})_{i,j}$ is bounded, and that

$$c_{i,j} \geq c_{i,j+1} \qquad \forall i,j \in \mathbb{N}.$$

Note that

$$\bigwedge_{j=1}^{\infty} c_{i,j} = l \qquad \forall i \in \mathbb{N}.$$

Set now

$$a_{i,j} = c_{i,j} - l \; \forall i,j \in \mathbb{N},$$

then

$$a_{i,j} \geq a_{i,j+1} \qquad \forall i,j \in \mathbb{N}$$

and

$$\bigwedge_{j=1}^{\infty} a_{i,j} = 0 \qquad \forall i \in \mathbb{N}.$$

Thus, $(a_{i,j})_{i,j}$ is a (D)-sequence. From this, by hypothesis and Proposition 2.4 again, we get:

$$l = \bigwedge_{\varphi \in \Phi} \left(\bigvee_{i=1}^{\infty} (a_{i,\varphi(i)} + l) \right) = \left[\bigwedge_{\varphi \in \Phi} \left(\bigvee_{i=1}^{\infty} c_{i,\varphi(i)} \right) \right] = v,$$

that is the assertion.

We recall that, in the context of Riesz spaces, which we will introduce in the next paragraph, weak σ-distributivity of R is equivalent to the property that, for every nonempty set X and for each algebra $\mathcal{F}$, every order σ-additive measure m, defined on $\mathcal{F}$, has an order σ-additive extension to the σ-algebra $\mathcal{A}(\mathcal{F})$ (see [278]). We note that, even if R is not weakly σ-distributive, it may happen that there exists still some algebra Σ such that every order σ-additive measure m, defined on Σ, has an order σ-additive extension to the σ-algebra $\mathcal{A}(\Sigma)$: for example, this is the case of the algebra of all clopen sets of a compact totally disconnected topological space and the Stone extension, which exists for *every* Dedekind complete l-group R (see [16]).

The following fundamental relation between (O)- and (D)-convergence in l-groups, related to weak σ-distributivity, holds (see [99], Proposition 1, pp. 134-135):

Proposition 2.22 If $(O)\lim_n r_n = r$, then $(D)\lim_n r_n = r$. Moreover, if R is weakly σ-distributive, then

$$[(D)\lim_n r_n = r] \Rightarrow [(O)\lim_n r_n = r].$$

Proof: We begin with the first part. By hypothesis, there exists an (O)-sequence $(a_n)_n$ such that

$$|r_n - r| \le a_n \qquad \forall n \in \mathbb{N}.$$

For every $j \in \mathbb{N}$, put

$$a_{i,j} = a_j \qquad \forall i \in \mathbb{N}.$$

Let $\varphi \in \Phi$. If we put $n_0 = \min\{\varphi(n) : n \in \mathbb{N}\}$, then for every $n \ge n_0$ we have:

$$|r_n - r| \le a_n \le a_{n_0} \le \bigvee_{n=1}^{\infty} a_{\varphi(n)} = \bigvee_{i=1}^{\infty} a_{i,\varphi(i)}.$$

Hence, $(D)\lim_n r_n = r$.

We now turn to the second part. Suppose that R is a weakly σ-distributive l-group and $(D)\lim_n r_n = r$. Put

$$a_n = \bigvee_{i=n}^{\infty} |r_i - r| \qquad \forall n \in \mathbb{N}.$$

Then $a_n \ge a_{n+1} \ge 0$ and $|r_n - r| \le a_n \; \forall n \in \mathbb{N}$. By definition, for every $\varphi \in \Phi$ there exists $n_0 \in \mathbb{N}$ such that

$$|r_n - r| \le \bigvee_{i=1}^{\infty} a_{i,\varphi(i)}$$

for every $n \ge n_0$. It follows

$$a_{n_0} = \bigvee_{n=n_0}^{\infty} |r_n - r| \le \bigvee_{i=1}^{\infty} a_{i,\varphi(i)}.$$

Moreover we have:

$$\bigwedge_{n=1}^{\infty} a_n \le a_{n_0} \le \bigvee_{i=1}^{\infty} a_{i,\varphi(i)}.$$

Since

$$\bigwedge_{n=1}^{\infty} a_n \le \bigvee_{i=1}^{\infty} a_{i,\varphi(i)} \qquad \forall \varphi \in \Phi$$

and R is weakly σ-distributive, then

$$0 \leq \bigwedge_{n=1}^{\infty} a_n \leq \bigwedge_{\varphi \in \Phi} \left(\bigvee_{i=1}^{\infty} a_{i,\varphi(i)} \right) = 0.$$

Thus, the sequence $(a_n)_n$ is an (O)-sequence and the assertion follows.

The (D)-sequences $(a_{i,j})_{i,j}$ play the role as $\varepsilon > 0$ in the real-valued case (see also [227], p. 41 and Chapter 3). Indeed, if $(a_{i,j})_{i,j}$ is a (D)-sequence in R, then for every $\varepsilon > 0$ and $i \in \mathbb{N}$ there exists $\varphi(i)$ such that $a_{i,\varphi(i)} \leq \varepsilon$. Thus

$$\bigvee_{i=1}^{\infty} a_{i,\varphi(i)} \leq \varepsilon.$$

We now investigate some properties of (D)-convergence both when the (D)-limit is unique and when there exist "several" (D)-limits (the (D)-limit is *always* unique if and only if the involved l-group R is weakly σ-distributive).

Let R be a Dedekind complete l-group, R not weakly σ-distributive (such groups do exist, see also [278]). There exist a (D)-sequence $(a_{i,j})_{i,j}$ and a positive element $b \in R$, for which

$$\bigwedge_{\varphi \in \Phi} \left(\bigvee_{i=1}^{\infty} a_{i,\varphi(i)} \right) = b \neq 0.$$

Now we can see that the null sequence $(0)(D)$-converges to b.

We shall characterize the set of all (D)-limits of sequences $(r_n)_n$, after the next propositions (see also [18]).

Proposition 2.23   *If $(r_n)_n$ (D)-converges to r and $(s_n)_n$ (D)-converges to s, then $(r_n + s_n)_n$ $[(r_n - s_n)_n]$ (D)-converges to $r + s$ $[r - s]$.*

Moreover, if

$$A_1 := \{r : r = (D)\lim_n r_n\}, \quad A_2 := \{s : s = (D)\lim_n s_n\}, \quad A_3 := \{t : t = (D)\lim_n (r_n + s_n)\},$$

then $A_1 + A_2 = A_3$.

Proof: Let $(a_{i,j})_{i,j}$ [resp. $(b_{i,j})_{i,j}$] be a (D)-sequence; moreover, $\forall \varphi \in \Phi$, let $n_0(\varphi)[n_1(\varphi)] \in \mathbb{N}$, satisfying the conditions of (D)-convergence, relatively to the sequence $(r_n)_n$ [resp. $(s_n)_n$]. Put $c_{i,j} := 2(a_{i,j} + b_{i,j})$ for all $i, j \in \mathbb{N}$. It is easy to check that $(c_{i,j})_{i,j}$ is a (D)-sequence. Let $\varphi \in \Phi$, and set $n_2(\varphi) := \max\{n_0(\varphi), n_1(\varphi)\}$. For each $n \geq n_2$, it holds:

$$|r_n + s_n - (r+s)| [|r_n - s_n - (r-s)|] \leq |r_n - r| + |s_n - s| \leq$$

$$\leq \bigvee_{i=1}^{\infty} a_{i,\varphi(i)} + \bigvee_{i=1}^{\infty} b_{i,\varphi(i)} \leq \bigvee_{i=1}^{\infty} (a_{i,\varphi(i)} + b_{i,\varphi(i)}) + \bigvee_{i=1}^{\infty} (a_{i,\varphi(i)} + b_{i,\varphi(i)}) =$$

$$= 2 \bigvee_{i=1}^{\infty} (a_{i,\varphi(i)} + b_{i,\varphi(i)}) = \bigvee_{i=1}^{\infty} 2(a_{i,\varphi(i)} + b_{i,\varphi(i)}) = \bigvee_{i=1}^{\infty} c_{i,\varphi(i)}.$$

Now, let $r \in A_1, s \in A_2$. We have:

$$r + s = (D) \lim_n (r_n + s_n),$$

and so

$$A_1 + A_2 \subset A_3.$$

Conversely, choose $t \in A_3$. By hypothesis, there exists at least an element $\bar{r} \in A_1$. But $t = \bar{r} + (t - \bar{r})$, and $t - \bar{r}$ is a (D)-limit of the sequence $((r_n + s_n) - r_n)_n = (s_n)_n$. Thus the assertion follows.

Analogously as above, it is easy to prove the following properties:

Proposition 2.24 *If* $(r_n)_n$ (D)-*converges to* r *and* $(s_n)_n$ (D)-*converges to* s, *then* $(r_n \vee s_n)_n$, $(r_n \wedge s_n)_n$, $(|r_n|)_n$ (D)-*converge to* $r \vee s$, $r \wedge s$ *and* $|r|$ *respectively.*

Now, we give a characterization of all (D)-limits of a sequence $(r_n)_n$.

Proposition 2.25 *Let* B *be the set of all* (D)-*limits of the null sequence* (0). *Then* B *is a subgroup of* R *such that, whenever* $a, b \in B$ *and* $c \in R$, *with* $a \leq c \leq b$, *then* $c \in B$.

Moreover, given a (D)-*convergent sequence* $(r_n)_n$, *the set of all* (D)-*limits of* $(r_n)_n$ *is* $r + B$, *where* r *is any fixed* (D)-*limit of* $(r_n)_n$. *Finally,* $B = \{0\}$ *if and only if* R *is weakly* σ-*distributive.*

The first property is an easy consequence of 2.23 and 2.24. Let now $(r_n)_n$ be any sequence, (D)-convergent to r. If $b \in B$, then $(D) \lim_n (0) = b$. Hence, $(D) \lim_n (r_n + 0) = r + b$, that is, $r + b$ is one of the (D)-limits of $(r_n)_n$. Moreover, if $(D) \lim_n r_n = r$ and $(D) \lim_n r_n = s$, then $(D) \lim_n (r_n - r_n) = r - s$, that is $r - s \in B$.

Finally, if $B = \{0\}$, there exist no positive non-trivial elements $b \in R$, such that

$$b = \bigwedge_{\varphi \in \Phi} \left(\bigvee_{i=1}^{\infty} a_{i, \varphi(i)} \right)$$

for some (D)-sequence $(a_{i,j})_{i,j}$ (such an element would be a (D)-limit of (0)). Conversely, if R is weakly σ-distributive, then $\bigwedge_{\varphi \in \Phi} \left(\bigvee_{i=1}^{\infty} a_{i, \varphi(i)} \right)$ is always equal to 0; hence $B = \{0\}$, and the (D)-limit is unique.

We now investigate some other properties of (D)-convergence.

Definition 2.26 Let R be an l-group. A sequence $(r_n)_n$ in R is said to be (D)-*Cauchy* if there exists a (D)-sequence $(a_{i,j})_{i,j}$ in R, such that, $\forall \varphi \in \Phi, \exists n_0 \in N$ such that

$$|r_n - r_m| \leq \bigvee_{i=1}^{\infty} a_{i, \varphi(i)}, \forall n, m \in \mathbb{N}, \ n, m \geq n_0.$$

Definition 2.27 Let Λ be any nonempty set, $(r_n^\lambda)_n$ a family of sequences in R, $r^\lambda, \lambda \in \Lambda$, be a family of elements of R. We say that the family $(r_n^\lambda)_n$ converges to r^λ *uniformly with respect to* (the parameter) $\lambda \in \Lambda$, or briefly, $\Lambda-(D)$- *converges to* r^λ, and we write $\Lambda-(D)\lim_n r_n^\lambda = r^\lambda$ [resp. *is Cauchy uniformly with respect to* $\lambda \in \Lambda$, *or is* $\Lambda-(D)$- *Cauchy*], if there exists a (D)-sequence $(a_{i,j})_{i,j}$ in R, such that, $\forall\, \varphi \in \Phi, \exists\, n_0(\varphi) \in \mathbb{N}$:

$$|r_n^\lambda - r^\lambda| \leq \bigvee_{i=1}^{\infty} a_{i,\varphi(i)}, \quad \forall\, n \geq n_0, \forall\, \lambda \in \Lambda$$

[resp.

$$|r_n^\lambda - r_m^\lambda| \leq \bigvee_{i=1}^{\infty} a_{i,\varphi(i)}, \quad \forall\, n,m \geq n_0, \forall\, \lambda \in \Lambda].$$

Analogously as above, we can prove the following:

Proposition 2.28   Let Λ be any nonempty set. If $(r_n)_n^\lambda$ $\Lambda-(D)$-converges to r^λ, and $(s_n)_n^\lambda$ $\Lambda-(D)$-converges to s^λ, then $(r_n^\lambda + s_n^\lambda)_n$ $[(r_n^\lambda - s_n^\lambda)_n, (|r_n^\lambda|)_n]$ $\Lambda-(D)$-converges to $r^\lambda + s^\lambda$ $[r^\lambda - s^\lambda, |r^\lambda|]$. Moreover, if

$$A := \{r^\lambda : r^\lambda = \Lambda-(D)\lim_n r_n^\lambda\}, B := \{s^\lambda : s^\lambda = \Lambda-(D)\lim_n s_n^\lambda\},$$
$$C := \{t^\lambda : t^\lambda = \Lambda-(D)\lim_n (r_n^\lambda + s_n^\lambda)\},$$

then $A + B = C$. Moreover, for each $\Lambda-(D)$-convergent sequence $(r_n)_n^\lambda$ and for every fixed $\Lambda-(D)$-limit r^λ of $(r_n^\lambda)_n$, the set of all $\Lambda-(D)$-limits of $(r_n^\lambda)_n$ is of the type

$$r^\lambda + B^*,$$

where B^* is the set of all $\Lambda-(D)$-limits of the null sequence (0).

Definition 2.29   An l-group R is called (D)-*complete* (or complete with respect to (D)-convergence) if every (D)-Cauchy sequence is (D)-convergent.

Now, we prove that every Dedekind complete l-group is (D)-complete; more precisely, we prove a more general theorem.

Definition 2.30   Let $\Lambda \neq \varnothing$ be any set. An l-group R is said to be $\Lambda-(D)$-*complete* if every family of sequences, which is (D)-Cauchy uniformly with respect to $\lambda \in \Lambda$, is (D)-convergent uniformly with respect to $\lambda \in \Lambda$.

Theorem 2.31   Let Λ be an arbitrary nonempty set, and R be a Dedekind complete l-group. Then, R is $\Lambda-(D)$-complete.

Proof: (see also [18])   By hypothesis, there exists a (D)-sequence $(a_{i,j})_{i,j}$, such that, $\forall\, \varphi \in \Phi, \exists\, s(\varphi) \in \mathbb{N}$:

$$|r_n^\lambda - r_m^\lambda| \leq \bigvee_{i=1}^{\infty} a_{i,\varphi(i)} \quad \forall\, n,m \geq s, \forall\, \lambda \in \Lambda.$$

Put $K := \{s \in \mathbb{N} : \exists \varphi = \varphi_s \in \Phi : s = s(\varphi)\}$. Fix $s \in K$ and $\lambda \in \Lambda$. Then, $\forall n \geq s$,

$$| r_n^\lambda | \leq | r_n^\lambda - r_s^\lambda | + | r_s^\lambda | \leq \bigvee_{i=1}^\infty a_{i,\varphi_s(i)} + | r_s^\lambda | \leq \bigvee_{i,j=1}^\infty a_{i,j} + | r_s^\lambda |.$$

For each $s \in \mathbb{N}$, $\forall \lambda \in \Lambda$, set $B_s := \{n \in \mathbb{N} : n \geq s\}$, $D_s^\lambda := \{r_n^\lambda : n \in B_s\}$. We have that $\forall s \in K$, $\forall \lambda \in \Lambda$, D_s^λ is bounded, and so there exist in R the quantities

$$a_s^\lambda := \inf D_s^\lambda, \; b_s^\lambda := \sup D_s^\lambda.$$

Fix arbitrarily s_1 and $s_2 \in K$, and $\lambda \in \Lambda$. Let $s := \max\{s_1, s_2\}$. Then $B_s \subset B_{s_1} \cap B_{s_2}$, and hence D_s^λ is bounded. So, we have:

$$a_{s_1}^\lambda \leq a_s^\lambda \leq b_s^\lambda \leq b_{s_2}^\lambda.$$

Therefore, there exists $r^\lambda \in R$, such that $\sup_{s_1 \in K} a_{s_1}^\lambda \leq r^\lambda \leq \inf_{s_2 \in K} b_{s_2}^\lambda$.

Choose arbitrarily $\varphi \in \Phi$. Then, there exists $s = s(\varphi) \in \mathbb{N}$, such that

$$r_n^\lambda \leq r_m^\lambda + \bigvee_{i=1}^\infty a_{i,\varphi(i)}, \quad \forall m, n \geq s. \tag{2.5}$$

By fixing m and taking the supremum with respect to n, we get:

$$b_{s(\varphi)}^\lambda \leq r_m^\lambda + \bigvee_{i=1}^\infty a_{i,\varphi(i)};$$

taking the infimum w. r. to m, we obtain

$$b_{s(\varphi)}^\lambda - a_{s(\varphi)}^\lambda \leq \bigvee_{i=1}^\infty a_{i,\varphi(i)}.$$

Thus, thanks to weak σ-distributivity of R,

$$0 \leq \inf_{s_2 \in K} b_{s_2}^\lambda - \sup_{s_1 \in K} a_{s_1}^\lambda = \inf_{\varphi \in \Phi} b_{s(\varphi)}^\lambda - \sup_{\varphi \in \Phi} a_{s(\varphi)}^\lambda \leq \bigwedge_{\varphi \in \Phi} \left(\bigvee_{i=1}^\infty a_{i,\varphi(i)} \right) = 0,$$

and hence

$$\inf_{\varphi \in \Phi} b_{s(\varphi)}^\lambda = \sup_{\varphi \in \Phi} a_{s(\varphi)}^\lambda = x^\lambda.$$

So, if $s = s(\varphi)$ is as in (24), then for all $n \geq s(\varphi)$ and $\lambda \in \Lambda$ we get

$$r^\lambda - r_n^\lambda \leq b_{s(\varphi)}^\lambda - r_n^\lambda \leq \bigvee_{i=1}^\infty a_{i,\varphi(i)};$$

similarly,

$$r_n^\lambda - r^\lambda \leq r_n^\lambda - a_{s(\varphi)}^\lambda \leq \bigvee_{i=1}^\infty a_{i,\varphi(i)}.$$

An immediate consequence of Theorem 2.31 is the following:

Corollary 2.32 *Every Dedekind complete l-group is (D)-complete.*

We now will give a characterization of weakly σ-distributive l-groups in terms of Egorov property (see also [175], pp. 458, 467 and [21]). More precisely, we prove the following:

Theorem 2.33 *If R is super Dedekind complete and weakly σ-distributive, then R has the Egorov property. Conversely, if R is Dedekind complete and has the Egorov property, then R is weakly σ-distributive.*

Proof: We begin with the first part. Let $(a_{i,j})_{i,j}$ be a (D)-sequence in R. By weak σ-distributivity of R, we have:

$$0 = \inf \left\{ \bigvee_{i=1}^{\infty} a_{i,\varphi(i)} : \varphi \in \Phi \right\}.$$

By super-Dedekind completeness of R, there exists a sequence $(\varphi_n)_n$ in Φ, such that

$$0 = \inf \left\{ \bigvee_{i=1}^{\infty} a_{i,\varphi_n(i)} : n \in \mathbb{N} \right\}.$$

Without loss of generality, we may assume $\varphi_n \leq \varphi_{n+1}$ for each $n \in \mathbb{N}$, and hence, by setting

$$b_n := \bigvee_{i=1}^{\infty} a_{i,\varphi_n(i)}$$

for every $n \in \mathbb{N}$, we get that $(b_n)_n$ is an (O)-sequence. Put now $\varphi(n) = \varphi_n(n)$ for each n. We have:

$$a_{n,\varphi(n)} \leq \bigvee_{i=1}^{\infty} a_{i,\varphi_n(i)} = b_n,$$

for all n, and thus the first part of the proposition is proved, by virtue of 2.19.

We now turn to the second part. Fix any (D)-sequence $(a_{i,j})_{i,j}$ in R. By hypothesis, we know that there exist an (O)-sequence $(b_m)_m$ and a sequence $(\varphi_m)_m$ of elements of Φ such that

$$\bigvee_{i=1}^{\infty} a_{i,\varphi_m(i)} \leq b_m \qquad \forall m \in \mathbb{N}. \qquad (2,6)$$

From (2.6) we get:

$$0 \leq \bigwedge_{\varphi \in \Phi} \left(\bigvee_{i=1}^{\infty} a_{i,\varphi(i)} \right) \leq \bigvee_{i=1}^{\infty} a_{i,\varphi_m(i)} \leq b_m \qquad \forall m \in \mathbb{N}. \qquad (2.7)$$

The assertion follows from (2.7) and arbitrariness of m; indeed, it is enough to observe that $(b_m)_m$ is an (O)-sequence.

Remark 2.34 In [175], pp. 466, 503, it is proved that, if the continuum hypothesis holds, then every Dedekind complete Riesz space with the Egorov property is super Dedekind complete. Moreover, the space $\mathbb{R}^{\mathbb{R}}$ of all real-valued functions defined on $\mathbb{R}$, endowed with pointwise ordering and convergence, does not have the Egorov property (see [175], Example 67.6 (v), pp. 465-466 and Theorem 75.1 (i) and (iii), p. 499). However, in $\mathbb{R}^{\mathbb{R}}$, the topology corresponding to pointwise convergence is locally convex, Hausdorff and σ-compatible (this means that any monotone increasing sequence $(b_n)_n$ with least upper bound b converges to b with respect to this topology). Thus, thanks to Corollary J of [278], we deduce that $\mathbb{R}^{\mathbb{R}}$ is weakly σ-distributive, though it is not super Dedekind complete.

Remark 2.35 Since $L^0(X, \mathcal{B}, \mu)$, where $(X, \mathcal{B}, \mu)$ is as above, is Dedekind complete and has the Egorov property, then it is weakly σ-distributive, thanks to Theorem 2.33.

2.3 Riesz Spaces

Definition 2.36 A *Riesz space* or *vector lattice* $R = (R, +, \cdot, \leq)$ is an l-group in which an application $\cdot : \mathbb{R} \times R \to R$ is defined, such that R is a real linear space and $\alpha x \leq \alpha y$ whenever $x, y \in R$, $x \leq y$ and $\alpha \in \mathbb{R}$, $\alpha \geq 0$.

Definition 2.37 A nonempty subset K of a Riesz space R is said to be *order dense* in R if, for each $r \in R$ such that $r \geq 0$ and $r \neq 0$, there exists $k \in K$ such that $0 \leq k \leq r$ and $k \neq 0$. We say that $\emptyset \neq K \subset R$ is *solid* in R if, whenever a, b, c are three elements of R such that $a, c \in K$ and $a \leq b \leq c$, then $b \in K$. A subset S of a Riesz space R is said to be an *ideal* (of R) if it is a Riesz space and a solid subset of R.

Definition 2.38 Let R be a Riesz space. A function $f : R \to \mathbb{R}$ is called a *linear functional* if

$$f(\alpha r + \beta s) = \alpha f(r) + \beta f(s) \qquad \forall \alpha, \beta \in \mathbb{R}, \quad \forall r, s \in R.$$

We say that $f : R \to \mathbb{R}$ is *increasing* [*decreasing*] if $f(r) \leq f(s)$ [$f(r) \geq f(s)$] whenever $r, s \in R$, $r \leq s$. We say that $f : R \to \mathbb{R}$ is *positive* if $f(r) \geq 0$ whenever $r \in R$, $r \geq 0$. It is readily seen that a linear functional f is positive if and only if it is increasing.

Definition 2.39 A linear functional $f : R \to \mathbb{R}$ is said to be *order bounded* if it can be expressed as the difference of two increasing linear functionals.

The following result holds (see also [289], pp. 99-101).

Proposition 2.40 *A linear functional is order bounded if and only if it maps bounded sets into bounded sets.*

We shall denote by R^+ the set of all order bounded linear functionals $f : R \to \mathbb{R}$. We observe that R^+, endowed with the "natural" sum and product for a scalar number, and with the following order:

$$f_1 \leq f_2 \iff f_1(r) \leq f_2(r) \qquad \forall r \in R, r \geq 0,$$

is a Dedekind complete Riesz space (see also [129]).

Definition 2.41 Let R be a Riesz space. An increasing function $f : R \to \mathbb{R}$ is said to be *order continuous* if $\sup_{b \in B} f(b) = f(a)$ whenever $B \subset R$ is a directed upwards set, such that $\sup B = a$, and $\inf_{b \in B} f(b) = f(a)$ whenever $B \subset R$ is a directed downwards set, such that $\inf B = a$. A linear functional $f : R \to \mathbb{R}$ is said to be *order continuous* if it can be expressed as the difference of two linear increasing order continuous functions.

We denote by $R^\times$ the set of all order continuous linear functionals $f : R \to \mathbb{R}$. We observe that even $R^\times$, endowed with the "natural" sum and product for a scalar number, and the order

$$f_1 \leq f_2 \iff f_1(r) \leq f_2(r) \qquad \forall r \in R, r \geq 0,$$

is a Dedekind complete Riesz space (see also [129]).

Let now $c : R \to R^{++}$ $[R^{\times\times}]$ the *evaluation map* or *canonical embedding,* defined by setting $c(r)(y) = y(r)$, $\forall r \in R, \forall y \in R^+$ $[R^\times]$. We say that a Riesz space R is a $+$-*space* [π-*space*] if for every $y \in R$, with $y \neq 0$, there exists a functional $f \in R^+[R^\times]$ such that $f(y) \neq 0$, that is if the evaluation map $c: R \to R^{++}[R^{\times\times}]$ is one-to-one. A Riesz space is said to be *perfect* is the evaluation map $c: R \to R^{\times\times}$ is bijective.

The following results hold (see [4, 121, 129, 174, 278]):

i)   If R is *any* Riesz space, then R^+ and $R^\times$ are Dedekind complete perfect Riesz spaces.

ii)   Every perfect Riesz space is a Dedekind complete π-space.

iii)   There exist Dedekind complete π-spaces which are not perfect (for example the space c_0 of all real-valued sequences $(x_n)_n$, such that $\lim_n x_n = 0$).

iv)   There are some π-spaces, which are not Dedekind complete, but every π-space is Archimedean.

v)   There exist Dedekind complete Riesz spaces, which are not π-spaces; there exist even Dedekind complete Riesz spaces R such that $R^+ = \{0\}$.

vi)   If Ω is any arbitrary nonempty set, then the space $\mathbb{R}^\Omega$ of all real-valued functions defined in Ω is a Dedekind complete π-space.

vii)   Every π-space is Archimedean, each ideal of a super Dedekind complete Riesz space is super Dedekind complete, and every ideal of any π-space is a π-space.

viii)   Every π-space is weakly σ-distributive.

Let now $\mathcal{F}$ be the class of all order dense ideals of R, and set $\Psi = \bigcup_{I \in \mathcal{F}} I^\times$. Then, a function $\psi \in \mathbb{R}^R$ belongs to Ψ if there exists an order dense ideal I of R such that ψ is an order continuous linear functional on I.

On Ψ, we define the following relation:

$\psi_1 \sim \psi_2$ when the set $\{r \in R : \psi_1(r) = \psi_2(r)\}$ contains an order dense ideal of R (see also [174]). As the intersection of a finite number of order dense ideals is still an order dense ideal, then $\sim$ is an equivalence relation.

Now, put $R^\rho := \Psi / \sim$. We define a sum, a product and an order relation on R^ρ in the following way:

$[\psi_1] + [\psi_2] = [\psi_3]$ when there exist $\psi'_1 \in [\psi_1]$, $\psi'_2 \in [\psi_2]$, $\psi'_3 \in [\psi_3]$, such that the set $\{r : \psi'_1(r) + \psi'_2(r) = \psi'_3(r)\}$ contains an order dense ideal (it is easy to check that this definition does not depend on $\psi'_1, \psi'_2, \psi'_3$, and that $[\psi_1 + \psi_2] = [\psi_1] + [\psi_2]$).

Analogously we define $\alpha \cdot [\psi]$, for every $[\psi] \in R^\rho$ and for all $\alpha \in \mathbb{R}$.

Furthermore, we will say that $[\psi] \geq 0$ if there exists $\psi' \in [\psi]$ such that the set $\{r \in R : \psi'(r) \geq 0\}$ contains an order dense ideal of R, and $[\psi_1] \leq [\psi_2]$ if $[\psi_2] - [\psi_1] \geq 0$.

Thus, R^ρ is a Dedekind complete Riesz space (see [174]).

Given an Archimedean Riesz space R, we define the evaluation map $c: R \to R^{\rho\rho}$ in the following way (see [174]). For every $r \in R$, $r \geq 0$, set

$$I_r = \{[\psi] \in R^\rho : r \in D_\psi, \psi \in [\psi]\},$$

where, for each ψ such that $[\psi] \in R^\rho$, D_ψ is the greatest order dense ideal of R on which $|\psi|$ can be extended as a real-valued map.

(*) Let $c(r)([\psi]) = \psi(r)$, $\forall r \geq 0$ and for all ψ such that $[\psi] \in I_r$, and put $c(r) = c(r^+) - c(r^-)$, $\forall r \in R$.

It is easy to prove that I_r is an order dense ideal of R^ρ and that c is well-defined.

Definition 2.42 An Archimedean Riesz space R is called a ρ-*space* if the evaluation map c defined in (*) is one-to-one.

We note that every π-space is a ρ-space, but there exist Dedekind complete ρ-spaces which are not π-spaces (see also [121]).

Now, we will introduce the very famous Maeda-Ogasawara-Vulikh representation theorem for Archimedean Riesz spaces (see [121]). We recall that the Maeda-Ogasawara-Vulikh theorem was generalized by Bernau (see [11]) also to the case of l-groups.

We begin with the following notations (see [121]). Let $\tilde{\mathbb{R}}$ be the set of all extended real numbers. For a topological Hausdorff space Ω, we set:

$\mathcal{C}(\Omega) = \{f : \Omega \to \mathbb{R}, f \text{ is continuous}\}$,

$\mathcal{C}_b(\Omega) = \{f : \Omega \to \mathbb{R}, f \text{ is continuous and bounded}\}$,

$\mathcal{C}_c(\Omega) = \{f : \Omega \to \mathbb{R}, f \text{ is continuous and of bounded support}\}$,

$\mathcal{C}_\infty(\Omega) = \{f : \Omega \to \tilde{\mathbb{R}}, f \text{ is continuous and } \{\omega \in \Omega : |f(\omega)| \neq +\infty\} \text{ is dense in } \Omega\}$.

We have that $\mathcal{C}(\Omega)$, $\mathcal{C}_b(\Omega)$ and $\mathcal{C}_c(\Omega)$ are Riesz spaces under the natural (that is pointwise) ordering.

The following result holds (see also [121]):

Proposition 2.43 In a locally compact Hausdorff space Ω, the following are equivalent:

a) Ω is Stonian, that is the (topological) closure of every open subset of Ω is still an open subset of Ω;

b) $\mathcal{C}(\Omega)$ is Dedekind complete;

c) $\mathcal{C}_b(\Omega)$ is Dedekind complete.

Remark 2.44 We observe that $\mathcal{C}([0,1])$, where $[0,1]$ is endowed with the usual Euclidean topology, is not Dedekind complete (see also [175]).

The following proposition holds (see [121]):

Proposition 2.45 *Let Ω be a Stonian topological space. Then $C_\infty(\Omega)$ is a universally complete Riesz space, where the addition is defined as follows: for $f, g \in C_\infty(\Omega)$ there exists exactly one $h \in C_\infty(\Omega)$, such that*

$$f(\omega) + g(\omega) = h(\omega) \quad \forall \omega \in \{\omega \in \Omega : |f(\omega)| \neq +\infty\} \cap \{\omega \in \Omega : |g(\omega)| \neq +\infty\};$$

we put $h := f + g$ (Similarly the product of scalar numbers is defined.)

We now turn to the very famous Maeda-Ogasawara-Vulikh representation theorem for Archimedean Riesz spaces.

Theorem 2.46 *Given an Archimedean Riesz space R, there exists a compact Hausdorff Stonian topological space Ω, unique up to homeomorphisms, such that R can be embedded as an order dense subspace of $C_\infty(\Omega)$. Moreover, if $(a_\lambda)_{\lambda \in \Lambda}$ is any net such that $a_\lambda \in R \ \forall \lambda$, and $a = \sup_\lambda a_\lambda \in R$ [where the supremum is taken with respect to R], then $a = \sup_\lambda a_\lambda$ with respect to $C_\infty(\Omega)$, and the set $\{\omega \in \Omega : (\sup_\lambda a_\lambda)(\omega) \neq \sup_\lambda a_\lambda(\omega)\}$ is meager in Ω.*

For further results related to Theorem 2.46, see also [39].

In the context of Riesz spaces, it is possible to define a "product" as a map which "links" three Riesz spaces R_1, R_2 and R_3. More precisely, a *product* is a map $\cdot : R_1 \times R_2 \to R_3$ satisfying the following axioms:

- $(r_1 + s_1) \cdot r_2 = r_1 \cdot r_2 + s_1 \cdot r_2$,

- $r_1 \cdot (r_2 + s_2) = r_1 \cdot r_2 + r_1 \cdot s_2$,

- $[r_1 \geq s_1, r_2 \geq 0] \Rightarrow [r_1 \cdot r_2 \geq s_1 \cdot r_2]$;

  $[r_1 \geq 0, r_2 \geq s_2] \Rightarrow [r_1 \cdot r_2 \geq r_1 \cdot s_2], \forall r_j, s_j \in R_j, j = 1, 2$;

- $[a_n \downarrow 0, b \geq 0] \Rightarrow [b \cdot a_n \downarrow 0], \quad \forall b \in R_1, \forall a_n \in R_2$;

- $[a_n \downarrow 0, b \geq 0] \Rightarrow [a_n \cdot b \downarrow 0], \quad \forall a_n \in R_1, \forall b \in R_2$.

There are several situations in which such products arise naturally: we now give some examples.

i) $R_1 = R_3, R_2 = \mathbb{R}$, and the product is the "product for scalar numbers".

ii) $R_1 = R_2 = R_3 = C_\infty(\Omega)$. The fourth property is fulfilled: indeed, if a_n, $n \in \mathbb{N}$, and b are as above, we have, up to the complement of a meager set:

$$0 = [\inf_n a_n](\omega) = \inf_n [a_n(\omega)] = \inf_n [a_n(\omega) \cdot b(\omega)] = [\inf_n a_n \cdot b](\omega).$$

As the complement of a meager set in Ω is dense in Ω, it follows that

$$\inf_n (a_n \cdot b) = 0.$$

Analogously, one can check the last property.

iii) $R_1 = X$, $R_2 = L(X,Y)$, $R_3 = Y$, where X and Y are two Dedekind complete Riesz spaces, and $L(X,Y)$ is the space of all order continuous linear operators, defined on X and taking values in Y (the definition of order continuous linear operator is analogous to the one of order continuous linear functional, see also [129,289]). Put $x \cdot f := f(x)$, $\forall x \in X$, $\forall f \in R_2$. The fourth property of the product follows from the fact that $[\inf_n f_n](x) = \inf_n [f_n(x)]$, $\forall x \in X$, $x \geq 0$. Moreover, if $f \in L(X,Y)$, $f \geq 0$ and $x_n \in X$, $x_n \downarrow 0$, then $f(x_n) \downarrow 0$, and so the last property of the product holds (for some references about definitions of product in Riesz spaces, see also [271, 272, 273]).

3 Kurzweil - Henstock Integral with Values in Riesz Spaces

Abstract: In this chapter we present the basic properties and results on the Kurzweil-Henstock integral for Riesz space-valued functions, defined on a bounded subinterval of the real line. We prove the uniform convergence theorem, and introduce also the Kurzweil-Stieltjes integral and some of its elementary properties.

3.1 Definition and Elementary Properties

One of the first problem in any integration theory with values in ordered spaces is impossibility to use the so-called ε-technique. Namely, in the real-valued case, if

$$s = \sup A$$

and $\varepsilon > 0$, then $s - \varepsilon < s$, and there exists $a \in A$ such that

$$s - \varepsilon < a.$$

This is not true in partially ordered sets. E.g. let R be the space of all real functions defined on $[0,1]$ with the usual ordering. Put

$$A = \{f_n : n \in \mathbb{N}\},$$

where $f_n : [0,1] \to \mathbb{R}$, $n \in \mathbb{N}$, is defined by setting

$$f_n(x) = 1 - x^n, \quad n \in \mathbb{N}, \quad x \in [0,1].$$

Then

$$\sup A = f,$$

where

$$f(x) = \begin{cases} 0, & \text{if } x = 1, \\ 1, & \text{if } x < 1. \end{cases}$$

If ε is considered as a constant function, then $g = f - \varepsilon$ is the function defined by

$$g(x) = \begin{cases} -\varepsilon, & \text{if } x = 1, \\ 1 - \varepsilon, & \text{if } x < 1, \end{cases}$$

and there exists no element $h \in A$ such that

$$f - \varepsilon = g \leq h.$$

Of course, instead of ε-technique, the double sequence technique proposed by D. H. Fremlin ([130]) can be used. Consider first the classical case of the real line. Let $(a_{i,j})_{i,j}$ be a bounded double sequence of real numbers such that

$$a_{i,j} \downarrow 0 (j \to +\infty, i \in \mathbb{N}),$$

Antonio Boccuto / Beloslav Riečan / Marta Vrábelová

that is a (D)-sequence or regulator in $\mathbb{R}$ (see Chapter 2). Then to any $i \in \mathbb{N}$ and $\varepsilon > 0$ there exists $\varphi(i)$ such that for any $j \geq \varphi(i)$ we have $a_{i,j} \leq \varepsilon$. Since the inequality $a_{i,\varphi(i)} \leq \varepsilon$ holds for any $i \in \mathbb{N}$ we have also

$$\bigvee_{i=1}^{\infty} a_{i,\varphi(i)} \leq \varepsilon.$$

Hence instead of $\varepsilon > 0$ we can use the entity $\bigvee_{i=1}^{\infty} a_{i,\varphi(i)}$.

As an illustration we show the definition of continuity of a function $f:(a,b) \to R$ in a point $x_0 \in (a,b)$. From now on, let Φ be as in the previous chapter. The function f is continuous at x_0 if and only if there exists a regulator $(a_{i,j})_{i,j}$ in R such that to any $\varphi \in \Phi$ there exists $\delta > 0$ such that

$$|f(x) - f(x_0)| \leq \bigvee_{i=1}^{\infty} a_{i,\varphi(i)}$$

whenever $x \in (x_0 - \delta, x_0 + \delta)$.

Since for the definition of the Kurzweil - Henstock integral the $\varepsilon - \delta$ approach is advisable, we shall use the double sequence technique.

The second problem is in the absence of the equality $\sum_{i=1}^{\infty} \dfrac{\varepsilon}{2^i} = \varepsilon$. It will be substitute by the famous Fremlin lemma:

Theorem 3.1 *Let R be a Dedekind complete Riesz space, $(a_{i,j}^{(n)})_{i,j}$, $n \in \mathbb{N}$, be a sequence of regulators in R. Then to every $a \in R$, $a \geq 0$, there corresponds a regulator $(a_{i,j})_{i,j}$ such that*

$$a \wedge \left(\sum_{n=1}^{k} \left(\bigvee_{i=1}^{\infty} a_{i,\varphi(i+n)}^{(n)} \right) \right) \leq \bigvee_{i=1}^{\infty} a_{i,\varphi(i)}$$

for any $k \in \mathbb{N}$ and $\varphi \in \Phi$.

Proof: Put $b_{i,j} = \bigvee_{r=1}^{i-1} 2^r a_{i,j}^{(r)}$, $a_{i,j} = a \wedge b_{i,j}$, $\forall i, j \in \mathbb{N}$. Evidently $(a_{i,j})_{i,j}$ is a (D)-sequence, and

$$a_{i,j}^{(n)} \leq 2^{-n} b_{i+n,j}.$$

Hence, $\forall k \in \mathbb{N}$ we get:

$$\sum_{n=1}^{k} \left(\bigvee_{i=1}^{\infty} a_{i,\varphi(i+n)}^{(n)} \right) \leq \sum_{n=1}^{k} 2^{-n} \left(\bigvee_{i=1}^{\infty} b_{i+n,\varphi(i+n)} \right) \leq \sum_{n=1}^{k} 2^{-n} \left(\bigvee_{t=1}^{\infty} b_{t,\varphi(t)} \right)$$

$$= \left(\sum_{n=1}^{k} 2^{-n} \right) \bigvee_{i=1}^{\infty} b_{i,\varphi(i)} \leq \bigvee_{i=1}^{\infty} b_{i,\varphi(i)}.$$

By the distributive law

$$a \wedge \left(\sum_{n=1}^{k} \left(\bigvee_{i=1}^{\infty} a_{i,\varphi(i+n)}^{(n)} \right) \right) \leq a \wedge \left(\bigvee_{i=1}^{\infty} b_{i,\varphi(i)} \right)$$

$$= \bigvee_{i=1}^{\infty} (a \wedge b_{i,\varphi(i)}) = \bigvee_{i=1}^{\infty} a_{i,\varphi(i)}.$$

This concludes the proof. 🍎

Recall that there exists an l-group-valued version of the Fremlin lemma (see [130,228]). For two (D)-sequences the corresponding assertion is straightforward:

Proposition 3.2 *Let* $(a_{i,j})_{i,j}$, $(b_{i,j})_{i,j}$ *be two* (D)-*sequences. Then there exists a* (D)-*sequence* $(c_{i,j})_{i,j}$, *such that*

$$\bigvee_{i=1}^{\infty} a_{i,\varphi(i)} + \bigvee_{i=1}^{\infty} b_{i,\varphi(i)} \leq \bigvee_{i=1}^{\infty} c_{i,\varphi(i)}$$

for any $\varphi \in \Phi$.

Proof: It is sufficient to put $c_{i,j} = 2(a_{i,j} \vee b_{i,j})$ $\forall i,j \in \mathbb{N}$. 🍎

Assumption 3.3 We shall assume that R is a Dedekind complete weakly σ-distributive Riesz space.

Definition 3.4 Let $f:[a,b] \to R$. If $\Pi = \{([x_{i-1},x_i],t_i): i=1,\ldots,n\}$ is a partition of $[a,b]$, then the *integral sum* or *Riemann sum* is defined by the formula

$$\sum_{\Pi} f = \sum_{i=1}^{n} (x_i - x_{i-1}) f(t_i).$$

Let $\delta:[a,b] \to \mathbb{R}^+$ any map. A partition $\Pi = \{([x_{i-1},x_i],t_i): i=1,\ldots,n\}$ of $[a,b]$ is said to be δ-*fine* if $[x_{i-1},x_i] \subset (t_i - \delta(t_i), t_i + \delta(t_i))$ $\forall i=1,\ldots,n$.

The function $f:[a,b] \to R$ is *integrable* (in the (KH)-sense) or (KH)-*integrable* on $[a,b]$, if there exist $I \in R$ and a (D)-sequence $(a_{i,j})_{i,j}$ such that to any $\varphi \in \Phi$ there exists a map $\delta:[a,b] \to \mathbb{R}^+$ such that

$$\left| \sum_{\Pi} f - I \right| \leq \bigvee_{i=1}^{\infty} a_{i,\varphi(i)}$$

for any δ-fine partition Π.

Lemma 3.5 *The element* I *from Definition 3.4 is determined uniquely.*

Proof: Let I,J be such elements, $(a_{i,j})_{i,j},(b_{i,j})_{i,j}$ be corresponding (D)-sequences, $\varphi \in \Phi$, and δ_1,δ_2 be $\mathbb{R}^+$-valued maps, defined on $[a,b]$, such that

$$\left| \sum_{\Pi_1} f - I \right| \leq \bigvee_{i=1}^{\infty} a_{i,\varphi(i)}, \quad \left| \sum_{\Pi_2} f - J \right| \leq \bigvee_{i=1}^{\infty} b_{i,\varphi(i)},$$

for any δ_i-fine partition of $[a,b]$ Π_i, $i=1,2$. Put $\delta = \min\{\delta_1,\delta_2\}$, and $(c_{i,j})_{i,j}$ according to Proposition 3.2. Then for any δ-fine partition Π of $[a,b]$ we have:

$$|I-J| \leq |I-\sum_{\Pi} f| + |\sum_{\Pi} f - J| \leq \bigvee_{i=1}^{\infty} c_{i,\varphi(i)}.$$

Since R is weakly σ-distributive, we obtain

$$0 \leq |I-J| \leq \bigwedge_{\varphi \in \Phi} \left(\bigvee_{i=1}^{\infty} c_{i,\varphi(i)} \right) = 0.$$
🍎

Notation 3.6 The unique element I from Definition 3.4 will be denoted by $\int f$, or $\int_a^b f$.

Proposition 3.7 *If f, g are integrable on $[a,b]$ and $\alpha \in \mathbb{R}$, then $f+g$, αf are integrable on $[a,b]$ too, and*

$$\int (f+g) = \int f + \int g, \quad \int \alpha f = \alpha \int f.$$

Proof: It can be obtained by equalities

$$\sum_{\Pi} (f+g) = \sum_{\Pi} f + \sum_{\Pi} g, \quad \sum_{\Pi} (\alpha f) = \alpha \sum_{\Pi} f.$$

Proposition 3.8 Let f, g be integrable on $[a,b]$, $f(t) \leq g(t)$ for any $t \in [a,b]$. Then

$$\int f \leq \int g.$$

Proof: Consider first an integrable mapping $h : [a,b] \to R$ such that $h(t) \geq 0$ for any $t \in [a,b]$. Then $\sum_{\Pi} h \geq 0$ for any partition Π. If $\varphi \in \Phi$ and $\delta : [a,b] \to \mathbb{R}^+$ is a corresponding map, then

$$-\int h \leq \sum_{\Pi} h - \int h \leq \left| \sum_{\Pi} h - \int h \right| \leq \bigvee_{i=1}^{\infty} a_{i,\varphi(i)}$$

for any δ-fine partition Π. Since R is weakly σ-distributive, we get

$$-\int h \leq \bigwedge_{\varphi \in \Phi} \left(\bigvee_{i=1}^{\infty} a_{i,\varphi(i)} \right) = 0,$$

hence $\int h \geq 0$. Put $h = g - f$. Then

$$0 \leq \int h = \int g - \int f.$$
🍎

Proposition 3.9 *(Bolzano-Cauchy condition). A mapping $f : [a,b] \to R$ is integrable if and only if the following condition is satisfied:*

There exists a (D)-sequence $(a_{i,j})_{i,j}$ such that, for every $\varphi \in \Phi$, there is a map $\delta : [a,b] \to \mathbb{R}^+$ so that

$$\left| \sum_{\Pi_1} f - \sum_{\Pi_2} f \right| \leq \bigvee_{i=1}^{\infty} a_{i,\varphi(i)}$$

whenever Π_1, Π_2 are δ-fine partitions of $[a,b]$.

Proof: Evidently integrability implies the Bolzano-Cauchy condition by the triangle inequality.

Let the Bolzano-Cauchy condition be satisfied, and $\varphi \in \Phi$. Put

$$M_\varphi = \{\delta : [a,b] \to \mathbb{R}^+ : \Pi_1, \Pi_2 \text{ are } \delta - \text{fine partitions} \Rightarrow |\sum_{\Pi_1} f - \sum_{\Pi_2} f| \le \bigvee_{i=1}^{\infty} a_{i,\varphi(i)}\},$$

$$a_\varphi = \bigwedge \{\sum_{\Pi} f : \Pi \text{ is } \delta \text{ fine}, \delta \in M_\varphi\},$$

$$b_\varphi = \bigvee \{\sum_{\Pi} f : \Pi \text{ is } \delta \text{ fine}, \delta \in M_\varphi\}.$$

Let $\delta_1 \in M_\varphi, \delta_2 \in M_\psi$. Put $\delta = \min\{\delta_1, \delta_2\} \in M_\varphi \cap M_\psi$. If Π is δ-fine, then Π is δ_1-fine, hence $a_\varphi \le \sum_\Pi f$, and Π is also δ_2-fine, hence $\sum_\Pi f \le b_\psi$. It follows that $a_\varphi \le b_\psi$ for any $\varphi, \psi \in \Phi$, and hence $\bigvee_{\varphi \in \Phi} a_\varphi \le \bigwedge_{\varphi \in \Phi} b_\varphi$. On the other hand

$$\sum_{\Pi_1} f \le \sum_{\Pi_2} f + \bigvee_{i=1}^{\infty} a_{i,\varphi(i)}$$

for all δ-fine partitions Π_1, Π_2, with $\delta \in M_\varphi$. By this inequality we obtain

$$b_\varphi \le a_\varphi + \bigvee_{i=1}^{\infty} a_{i,\varphi(i)},$$

hence, by weak σ-distributivity, we get

$$\bigwedge_{\varphi \in \Phi} b_\varphi - \bigvee_{\varphi \in \Phi} a_\varphi = \bigwedge_{\varphi \in \Phi} (b_\varphi - a_\varphi) \le \bigwedge_{\varphi \in \Phi} \left(\bigvee_{i=1}^{\infty} a_{i,\varphi(i)} \right) = 0.$$

We have obtained the equality

$$\bigvee_{\varphi \in \Phi} a_\varphi = \bigwedge_{\varphi \in \Phi} b_\varphi.$$

Denote by I the common value, and consider again the double sequence $(a_{i,j})_{i,j}$. To any $\varphi \in \Phi$ choose $\delta \in M_\varphi$. Then for any δ-fine partition Π we obtain

$$\sum_{\Pi} f - I \le b_\varphi - a_\varphi \le \bigvee_{i=1}^{\infty} a_{i,\varphi(i)},$$

$$I - \sum_{\Pi} f \le b_\varphi - a_\varphi \le \bigvee_{i=1}^{\infty} a_{i,\varphi(i)},$$

hence

$$|\sum_{\Pi} f - I| \le \bigvee_{i=1}^{\infty} a_{i,\varphi(i)}. \quad \text{🍎}$$

Proposition 3.10 If f is integrable on $[a,b]$, and $[c,d] \subset [a,b]$, then f is integrable on $[c,d]$ too.

Proof: Since f is integrable on $[a,b]$, there exists a (D)-sequence $(a_{i,j})_{i,j}$ such that for any $\varphi \in \Phi$ there exists a map $\delta : [a,b] \to \mathbb{R}^+$ such that

$$\left| \sum_{\Pi_1} f - \sum_{\Pi_2} f \right| \le \bigvee_{i=1}^{\infty} a_{i,\varphi(i)}$$

whenever Π_1, Π_2 are δ-fine partitions of $[a,b]$. In the interval $[c,d]$ consider the same (D)-sequence $(a_{i,j})_{i,j}$, but the map $\delta' = \delta|_{[c,d]}$. Let Π, Π' be δ'-fine partitions of $[c,d]$. Choose any $\delta|_{[a,c]}$- and $\delta|_{[d,b]}$-fine partitions Π_0 and Π'_0 of $[a,c]$ and $[d,b]$ respectively, and put $\Pi_1 = \Pi \cup \Pi_0 \cup \Pi_0, \Pi_2 = \Pi' \cup \Pi_0 \cup \Pi_0$. Then Π_1, Π_2 are δ-fine partitions of $[a,b]$. Therefore

$$\left| \sum_{\Pi} f - \sum_{\Pi'} f \right| =$$

$$\left| \sum_{\Pi} f + \sum_{\Pi_0} f + \sum_{\Pi_0} f - \sum_{\Pi'} f - \sum_{\Pi_0} f - \sum_{\Pi_0} f \right| =$$

$$\left| \sum_{\Pi_1} f - \sum_{\Pi_2} f \right| \le \bigvee_{i=1}^{\infty} a_{i,\varphi(i)},$$

and f is integrable on $[c,d]$ by the Bolzano-Cauchy condition (Proposition 3.9).

Proposition 3.11 *Let* $c \in (a,b)$, $f:[a,b] \to R$ *be integrable both on* $[a,c]$ *and on* $[c,b]$. *Then* f *is integrable on* $[a,b]$ *too, and*

$$\int_a^b f = \int_a^c f + \int_c^b f.$$

Proof: Since f is integrable on $[a,c]$ and on $[c,b]$, there are regulators $(a_{i,j})_{i,j}$, $(b_{i,j})_{i,j}$ such that for any $\varphi \in \Phi$ there exist maps $\delta_1 : [a,c] \to \mathbb{R}^+$, $\delta_2 : [c,b] \to \mathbb{R}^+$ such that

$$\left| \sum_{\Pi_1} f - \int_a^c f \right| \le \bigvee_{i=1}^{\infty} a_{i,\varphi(i)}, \quad \left| \sum_{\Pi_2} f - \int_c^b f \right| \le \bigvee_{i=1}^{\infty} b_{i,\varphi(i)}$$

whenever Π_1 and Π_2 are δ_1- and δ_2-fine partitions of $[a,b]$ respectively. Construct $(c_{i,j})_{i,j}$ according with Proposition 3.2 and $\delta : [a,b] \to \mathbb{R}^+$ in the following way:

$$\delta(x) = \begin{cases} \min\{\delta_1(x), c-x\}, & \text{if } x < c; \\ \min\{\delta_1(x), \delta_2(x)\}, & \text{if } x = c; \\ \min\{x-c, \delta_2(x)\}, & \text{if } x > c. \end{cases}$$

Then to any δ-fine partition Π of $[a,b]$ there are partitions Π_1 of $[a,c]$, Π_2 of $[c,b]$, δ_1- and δ_2-fine respectively, such that

$$\sum_{\Pi} f = \sum_{\Pi_1} f + \sum_{\Pi_2} f.$$

Therefore

$$\left| \sum_{\Pi} f - \int_a^c f - \int_c^b f \right| \le$$

$$\le \left| \sum_{\Pi_1} f - \int_a^c f \right| + \left| \sum_{\Pi_2} f - \int_c^b f \right| \le$$

$$\le \bigvee_{i=1}^{\infty} a_{i,\varphi(i)} + \bigvee_{i=1}^{\infty} b_{i,\varphi(i)} \le \bigvee_{i=1}^{\infty} c_{i,\varphi(i)}.$$

We have obtained that f is integrable on $[a,b]$, and

$$\int_a^b f = \int_a^c f + \int_c^b f. \quad \text{🍎}$$

3.2 Uniform Convergence Theorem

Although a special chapter will be devoted to convergence theorems (Chapter 6), we present here a simple example.

Definition 3.12 A sequence $(f_n)_n$ of R-valued functions, defined on $[a,b]$, *converges uniformly* to a function f on $[a,b]$, if there exists a (D)-sequence $(a_{i,j})_{i,j}$ such that for every $\varphi \in \Phi$ there exists $n_0 \in \mathbb{N}$ such that for any $n \geq n_0$ and each $x \in [a,b]$ there holds

$$| f_n(x) - f(x)| \leq \bigvee_{i=1}^{\infty} a_{i,\varphi(i)}.$$

Theorem 3.13 *If $(f_n : [a,b] \to R)_n$ is a sequence of integrable functions uniformly converging to a bounded function $f : [a,b] \to R$, then f is integrable, and*

$$(D)\lim_n \int_a^b f_n = \int_a^b f.$$

Proof: First mention that to any $\varphi \in \Phi$ there exists $n_0 \in \mathbb{N}$ such that

$$|\sum_{\Pi} f - \sum_{\Pi} f_n| \leq (b-a)\bigvee_{i=1}^{\infty} a_{i,\varphi(i)}$$

for every $n \geq n_0$ and each partition Π of $[a,b]$, $\Pi = \{([x_{i-1},x_i],t_i) : i = 1,\ldots,k\}$. Indeed,

$$|\sum_{\Pi} f - \sum_{\Pi} f_n| = |\sum_{i=1}^{k}(x_i - x_{i-1})f(t_i) - \sum_{i=1}^{k}(x_i - x_{i-1})f_n(t_i)|$$

$$\leq \sum_{i=1}^{k}(x_i - x_{i-1})|f(t_i) - f_n(t_i)| \qquad (3.1)$$

$$\leq \sum_{i=1}^{k}(x_i - x_{i-1})\bigvee_{i=1}^{\infty} a_{i,\varphi(i)} = (b-a)\bigvee_{i=1}^{\infty} a_{i,\varphi(i)}.$$

Moreover, proceeding analogously as in (3.1), since f is bounded, for every two partitions Π_1 and Π_2 of $[a,b]$ we get:

$$\left| \sum_{\Pi_1} f - \sum_{\Pi_2} f \right| \leq 2(b-a)M,$$

where

$$M = \sup_{x \in [a,b]} | f(x)|.$$

Let now $n \in \mathbb{N}$. Since f_n is integrable on $[a,b]$, there exists a (D)-sequence $(b_{i,j}^{(n)})_{i,j}$ such that to any $\varphi \in \Phi$ there exists a mapping $\delta_n : [a,b] \to \mathbb{R}^+$ such that

$$\left|\sum_{\Pi_1} f_n - \sum_{\Pi_2} f_n\right| \leq \bigvee_{i=1}^{\infty} b_{i,\varphi(i+n)}^{(n)}$$

for any two δ_n-fine partitions Π_1, Π_2 of $[a,b]$. Then, for all δ_{n_0}-fine partitions Π_1, Π_2 of $[a,b]$, we have:

$$\left|\sum_{\Pi_1} f - \sum_{\Pi_2} f\right|$$

$$\leq (2(b-a)M) \wedge \left(\left|\sum_{\Pi_1} f - \sum_{\Pi_1} f_{n_0}\right| + \left|\sum_{\Pi_1} f_{n_0} - \sum_{\Pi_2} f_{n_0}\right| + \left|\sum_{\Pi_2} f_{n_0} - \sum_{\Pi_2} f\right|\right)$$

$$\leq (2(b-a)M) \wedge \left(2(b-a)\bigvee_{i=1}^{\infty} a_{i,\varphi(i)} + \bigvee_{i=1}^{\infty} b_{i,\varphi(i+n_0)}^{(n_0)}\right) \leq \bigvee_{i=1}^{\infty} c_{i,\varphi(i)},$$

where $(c_{i,j})_{i,j}$ is a suitable regulator, existing by virtue of the Fremlin Lemma (Theorem 3.1). From the Bolzano-Cauchy condition (Proposition 3.9) we obtain that f is integrable on $[a,b]$. We have to prove

$$(D)\lim_n \int_a^b f_n = \int_a^b f. \qquad (3.2)$$

Fix again $n \geq n_0$. We have

$$-\bigvee_{i=1}^{\infty} a_{i,\varphi(i)} \leq f_n(x) - f(x) \leq \bigvee_{i=1}^{\infty} a_{i,\varphi(i)} \quad \forall x \in [a,b].$$

Since f_n and f are integrable on $[a,b]$, by taking the integrals we get

$$-(b-a)\bigvee_{i=1}^{\infty} a_{i,\varphi(i)} \leq \int_a^b f_n - \int_a^b f \leq (b-a)\bigvee_{i=1}^{\infty} a_{i,\varphi(i)},$$

because the (KH)-integral is a monotone linear functional. Thus the equality (3.2) follows. This completes the proof. $\blacksquare$

3.3 Kurzweil - Stieltjes Integral

Usually in the Stieltjes-type integral $\int f\, dF$ the Stieltjes integral sums

$$\sum_{\Pi}^{(F)} f = \sum_{i=1}^{n} (F(x_i) - F(x_{i-1})) f(t_i)$$

are considered instead of the Riemann ones

$$\sum_{\Pi} f = \sum_{i=1}^{n} (x_i - x_{i-1}) f(t_i).$$

Definition 3.14 Let $f:[a,b] \to R$, $F:[a,b] \to \mathbb{R}$. We say that the (KH)-*Stieltjes integral* of f with respect to F exists in R (on $[a,b]$) if there exist $I \in R$ and a (D)-sequence $(a_{i,j})_{i,j}$ such that to any $\varphi \in \Phi$ there exists a map $\delta:[a,b] \to \mathbb{R}^+$, such that

$$\left|\sum_{\Pi}^{(F)} f - I\right| \leq \bigvee_{i=1}^{\infty} a_{i,\varphi(i)}$$

whenever Π is a δ-fine partition of $[a,b]$.

As before, by weak σ-distributivity of R, the element I is determined uniquely; it will be denoted by the symbol

$$\int_a^b f\, dF.$$

Of course, it is possible to integrate a real-valued function $F:[a,b]\to\mathbb{R}$ with respect to a vector-valued map $f:[a,b]\to R$, i.e. to work with the integral sums

$$\sum_{\Pi}^{(f)} F = \sum_{i=1}^{n} F(t_i)(f(x_i)-f(x_{i-1})).$$

Definition 3.15 We say that the (KH)-*Stieltjes integral* of F with respect to f exists in R (on $[a,b]$), if there exist $I\in R$ and a (D)-sequence $(a_{i,j})_{i,j}$ such that to any $\varphi\in\Phi$ there exists a map $\delta:[a,b]\to\mathbb{R}^+$, such that

$$\left|\sum_{\Pi}^{(f)} F - I\right| \leq \bigvee_{i=1}^{\infty} a_{i,\varphi(i)}$$

whenever Π is any δ-fine partition of $[a,b]$.

Again the entity I is determined uniquely. We shall denote it by

$$I = \int_a^b F\, df.$$

Definition 3.16 A function $f:[a,b]\to R$ is *continuous at a point* $t_0\in[a,b]$ with respect to the regulator $(a_{i,j})_{i,j}$, if, to every $\varphi\in\Phi$, there exists $\delta>0$ such that

$$|f(t)-f(t_0)| \leq \bigvee_{i=1}^{\infty} a_{i,\varphi(i)}$$

whenever $t\in[a,b]$, $|t-t_0|<\delta$.

Definition 3.17 Let $g:[a,b]\to\mathbb{R}$, and let $(a_{i,j})_{i,j}$ be a (D)-sequence in R. We say that $g:[a,b]\to\mathbb{R}$ is *of bounded semivariation* with respect to $(a_{i,j})_{i,j}$ if there exists a (D)-sequence $(b_{i,j})_{i,j}$ in R such that, for every finite collection $\{[t_{i-1},t_i]:i=1,\ldots,n\}$ of pairwise nonoverlapping subintervals of $[a,b]$, we have, for each $\varphi\in\Phi$,

$$\left|\sum_{i=1}^{n} (g(t_i)-g(t_{i-1}))x_i\right| \leq \bigvee_{j=1}^{\infty} b_{j,\varphi(j)}$$

whenever $x_i\in R$, $|x_i|\leq\bigvee_{j=1}^{\infty} a_{j,\varphi(j)}$ $(i=1,\ldots,n)$.

Theorem 3.18 *Let* $(a_{i,j})_{i,j}$ *be a* (D)*-sequence, and let* $f:[a,b] \to R$ *be continuous at every point of* $[a,b]$ *with respect to the same regulator* $(a_{i,j})_{i,j}$. *Let* $F:[a,b] \to \mathbb{R}$ *have bounded semivariation with respect to* $(a_{i,j})_{i,j}$. *Then the* (KH)*-Stieltjes integral* $\int_a^b f \, dF$ *exists.*

Proof: By continuity of f with respect to the same regulator $(a_{i,j})_{i,j}$, to every $\varphi \in \Phi$ there exists a function $\delta:[a,b] \to \mathbb{R}^+$ such that $|u-u_0| < \delta(u_0)$, $u,u_0 \in [a,b]$ implies

$$|f(u)-f(u_0)| \le \bigvee_{i=1}^{\infty} a_{i,\varphi(i)}.$$

Let Π_1, Π_2 be δ-fine partitions of $[a,b]$. Without loss of generality, we can suppose that Π_2 is a refinement of Π_1. Let $\Pi_1 = \{([t_{i-1},t_i],s_i) : i=1,\ldots,k\}$ and $\Pi_{2,i}$ the subpartition of $[t_{i-1},t_i]$ given by Π_2, $\Pi_{2,i} = \{([t_{i_{j-1}},t_{i_j}],s_{i,j}) : j=1,\ldots,k_i\}$. Then we have

$$\left| \sum_{\Pi_1}^{(F)} f - \sum_{\Pi_2}^{(F)} f \right|$$

$$= \sum_{i=1}^{k} (f(s_i)(F(t_i)-F(t_{i-1})) - \sum_{j=1}^{k_i} f(s_{i,j})(F(t_{i,j})-F(t_{i,j-1})).$$

As $t_{i,0}=t_{i-1}, t_{i,k_i}=t_i$, we have

$$\left| \sum_{\Pi_1}^{(F)} f - \sum_{\Pi_2}^{(F)} f \right| = \left| \sum_{i=1}^{k} \sum_{j=1}^{k_i} (f(s_i)-f(s_{i,j}))(F(t_{i,j})-F(t_{i,j-1})) \right| \le \bigvee_{i=1}^{\infty} b_{i,\varphi(i)},$$

since $|s_i - s_{i,j}| \le \delta(s_i)$ and hence $|f(s_i)-f(s_{i,j})| \le \bigvee_{i=1}^{\infty} a_{i,\varphi(i)}$ (Here, $(b_{i,j})_{i,j}$ is a regulator in R, existing because F is of bounded semivariation with respect to $(a_{i,j})_{i,j}$). ♣

4 Double Integrals

Abstract: We introduce the theory of the double integrals for Riesz space-valued mappings, defined on a bounded subrectangle of the Euclidean plane, and prove some versions of the Fubini theorems. We deal also with some concepts of continuity for Riesz space-valued functions, related with these kinds of results.

In this chapter we define the Kurzweil-Henstock integral for Riesz space-valued functions defined on bounded intervals in $\mathbb{R}^2$ and some types of the Fubini theorem are presented (see also [267]). Similar theorems in the real case are obtained in [170, 188, 254].

Let $E = [a,b] \times [c,d]$ be a bounded 2-dimensional interval of the set of the real numbers. A *division* D of the interval E is a set of non-overlapping bounded intervals E_i, $i = 1, 2, ..., n$ with $\bigcup_{i=1}^{n} E_i = E$.

Let $\delta : [a,b] \times [c,d] \to \mathbb{R}^+ \times \mathbb{R}^+$, $\delta(x,y) = (\delta_1(x,y), \delta_2(x,y))$. A δ-*fine partition* of the interval $[a,b] \times [c,d]$ is a set of the type

$$P = \{([a_i, b_i] \times [c_i, d_i], (t_{i1}, t_{i2})) = (E_i, t_i) : i = 1, 2, ..., n\},$$

where $t_i \in E_i, E_i \subset [a,b] \times [c,d]$,

$$E_i \subset (t_{i1} - \delta_1(t_{i1}, t_{i2}), t_{i1} + \delta_1(t_{i1}, t_{i2})) \times (t_{i2} - \delta_2(t_{i1}, t_{i2}), t_{i2} + \delta_2(t_{i1}, t_{i2}))$$

for $i = 1, 2, ..., n$, E_i are non-overlapping and $\bigcup_{i=1}^{n} E_i = [a,b] \times [c,d]$.

The existence of at least one $\delta = (\delta_1, \delta_2)$-fine partition P of $[a,b] \times [c,d]$ is guaranteed by the existence of a δ_1-fine partition of $[a,b]$ and a δ_2-fine partition of $[c,d]$ (see subsection "Compound partitions").

4.1 Definition of the Double Integral

Let R be a Dedekind complete weakly σ-distributive Riesz space, and $\Phi = \mathbb{N}^{\mathbb{N}}$.

Definition 4.1 A function $f : [a,b] \times [c,d] \to R$ is called (KH)-*integrable* (in the Kurzweil-Henstock sense) if there exist $I \in R$ and a regulator $(a_{i,j})_{i,j}$ in R such that for every $\varphi \in \Phi$ there is $\delta : [a,b] \times [c,d] \to \mathbb{R}^+ \times \mathbb{R}^+$ such that

$$\left| \sum_{i=1}^{n} f(t_{i1}, t_{i2})(b_i - a_i)(d_i - c_i) - I \right| \leq \bigvee_{i=1}^{\infty} a_{i,\varphi(i)}$$

for every δ-fine partition $P = \{([a_i, b_i] \times [c_i, d_i], (t_{i1}, t_{i2})) = (E_i, t_i) : i = 1, 2, ..., n\}$.

The element I we will called the *double integral* of f on $[a,b] \times [c,d]$, and we denote it by

$$(KH) \int_c^d \int_a^b f(x,y) \, dxdy$$

or more simply

Antonio Boccuto / Beloslav Riečan / Marta Vrábelová

$$(KH)\iint_E f(t)\,dt$$

We will omit (KH) in proofs and examples.

Furthermore, we denote

$m([a,b]\times[c,d]) = (b-a)(d-c) = m([a,b)\times[c,d)) = m((a,b]\times[c,d))$, and so on.

Remark 4.2 The proofs of uniqueness and of the elementary properties of the double (KH)-integral are similar as the ones in Section 3.

Example 4.3 Let $E = [a,b]\times[c,d]$, $H \subset E$ be a bounded interval, $a \in R, a \neq 0$. We show that the function $f = a\chi_H$, where χ_H is the characteristic function of H, is (KH)-integrable on E and

$$\iint_E a\chi_H(t)\,dt = a\,m(H).$$

First we define the regulator $(a_{i,j})_{i,j}$ by setting

$$a_{i,j} = \frac{|a|}{j}, \quad i,j \in \mathbb{N}.$$

Let $\varphi \in \Phi$ be arbitrary. Take $\varepsilon = 1/\min_i \varphi(i)$. There exist an open interval $G \supset H$ and a closed interval $F \subset H$ such that $m(G) - m(F) < \varepsilon$.

For $(x,y) \in H$ define $\delta(x,y) = (\delta_1(x,y), \delta_2(x,y))$ in such a way that

$$(x-\delta_1(x,y), x+\delta_1(x,y)) \times (y-\delta_2(x,y), y+\delta_2(x,y)) \subset G.$$

For $(x,y) \in E \setminus H$ define $\delta(x,y) = (\delta_1(x,y), \delta_2(x,y))$ in such a way that

$$((x-\delta_1(x,y), x+\delta_1(x,y)) \times (y-\delta_2(x,y), y+\delta_2(x,y))) \cap F = \varnothing.$$

Take any arbitrary δ-fine partition $P = \{(E_i, t_i) : i = 1, 2, \ldots, m\}$ of E. Then

$$\sum_P a\chi_H = \sum_{t_i \in H} a\,m(E_i) = a\sum_{t_i \in H} m(E_i).$$

We have non-overlapping closed intervals E_i fulfilling $F \subset \bigcup_{t_i \in H} E_i \subset G$ and, from the finite additivity of m on the set of rectangles (we omit the proof of that fact in this place), we get:

$$a\,m(F) \leq \sum_P a\chi_H \leq a\,m(G).$$

Therefore

$$\left| \sum_P a\chi_H - a\,m(H) \right| \leq |a\,m(G) - a\,m(H)| \leq |a| |m(G) - m(F)|$$

$$\leq |a|\varepsilon = \frac{|a|}{\min_i \varphi(i)} = \bigvee_{i=1}^{\infty} a_{i,\varphi(i)}. \qquad \maltese$$

Example 4.4 Let $a_k \in R\ (k \in \mathbb{N})$ and $s = \sum_{k=1}^{\infty} a_k = \lim_n \sum_{k=1}^{n} a_k = \lim_n s_n \in R$. Hence, there is a regulator $(b_{i,j})_{i,j}$ in R such that for every $\varphi \in \Phi$ we have

$$|s - s_n| \le \bigvee_{i=1}^{\infty} b_{i,\varphi(i)} \tag{4.1}$$

and

$$|a_n| = |s_n - s_{n-1}| \le |s_n - s| + |s - s_{n-1}| \le 2 \bigvee_{i=1}^{\infty} b_{i,\varphi(i)} \tag{4.2}$$

whenever $n > n_0$. Denote $M = \sup\{|a_k| : k \in \mathbb{N}\}$. Then $M \in R$, since R is Dedekind complete.

Let $E = [0,1] \times [0,1]$. Put

$$K_j = \left[\frac{1}{2^{2j}}, \frac{1}{2^{2j-1}}\right] \times \left[\frac{1}{2^{2j}}, \frac{1}{2^{2j-1}}\right], \quad j \in \mathbb{N}$$

(see Figure 4.1).

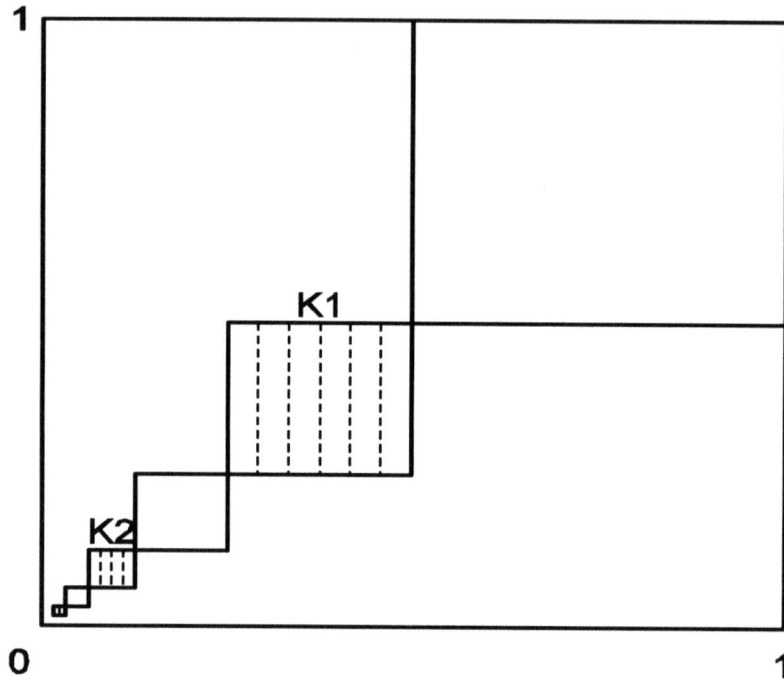

Fig. 4.1 The sets K_j

Define the function f on E by the formula

$$f = \sum_{j=1}^{\infty} a_j 2^{4j} \chi_{K_j}.$$

Hence $f(x,y) = a_j 2^{4j}$ if $(x,y) \in K_j (j \in \mathbb{N})$ and $f(x,y) = 0$ at the other points in E. We will show that f is (KH)-integrable on E and

$$\iint_E f(x,y)\,dxdy = \sum_{k=1}^{\infty} a_k.$$

To prove it, take any arbitrary $\varphi \in \Phi$ and put $\varepsilon = 1/\min_i \varphi(i)$. Then $\varepsilon \in R, 0 < \varepsilon \leq 1$. Further put

$$\alpha_j = \frac{1}{2^{2j}} - \frac{\varepsilon}{2^{6j}}, \beta_j = \frac{1}{2^{2j-1}} + \frac{\varepsilon}{2^{6j}}, O_j = (\alpha_j, \beta_j) \times (\alpha_j, \beta_j).$$

Then $O_j \supset K_j$ is an open interval with

$$m(O_j) - m(K_j) < \frac{\varepsilon}{2^{5j}}.$$

Define a gauge δ on E by putting:

$\delta(x,y) = (\min\{x - \alpha_j, \beta_j - x\}, \min\{y - \alpha_j, \beta_j - y\})$ when $(x,y) \in K_j (j \in \mathbb{N})$;

$\delta(x,y) = (\delta_1(x,y), \delta_2(x,y))$, where $\delta_1(x,y)$ and $\delta_2(x,y)$ are positive numbers such that

$(x - \delta_1(x,y), x + \delta_1(x,y)) \times (y - \delta_2(x,y), y + \delta_2(x,y)) \cap K_j = \varnothing$ for any j when

$(x,y) \notin K_j \ (j \in \mathbb{N})$ and $(x,y) \neq (0,0)$;

$\delta(0,0) = (\gamma, \gamma)$, where $\gamma = 1/(2^{2n_0})$.

Let $P = \{(E_i, t_i) : i = 0, 1, \ldots, m\}$ be a δ-fine partition of E. Then $t_0 = (0,0)$ and $E_0 \subset [0, \gamma) \times [0, \gamma)$. Let $q \in \mathbb{N}$ be such that K_q is not a subset of E_0 but $K_{q+1} \subset E_0$. Obviously $q > n_0$. We get

$$\sum_P f = \sum_{i=0}^m f(t_i) m(E_i) = \sum_{j=1}^q \sum_{t_i \in K_j} f(t_i) m(E_i)$$

$$= \sum_{j=1}^q a_j 2^{4j} \sum_{t_i \in K_j} m(E_i).$$

If $j < q$, then

$$\frac{1}{2^{4j}} = m(K_j) \leq \sum_{t_i \in K_j} m(E_i) \leq m(O_j) < m(K_j) + \frac{\varepsilon}{2^{5j}}.$$

Therefore

$$1 \leq 2^{4j} \sum_{t_i \in K_j} m(E_i) < 2^{4j} m(K_j) + \frac{\varepsilon}{2^j}$$

and hence

$$0 \leq 2^{4j} \sum_{t_i \in K_j} m(E_i) - 1 < \frac{\varepsilon}{2^j}. \tag{4.3}$$

If $j = q$,

$$\sum_{t_i \in K_q} m(E_i) < m(O_q) < \frac{1}{2^{4q}} + \frac{\varepsilon}{2^{5q}} \leq \frac{2}{2^{4q}},$$

so

$$\left|\sum_{t_i \in K_q} f(t_i)m(E_i)\right| = |a_q| 2^{4q} \sum_{t_i \in K_q} m(E_i) < 2|a_q| < 4\bigvee_{i=1}^{\infty} b_{i,\varphi(i)}. \qquad (4.4)$$

Now, by (4.3), (4.4), (4.2) and (4.1),

$$\left|\sum_P f - \sum_{j=1}^{\infty} a_j\right| = \left|\sum_{j=1}^{q} a_j 2^{4j} \sum_{t_i \in K_j} m(E_i) - \sum_{j=1}^{\infty} a_j\right|$$

$$\leq \left|\sum_{j=1}^{q-1} a_j \left(2^{4j} \sum_{t_i \in K_j} m(E_i) - 1\right)\right| + \left|a_q 2^{4j} \sum_{t_i \in K_q} m(E_i)\right| + |a_q|$$

$$+ \left|\sum_{j=q+1}^{\infty} a_j\right| \leq \sum_{j=1}^{q-1} |a_j| \frac{\varepsilon}{2^j} + 4\bigvee_{i=1}^{\infty} b_{i,\varphi(i)} + 2\bigvee_{i=1}^{\infty} b_{i,\varphi(i)} + \bigvee_{i=1}^{\infty} b_{i,\varphi(i)}$$

$$\leq M\frac{1}{\min_i \varphi(i)} + 7\bigvee_{i=1}^{\infty} b_{i,\varphi(i)}.$$

Put $c_{i,j} = M/j, a_{i,j} = 2(c_{i,j} + 7b_{i,j})$ for $i, j \in \mathbb{N}$. Then $(c_{i,j})_{i,j}$ is a regulator in R and

$$\bigvee_{i=1}^{\infty} c_{i,\varphi(i)} = \bigvee_{i=1}^{\infty} \frac{M}{\varphi(i)} = \frac{M}{\min_i \varphi(i)}.$$

We have found the regulator $(a_{i,j})_{i,j}$ in R, which is such that for every $\varphi \in \Phi$ there exists a gauge δ such that

$$\left|\sum_P f - \sum_{j=1}^{\infty} a_j\right| \leq \bigvee_{i=1}^{\infty} a_{i,\varphi(i)}$$

whenever P is $\delta-$fine. 🍎

4.2 Double Integral of Continuous Functions

Definition 4.5 We say that the function $f: K \to R$, $K \subset \mathbb{R}^2$, is *continuous with respect to a common regulator on* K if there is a regulator $(a_{i,j})_{i,j}$ in R such that for every $\varphi \in \Phi$ and for every $t \in K$ there exists an open interval $J_t \subset \mathbb{R}^2$ containing t such that $|f(t) - f(t')| \leq \bigvee_{i=1}^{\infty} a_{i,\varphi(i)}$ whenever $t' \in J_t \cap K$.

Definition 4.6 Let $K \subset \mathbb{R}^2$ and $f_n, f: K \to R$. Then $f_n \to f$ on K *with respect to a common regulator* if there is a regulator $(a_{i,j})_{i,j}$ in R such that for every $t \in K$ there exists $p(t) \in \mathbb{N}$ such that $|f_n(t) - f(t)| \leq \bigvee_{i=1}^{\infty} a_{i,\varphi(i)}$ when $n > p(t)$.

Lemma 4.7 Let $K \subset \mathbb{R}^2$ be compact and E be a compact 2-dimensional interval containing K. Let R be a Dedekind complete weakly σ-distributive Riesz space. If $f: K \to R$ is continuous with respect to a common regulator on K, $f \geq 0$ and f is bounded on K, and if f is extended to E

by setting $f(t) = 0$ *for* $t \in E \setminus K$, *then there is a sequence of step functions* $(s_n)_{n=1}^{\infty}, s_k \geq 0$ $(k \in \mathbb{N})$ *such that* $s_k \downarrow f$ *on* E *with respect to a common regulator.*

Proof: We split the interval E into pairwise disjoint intervals with diameter smaller then $1/k$. Denote this collection of subintervals of E by $D_k = \{E_i^k : i = 1, 2, \ldots, m_k\}, \bigcup_{i=1}^{m_k} E_i^k = E$. Take the sequence $(D_k)_{k=1}^{\infty}$ such that D_{k+1} is a refinement of D_k for all k (that is, for every E_i^{k+1} there is E_j^k with $E_i^{k+1} \subset E_j^k$).

The collection D_k determines the step function s_k. When $t \in E$, then for every k there is a unique E_i^k with $t \in E_i^k$. Define

$$s_k(t) = \bigvee \{f(t) : t \in E_i^k\} \text{ for } t \in E_i^k \, (i = 1, 2, \ldots, m_k).$$

Obviously $s_k(t) \geq s_{k+1}(t)$ for every $t \in E$ and all k. We show that $s_k(t) \downarrow f(t)$ for every $t \in E$. If $t \in E \setminus K$, then there is an open interval E_t such that $t \in E_t$ and $E_t \cap K = \varnothing$ and hence, for sufficiently large k, the subinterval of D_k containing t is a subset of E_t and $s_k(t) = 0$. Since $f(t) = 0$, then $s_k(t) \downarrow f(t)$. If $t \in K$, then, since f is continuous with respect to a common regulator, there is a regulator $(a_{i,j})_{i,j}$ in R and for any arbitrary fixed $\varphi \in \Phi$ we can find an open interval J_t with $t \in J_t$ and

$$|f(t) - f(t')| \leq \bigvee_{i=1}^{\infty} a_{i,\varphi(i)}$$

when $t' \in J_t \cap K$. So

$$0 \leq f(t') \leq f(t) + \bigvee_{i=1}^{\infty} a_{i,\varphi(i)}$$

for every $t' \in J_t$ (since $f(t') = 0$ when $t' \in E \setminus K$). Furthermore, there is $p(t) \in \mathbb{N}$ such that for every $k > p(t)$ there exists $E_i^k \subset J_t$ for which

$$s_k(t) = \sup_{t' \in E_i^k} f(t') \leq f(t) + \bigvee_{i=1}^{\infty} a_{i,\varphi(i)}.$$

Therefore $f(t) \leq s_k(t) \leq f(t) + \bigvee_{i=1}^{\infty} a_{i,\varphi(i)}$ and hence

$$|s_k(t) - f(t)| \leq \bigvee_{i=1}^{\infty} a_{i,\varphi(i)}$$

for every $k > p(t)$. 🍎

Theorem 4.8 *Suppose that* $K \subset \mathbb{R}^2$ *is a compact set,* R *is Dedekind complete,* $f : K \to R$ *is continuous with respect to a common regulator and bounded on* K *and* $f(t) = 0$ *for* $t \in E \setminus K$, *where* E *is a compact interval with* $K \subset E$. *Then* f *is* (KH)-*integrable on* E.

Proof: Consider $f \geq 0$. Then by Lemma 4.7 there is a sequence $(s_k)_{k=1}^{\infty}$ of step functions such that $s_k \downarrow f$ on E with respect to a common regulator. By the Levi monotone convergence theorem (see Chapter 1), f is (KH)-integrable.

Generally, $f = f^+ - f^-$, where $f^+ \geq 0, f^- \geq 0$ are (KH)-integrable and f is (KH)-integrable by linearity of the integral. ◖

Theorem 4.9 *Let $[a,b],[c,d]$ be bounded closed intervals in $\mathbb{R}$ and $E = [a,b] \times [c,d]$. Let $f : E \to R$ be a bounded function such that there exists a sequence of step functions $(s_k)_{k=1}^\infty, s_k = \sum_{i=1}^{m_k} a_i^k \chi_{E_i^k}$, where $E_i^k (i = 1,2,...,m_k)$ are pairwise disjoint subintervals of E. Let $s_k \to f$ on E with respect to a common regulator. Then*

    (i) $f(.,y)$ is integrable on $[a,b]$ for every $y \in [c,d]$;

    (ii) $h(y) = \int_a^b f(x,y)\,dx$ is integrable on $[c,d]$;

    (iii) f is integrable on E and

$$\iint_E f(x,y)\,dxdy = \int_c^d \left(\int_a^b f(x,y)\,dx \right) dy.$$

Proof: If $f = a\chi_{[s,t] \times [u,v]}$, then (i)–(iii) hold. Indeed,

$$\int_a^b f(x,y)\,dx = \int_a^b a\chi_{[s,t]}(x)\chi_{[u,v]}(y)\,dx = a\chi_{[u,v]}(y) l([a,b] \cap [s,t]),$$

$$\int_c^d \left(\int_a^b f(x,y)\,dx \right) dy = a l([a,b] \cap [s,t]) l([c,d] \cap [u,v]) = \iint_E f(x,y)\,dxdy$$

(l denotes the length of the interval in $\mathbb{R}$).

In the case f is a step function, then the assertions of the theorem follow from linearity of the integral.

In the general case there is $M \in R$, $M > 0$, with $|f(t)| \leq M$ for every $t \in E$. Take the function

$$s_k'(t) = (s_k(t) \wedge (-M)) \vee M$$

for every $t \in E$ instead of s_k. Then $s_k'(t) \leq M$ are integrable step functions for every $t \in E, s_k'$ $(k \in \mathbb{N})$, and

$$s_k'(t) \to (f(t) \wedge (-M)) \vee M = f(t)$$

for every $t \in E$ with respect to a common regulator. By the Lebesgue dominated convergence theorem (see Chapter 1), f is integrable on E and

$$\iint_E f(t)\,dt = \lim_{k \to \infty} \iint_E s_k(t)\,dt.$$

Now $s_k'(.,y)$ is (KH)-integrable on $[a,b]$, $s_k'(.,y) \to f(.,y)$ for every $y \in [c,d]$ with respect to a common regulator and again by the dominated convergence theorem, $f(.,y)$ is (KH)-integrable with

$$\int_a^b f(x,y)\,dx = \lim_k \int_a^b s_k'(x,y)\,dx.$$

The function $h_k(y) = \int_a^b s_k'(x,y)dx$ is also (KH)-integrable on $[c,d]$ by the first part of this proof and

$$h_k(y) \to \lim_k \int_a^b s_k'(x,y)dx = \int_a^b f(x,y)dx = h(y).$$

Since $|h(y)| \le M(b-a)$, by the dominated convergence theorem, $h(y)$ is integrable and

$$\int_a^b h(y)dy = \lim_k \int_c^d h_k(y)dy = \lim_k \int_c^d \left(\int_a^b s_k'(x,y)dx \right)dy$$

$$= \lim_k \iint_E s_k'(x,y)dxdy = \iint_E f(x,y)dxdy$$

On the other hand

$$\int_c^d h(y)dy = \int_c^d \left(\lim_k \int_a^b s_k'(x,y)dx \right)dy = \int_c^d \left(\int_a^b f(x,y)dx \right)dy.$$

Hence

$$\iint_E f(x,y)dxdy = \int_c^d \left(\int_a^b f(x,y)dx \right)dy,$$

and by symmetry we get

$$\iint_E f(x,y)dxdy = \int_c^d \left(\int_a^b f(x,y)dx \right)dy = \int_a^b \left(\int_c^d f(x,y)dy \right)dx \qquad \text{◆}$$

Corollary 4.10 *Let the function f fulfil the assumptions of Theorem 4.8. Then the assumptions of Theorem 4.9 hold.*

Proof: There are sequences of step functions $(s_n^+)_{n=1}^{\infty}$, $(s_n^-)_{n=1}^{\infty}$, such that $s_n^+ \downarrow f^+$ with a regulator $(a_{i,j})_{i,j}$ in R and $s_n^- \downarrow f^-$ with a regulator $(b_{i,j})_{i,j}$ in R. Then $s_n^+ - s_n^-$ is a step function and $(s_n^+ - s_n^-) \to f$ with the regulator $(2(a_{i,j} + b_{i,j}))_{i,j}$. ◆

Definition 4.11 Let $f : [a,b] \times [c,d] \to R$. We say that

$$h(y) = \int_a^b f(x,y)dx$$

exists uniformly with respect to $y \in [c,d]$ if there exists a regulator $(b_{i,j})_{i,j}$ in R such that for every $y \in [c,d]$ and every $\varphi \in \Phi$ there exists $\delta_y : [a,b] \to \mathbb{R}^+$ such that

$$\left| \sum_{i=1}^{n_y} f(t_{yi},y)(a_{yi} - a_{y(i-1)}) - h(y) \right| \le \bigvee_{i=1}^{\infty} b_{i,\varphi(i)}$$

for every δ-fine partition $P = \{([a_{y(i-1)}, a_{yi}], t_{yi}) : i = 1, 2, \ldots, n_y\}$ of $[a,b]$.

4.3 Compound Partitions

Let $\delta : [a,b] \times [c,d] \to \mathbb{R}^+ \times \mathbb{R}^+$, $\delta(x,y) = (\delta_1(x,y), \delta_2(x,y))$. Suppose that for every $y \in [c,d]$ the partition

$$P_y = \{([a_{y(i-1)}, a_{yi}], x_{yi}) : i = 1, 2, \ldots, n_y\}$$

of the interval $[a,b]$ is $\delta_1(.,y)$-fine. Define

$$\delta':[c,d]\to\mathbb{R}^+,\ \delta'(y)=\min\{\delta_2(x_{yi},y):i=1,2,...,n_y\}.$$

Let F be a δ'-fine partition of $[c,d]$,

$$F=\{([b_{j-1},b_j],y_j):j=1,2,...,n\}.$$

Now, we can construct a δ-fine compound partition of $[a,b]\times[c,d]$ by putting

$$P=\left\{\left([a_{y_j(i-1)},a_{y_ji}]\times[b_{j-1},b_j],(x_{y_ji},y_j)\right):j=1,2,...,n,i=1,2,...,n_{y_j}\right\}.$$

4.4 Fubini's Theorem

Theorem 4.12 *Let R be a Dedekind complete weakly σ-distributive Riesz space. Let $f:[a,b]\times[c,d]\to R$ be (KH)-integrable with integral $I\in R$. Suppose that*

$$h(y)=\int_a^b f(x,y)dx$$

exists uniformly with respect to $y\in[c,d]$. Then $h(y)$ is (KH)-integrable on $[c,d]$ and

$$(KH)\int_c^d h(y)dy=I.$$

Proof: By the assumptions of the theorem there is a regulator $(a_{i,j})_{i,j}$ in R such that for every $\varphi\in\Phi$ there exists

$$\delta:[a,b]\times[c,d]\to\mathbb{R}^+\times\mathbb{R}^+,\delta(x,y)=(\delta_1(x,y),\delta_2(x,y)),$$

such that

$$\left|\sum_{i=1}^n f(t_{i1},t_{i2})(b_i-a_i)(d_i-c_i)-I\right|\leq\bigvee_{i=1}^\infty a_{i,\varphi(i)}$$

for every δ-fine partition $P=\{([a_i,b_i]\times[c_i,d_i],(t_{i1},t_{i2})):i=1,2,...,n\}$.

Furthermore, there exists a regulator $(b_{i,j})_{i,j}$ in R such that for every $y\in[c,d]$ and for every $\varphi\in\Phi$ there exists $\delta_y:[a,b]\to\mathbb{R}^+$ such that

$$\left|\sum_{i=1}^{n_y} f(x_{yi},y)(a_{yi}-a_{y(i-1)})-\int_a^b f(x,y)dx\right|\leq\bigvee_{i=1}^\infty b_{i,\varphi(i)}$$

for every δ_y-fine partition $P_y=\left\{\left([a_{y(i-1)},a_{yi}],x_{yi}\right):i=1,2,...,n_y\right\}$.

Now, let $\varphi\in\Phi$ be arbitrary. Put

$$\delta'_y=\min\{\delta_y,\delta_1(.,y)\}.$$

Let $P_y = \left\{ \left([a_{y(i-1)}, a_{yi}], x_{yi} \right) : i = 1, 2, ..., n_y \right\}$ be a δ_y'-fine partition of $[a, b]$. Then P_y is δ_y-fine and $\delta_1(., y)$-fine. Define

$$\delta' : [c, d] \to \mathbb{R}^+, \delta'(y) = \min\{\delta_2(x_{yi}, y) : i = 1, 2, ..., n_y\}.$$

Let $F = \left\{ \left([b_{j-1}, b_j], y_j \right) : j = 1, 2, ..., n \right\}$ be an arbitrary δ'-fine partition of $[c, d]$.

Construct the compound partition from F and $P_{y_j}, y_j \in F$,

$$P = \left\{ \left([a_{y_j(i-1)}, a_{y_j i}] \times [b_{j-1}, b_j], (x_{y_j i}, y_j) \right) : j = 1, 2, ..., n, i = 1, 2, ..., n_{y_j} \right\}.$$

Then P is δ-fine and

$$\left| \sum_{j=1}^{n} h(y_j)(b_j - b_{j-1}) - I \right| = \left| \sum_{j=1}^{n} \int_a^b f(x, y_j) dx (b_j - b_{j-1}) - I \right|$$

$$\leq \left| \sum_{j=1}^{n} \int_a^b f(x, y_j) dx (b_j - b_{j-1}) - \sum_{j=1}^{n} \sum_{i=1}^{n_j} f(x_{y_j i}, y_j)(a_{y_j i} - a_{y_j(i-1)})(b_j - b_{j-1}) \right|$$

$$+ \left| \sum_{j=1}^{n} \sum_{i=1}^{n_j} f(x_{y_j i}, y_j)(a_{y_j i} - a_{y_j(i-1)})(b_j - b_{j-1}) - I \right|$$

$$\leq \sum_{j=1}^{n} (b_j - b_{j-1}) \left| \int_a^b f(x, y_j) dx - \sum_{i=1}^{n_j} f(x_{y_j i}, y_j)(a_{y_j i} - a_{y_j(i-1)}) \right| + \bigvee_{i=1}^{\infty} a_{i, \varphi(i)}$$

$$\leq \bigvee_{i=1}^{\infty} b_{i, \varphi(i)}(d - c) + \bigvee_{i=1}^{\infty} a_{i, \varphi(i)} = \bigvee_{i=1}^{\infty} c_{i, \varphi(i)} + \bigvee_{i=1}^{\infty} a_{i, \varphi(i)} \leq \bigvee_{i=1}^{\infty} d_{i, \varphi(i)},$$

where $c_{i,j} = b_{i,j}(d - c)$, $(c_{i,j})_{i,j}$ is a regulator in R and $(d_{i,j})_{i,j} = 2(c_{i,j} + a_{i,j})$. The sequence

$(d_{i,j})_{i,j}$ is a regulator in R such that, for every δ'-fine partition F of $[c, d]$, $F = \left\{ \left([b_{j-1}, b_j], y_j \right) : j = 1, 2, ..., n \right\}$, we have

$$\left| \sum_{j=1}^{n} h(y_j)(b_j - b_{j-1}) - I \right| \leq \bigvee_{i=1}^{\infty} d_{i, \varphi(i)}. \quad \text{🍎}$$

5 Kurzweil - Henstock Integral in Topological Spaces

Abstract: We deal with the Kurzweil-Henstock integral for functions defined in an abstract compact topological space and taking values in Riesz spaces.

We introduce the theory and the fundamental properties, and in particular we prove the monotone convergence theorem.

5.1 Elementary Properties

In the chapter we shall work with a compact topological space instead of a compact interval $[a,b] \subset \mathbb{R}$.

Assumption 5.1 We assume that $(T,\mathcal{T})$ is a Hausdorff compact topological space, and that there are: a family $\mathcal{B}$ of Borel subsets of T such that $T \in \mathcal{B}$ and closed under intersections, and a mapping $\mu : \mathcal{B} \to \mathbb{R}^+$.

Definition 5.2 We say that $\mu : \mathcal{B} \to \mathbb{R}^+$ is *additive,* if

$$\mu(E \cup F) + \mu(E \cap F) = \mu(E) + \mu(F)$$

whenever $E, F, E \cup F \in \mathcal{B}$.

Let $\mathcal{S}_0$ be the σ-algebra of all Borel subsets of T. We say that $\mu : \mathcal{S}_0 \to \mathbb{R}^+$ is *regular,* if to any $E \in \mathcal{S}_0$ and any $\varepsilon > 0$ there exist C compact and U open such that $C \subset E \subset U$ and $\mu(U) < \mu(C) + \varepsilon$.

Definition 5.3 A *gauge* on T is a mapping $\gamma : T \to \mathcal{T}$ assigning to each $t \in T$ an open neighborhood $\gamma(t)$.

Example 5.4 Let $T = [a,b]$, $\delta : T \to \mathbb{R}^+$ be a map. Let $\mathcal{T}$ be the usual topology on the real line. Put $\gamma(t) = (t - \delta(t), t + \delta(t))$. Then γ is a gauge in the sense of Definition 5.3.

Definition 5.5 A *partition* of T is a finite collection $\{(E_i, t_i) : i = 1, \dots, k\}$ of couples such that

(i) $\bigcup_{i=1}^{k} E_i = T$;

(ii) $t_i \in E_i, E_i \in \mathcal{B}$;

(iii) $\mu(E_i \cap E_j) = 0 \, (i \neq j)$.

A collection Π satisfying axioms (ii) and (iii), but not necessarily (i), is called *decomposition* of T. The partition or decomposition Π is γ-*fine* $(\Pi \prec \gamma)$, if $E_i \subset \gamma(t_i) (i = 1, 2, \dots, k)$.

Definition 5.6 We say that $\mathcal{B}$ is *separating,* if there exists a sequence $(\Pi_n)_n$ of partitions such that Π_{n+1} is a refinement of $\Pi_n (n \in \mathbb{N})$ and for any $x, y \in T, x \neq y$ there exist $n \in \mathbb{N}$ and $E \in \Pi_n$ such that, if $x \in \overline{E}$ for some $E \in \Pi_n$, then $y \in T \setminus \overline{E}$.

Antonio Boccuto / Beloslav Riečan / Marta Vrábelová

Example 5.7 The compact interval $[a,b]$ with the usual topology is a Hausdorff compact topological space. The family $\mathcal{B}$ of all compact subintervals (including singletons and $\varnothing$) is separating, and μ defined by $\mu([c,d]) = d - c$ is additive and regular.

Lemma 5.8 *If $\mathcal{B}$ is separating, then to any gauge $\gamma : T \to \mathcal{T}$ there exists a γ-fine partition.*

Proof: Let us consider Π_1. We want to prove that the space T is *decomposable,* i.e. that there exists a partition $\{(E_i, t_i) : i = 1, \ldots, n\}$ of T such that $E_i \subset \gamma(t_i)(i = 1, \ldots, n)$. If not, then one of the elements of Π_1 (denote it by C_1) is not decomposable. Similarly there is an indecomposable $C_2 \in \Pi_2, C_2 \subset C_1$, etc. Evidently $\bigcap_n \overline{C_n} \neq \varnothing$. Let $x \in \bigcap_n \overline{C_n}$. Since $\mathcal{B}$ is separating, $\bigcap_n \overline{C_n} = \{x\}$. Since $\bigcap_n \overline{C_n} \subset \gamma(x)$, there exists n such that $x \in \overline{C_n} \subset \gamma(x)$. But then the singleton $\{(C_n, x)\}$ is a partition of C_n. This is a contradiction with the assumption that C_n is indecomposable. 🍎

Remark 5.9 If (T, d) is a compact metric space, $\mathcal{T}$ is a semiring of subsets of T such that to any $x \in T$ and every neighborhood U of x there exists $E \in \mathcal{B}$ such that $x \in \operatorname{int} E \subset E \subset U$ (where, given any subset E of any topological space, $\operatorname{int} E$ denotes its interior in the topological sense) and $\mathcal{B}$ is separating, then in correspondence with any set $B \in \mathcal{B}$ there exists a $\gamma_{|B}$-fine partition (see also [217], Lemma 1.2., p. 154 and Proposition 1.7., p. 156).

Definition 5.10 (see also [217]) A function $f : T \to \mathbb{R}$ is *(KH)-integrable* (or, in short, *integrable*) if there exists $I \in \mathbb{R}$ such that to any $\varepsilon > 0$ there exist a gauge γ such that

$$\left| \sum_\Pi f - I \right| < \varepsilon$$

for every partition $\Pi \prec \gamma$. Here the entity

$$\sum_\Pi f = \sum_{i=1}^n f(t_i) \mu(E_i)$$

is called *Riemann sum* or *integral sum*.

Evidently the number I is determined uniquely. It will be denoted by

$$(KH)\int_T f \, d\mu, \quad \int_T f \, d\mu \quad \text{or} \quad \int f \, d\mu.$$

Proposition 5.11 *The integral is a linear positive functional.*

Proof: The linearity follows by the identity

$$\sum_\Pi (\alpha f + \beta g) = \alpha \sum_\Pi f + \beta \sum_\Pi g.$$

The positivity follows by the implication

$$f \geq 0 \Rightarrow \sum_\Pi f \geq 0.$$
🍎

Proposition 5.12 *(Bolzano-Cauchy condition) A function $f : T \to \mathbb{R}$ is integrable if and only if the following condition holds:*

To any $\varepsilon > 0$ *there exists a gauge γ such that*

$$\left|\sum_{\Pi_1} f - \sum_{\Pi_2} f\right| < \varepsilon$$

for every two partitions $\Pi_1, \Pi_2 \prec \gamma$.

Proof: The necessity of this condition is evident. We now turn to the sufficiency. If the condition is satisfied, we can put $\varepsilon = \frac{1}{n}$. So, there exists a gauge γ_n such that

$$\left|\sum_{\Pi_1} f - \sum_{\Pi_2} f\right| < \frac{1}{n}$$

for every two partitions $\Pi_1, \Pi_2 \prec \gamma$. Put

$$a_n = \inf\{\sum_{\Pi} f : \Pi \prec \gamma_n\},$$

$$b_n = \sup\{\sum_{\Pi} f : \Pi \; p \; \gamma_n\}.$$

Then there exists exactly one I that belongs to the set $\bigcap_{n=1}^{\infty}[a_n, b_n]$. For every $\varepsilon > 0$ choose n such that $\frac{1}{n} < \varepsilon$ and put $\gamma = \gamma_n$. Then for each partition $\Pi \prec \gamma$ we obtain

$$\left|\sum_{\Pi} f - I\right| \le b_n - a_n \le \frac{1}{n} < \varepsilon. \quad \text{\tiny }$$

Definition 5.13 A function $f : T \to \mathbb{R}$ is said to be *(KH)-integrable* (or, in short, *integrable*) on a set $E \in \mathcal{B}$ if there exists $I_E \in \mathbb{R}$ such that to any $\varepsilon > 0$ there exist a gauge γ such that

$$\left|\sum_{\Pi} f - I_E\right| < \varepsilon$$

for every partition $\Pi \prec \gamma$ of E.

The element I_E is denoted by $\int_E f \, d\mu$.

Proposition 5.14 *If f is integrable on $E \in \mathcal{B}$, $F \subset E$, $F \in \mathcal{B}$, and $E \setminus F \in \mathcal{B}$, then f is integrable on F too.*

Proof: We shall use the Bolzano-Cauchy condition. To any $\varepsilon > 0$ there exists a gauge $\gamma : E \to T$ such that

$$\left|\sum_{\Pi_1} f - \sum_{\Pi_2} f\right| < \varepsilon$$

for every $\Pi_1, \Pi_2 \prec \gamma$. Put $\gamma_0 = \gamma_{|F}$ and $\Pi, \Pi' \prec \gamma_0$. Choose $\Pi_0 \prec \gamma|_{(E \setminus F)}$ and define $\Pi_1 = \Pi \cup \Pi_0$, $\Pi_2 = \Pi' \cup \Pi_0$. Then $\Pi_1, \Pi_2 \prec \gamma$ so that

$$\left|\sum_{\Pi_1} f - \sum_{\Pi_2} f\right| < \varepsilon.$$

But

$$\sum_{\Pi_1} f = \sum_{\Pi} f + \sum_{\Pi_0} f,$$

$$\sum_{\Pi_2} f = \sum_{\Pi'} f + \sum_{\Pi_0} f,$$

so that

$$\left|\sum_{\Pi} f - \sum_{\Pi'} f\right| < \varepsilon$$

for all $\Pi, \Pi' \prec \gamma_0$. 🍎

Proposition 5.15 *If* $E, F, G \in \mathcal{B}$, $E = F \cup G$, $\mu(F \cap G) = 0$, *and* f *is integrable on* E, *then*

$$\int_E f \, d\mu = \int_F f \, d\mu + \int_G f \, d\mu$$

Proof: Since f is integrable on E, we have to any $\varepsilon > 0$ a gauge $\gamma : E \to \mathcal{T}$ such that

$$\left|\sum_{\Pi} f - \int_E f \, d\mu\right| < \varepsilon$$

whenever $\Pi \prec \gamma$. Choose $\gamma_1 : F \to \mathcal{T}$, $\gamma_1 \subset \gamma|_F$ and $\gamma_2 : G \to \mathcal{T}$, $\gamma_2 \subset \gamma|_G$ such that

$$\left|\sum_{\Pi_1} f - \int_F f \, d\mu\right| < \varepsilon, \quad \left|\sum_{\Pi} f - \int_G f \, d\mu\right| < \varepsilon$$

whenever $\Pi_1 \prec \gamma_1, \Pi_2 \prec \gamma_2$. Evidently $\Pi_1 \cup \Pi_2 \prec \gamma$, so that

$$\left|\int_E f \, d\mu - \int_F f \, d\mu - \int_G f \, d\mu\right|$$

$$\leq \left|\int_E f \, d\mu - \sum_{\Pi_1 \cup \Pi_2} f\right| + \left|\sum_{\Pi_1 \cup \Pi_2} f - \sum_{\Pi_1} f - \sum_{\Pi_2} f\right|$$

$$+ \left|\sum_{\Pi_1} f - \int_F f \, d\mu\right| + \left|\sum_{\Pi_2} f - \int_G f \, d\mu\right| < 3\varepsilon.$$ 🍎

By induction, we can prove the following:

Proposition 5.16 *If* $E \in \mathcal{B}$, $E_i \in \mathcal{B}$, $i = 1, ..., n$, *and* $E = \bigcup_{i=1}^{n} E_i$, $\mu(E_i \cap E_j) = 0$ *whenever* $i \neq j$ *and* f *is integrable on* E, *then* f *is integrable on* E_i *for every* i, *and*

$$\int_E f \, d\mu = \sum_{i=1}^{n} \int_{E_i} f \, d\mu.$$

We note that, if we want to get an integral corresponding with our intuition, then the value of the integral of the simple function $\sum_{i=1}^{n} \alpha_i \chi_{E_i}$ should be $\sum_{i=1}^{n} \alpha_i \mu(E_i)$. To do this, let us prove the following:

Theorem 5.17 *Let* $\mu : \mathcal{S}_0 \to \mathbb{R}^+$ *be additive and regular,* $E \in \mathcal{S}_0$. *Then* χ_E *is integrable, and*

$$\int \chi_E d\mu = \mu(E).$$

Proof: Consider the chain $C \subset E \subset U$ according to Definition 5.2. Choose $\gamma : T \to \mathcal{T}$ such that

$$\gamma(t) \cap C = \varnothing,$$

if $t \notin C$,

$$\gamma(t) \subset U,$$

if $t \in C$. Let $\Pi \prec \gamma, \Pi = \{(E_i, t_i) : i = 1, \ldots, k\}$ be a partition. Then

$$\mu(C) \leq \sum_{i=1}^{k} \chi_E(t_i) \mu(E_i) = \sum_{t_i \in E} \mu(E_i) =$$
$$= \mu(\bigcup_{t_i \in E} E_i) \leq \mu(U).$$

Since $\mu(C) \leq \mu(E) \leq \mu(U)$, and $\mu(C) \leq \sum_{\Pi} \chi_E \leq \mu(U)$, we obtain

$$\left| \sum_{\Pi} \chi_E - \mu(E) \right| \leq \mu(U) - \mu(C) < \varepsilon.$$

This concludes the proof. 🍎

5.2 Monotone Convergence Theorem

We shall prove the following theorem:

Theorem 5.18 *Let* $(f_n : T \to \mathbb{R})_n$ *be a sequence of integrable functions,* $f_n \leq f_{n+1} (n \in \mathbb{N})$, $f(x) = \lim_{n \to +\infty} f_n(x)$ *for every* $x \in T$ *and let the sequence* $\left(\int f_n d\mu \right)_n$ *be bounded. Then* f *is integrable, and*

$$\int f \, d\mu = \lim_{n \to +\infty} \int f_n \, d\mu.$$

Before the proof of the theorem we shall expose a variant of the Henstock lemma which will be used in the proof.

Lemma 5.19 *(Henstock) Let* g *be an integrable function and let* $\gamma : T \to \mathcal{T}$ *be a gauge such that*

$$\left| \sum_{\Pi'} g - \int g d\mu \right| < \varepsilon$$

for every partition $\Pi' \prec \gamma$. *If* $\Pi = \{(E_i, t_i) : i = 1, \ldots, n\}$ *is any decomposition of* T, *then*

$$\left| \sum_{i\in L} g(t_i)\mu(E_i) - \sum_{i\in L} \int_{E_i} g\, d\mu \right| < \varepsilon$$

for every $L \subset \{1,\dots,n\}, L \neq \varnothing$.

Proof: Let $\eta > 0$. Put $\gamma_i = \gamma|_{E_i}$ and choose $\Pi_i \prec \gamma_i$ such that

$$\left| \sum_{\Pi_i} g - \int_{E_i} g\, d\mu \right| < \frac{\eta}{2^i}, \quad i = 1,\dots,n.$$

Let

$$\Pi = \{(E_i, t_i): i \in L\} \cup \bigcup_{i\in L'} \Pi_i,$$

where $L' = \{1,\dots,n\} \setminus L$. Evidently Π is a partition of T such that $\Pi \prec \gamma$, hence

$$\left| \sum_{\Pi} g - \int g\, d\mu \right| < \varepsilon.$$

But

$$\sum_{\Pi} g = \sum_{i\in L} g(t_i)\mu(E_i) + \sum_{i\in L'} \sum_{\Pi_i} g.$$

Therefore

$$\left| \sum_{i\in L} g(t_i)\mu(E_i) - \sum_{i\in L} \int_{E_i} g\, d\mu \right| =$$

$$= \left| \sum_{\Pi} g - \sum_{i\in L'} \sum_{\Pi_i} g - \int g\, d\mu + \sum_{i\in L'} \int_{E_i} g\, d\mu \right| \le$$

$$\le \left| \sum_{\Pi} g - \int g\, d\mu \right| + \sum_{i\in L'} \left| \sum_{\Pi_i} g - \int_{E_i} g\, d\mu \right| <$$

$$< \varepsilon + \eta.$$

Since the last inequality holds for every $\eta > 0$, we obtain the desired property.

Now we are able to prove Theorem 5.18.

Proof of Theorem 5.18: Let ε be a positive real number. Since the sequence $\left(\int f_n d\mu \right)_n$ is convergent, there exists m_0 such that

$$\left| \int f_i\, d\mu - \int f_m\, d\mu \right| < \varepsilon \tag{5.1}$$

for every $i, m \ge m_0$. By the fact that f_p is integrable $\forall p \in \mathbb{N}$ there exists a gauge γ_p such that for every partition $\Pi \prec \gamma_p$ we have

$$\left| \sum_{\Pi} f_p - \int f_p d\mu \right| < \frac{\varepsilon}{2^p}. \tag{5.2}$$

Since $\lim_{n\to+\infty} f_n(x) = f(x)$ for every $x \in T$, there exists $p(x) \ge m_0$ such that

$$|f_i(x) - f(x)| < \varepsilon \tag{5.3}$$

for every $i \geq p(x)$. Put

$$\gamma(x) = \gamma_1(x) \cap \gamma_2(x) \cap ... \cap \gamma_{p(x)}(x)$$

and choose a partition $\Pi \prec \gamma$. Then for every $i \in \mathbb{N}$ we have:

$$\left| \sum_{\Pi} f_i - \int f_i \, d\mu \right| \leq \left| \sum_{p(x_j) \geq i} f_i(x_j) \mu(E_j) - \sum_{p(x_j) \geq i} \int_{E_j} f_i \, d\mu \right|$$

$$+ \left| \sum_{p(x_j) < i} f_i(x_j) \mu(E_j) - \sum_{p(x_j) < i} \int_{E_j} f_i \, d\mu \right|. \qquad (5.4)$$

Construct Π similarly as in the proof of the Henstock lemma (Lemma 5.19), i.e.

$$\Pi = \{(E_j, x_j) : i \leq p(x_j)\} \cup \bigcup_{p(x_j) < i} \Pi_j,$$

where Π_j is a sufficiently fine partition of E_j, in such a way that $\Pi \prec \gamma_j$. Then

$$\left| \sum_{\Pi} f_i - \int f_i \, d\mu \right| < \frac{\varepsilon}{2^i}.$$

Hence by Lemma 5.19 $(L = \{j : p(x_j) \geq i\})$ we obtain

$$\left| \sum_{p(x_j) \geq i} f_i(x_j) \mu(E_j) - \sum_{p(x_j) \geq i} \int_{E_j} f_i \, d\mu \right| \leq \frac{\varepsilon}{2^i}. \qquad (5.5)$$

We shall now estimate the second part of the right side of (5.4). Evidently

$$\sum_{p(x_j) < i} f_i(x_j) \mu(E_j) = \sum_{m=1}^{i-1} \sum_{p(x_j) = m} f_i(x_j) \mu(E_j).$$

Consider $\Pi = \{(E_i, x_i) : i = 1 ..., n\} \prec \gamma$ and choose j such that $p(x_j) < i$. Let $m = p(x_j)$ and $L = \{k : p(x_k) = m\}$ so that $j \in L$. Now we can construct $\Pi' \prec \gamma_m$ such that $(E_k, x_k) \in \Pi'$ for every $k \in L$. Since

$$\left| \sum_{\Pi'} f_m - \int f_m \, d\mu \right| < \frac{\varepsilon}{2^m},$$

we obtain by Lemma 5.19

$$\left| \sum_{k \in L} f_m(x_k) \mu(E_k) - \sum_{k \in L} \int_{E_k} f_m \, d\mu \right| \leq \frac{\varepsilon}{2^m},$$

i.e.

$$\left| \sum_{p(x_k) = m} f_{p(x_k)}(x_k) \mu(E_k) - \sum_{p(x_k) = m} \int_{E_k} f_{p(x_k)} \, d\mu \right| \leq \frac{\varepsilon}{2^m},$$

so that

$$\left| \sum_{m=1}^{i-1} \sum_{p(x_k)=m} f_{p(x_k)}(x_k) - \sum_{m=1}^{i-1} \sum_{p(x_k)=m} \int_{E_k} f_{p(x_k)} d\mu \right| < \varepsilon. \qquad (5.6)$$

Furthermore, from (5.3) we have

$$\left| \sum_{p(x_j)<i} f_i(x_j)\mu(E_j) - \sum_{p(x_j)<i} f_{p(x_j)}(x_j)\mu(E_j) \right|$$

$$\leq \sum_{m=1}^{i-1} \sum_{p(x_j)=m} |f_i(x_j) - f_{p(x_j)}(x_j)| \, \mu(E_j) < 2\varepsilon. \qquad (5.7)$$

From (5.6) and (5.7) it follows

$$\left| \sum_{p(x_j)<i} f_i(x_j)\mu(E_j) - \sum_{p(x_j)<i} \int_{E_j} f_{p(x_j)} d\mu \right|$$

$$\leq \left| \sum_{p(x_j)<i} f_i(x_j)\mu(E_j) - \sum_{p(x_j)<i} f_{p(x_j)}(x_j)\mu(E_j) \right| \qquad (5.8)$$

$$+ \left| \sum_{p(x_j)<i} f_{p(x_j)}(x_j)\mu(E_j) - \sum_{p(x_j)<i} \int_{E_j} f_{p(x_j)} d\mu \right| < 3\varepsilon.$$

Since the sequence $(f_i)_i$ is nondecreasing, we have $f_i - f_{m_0} \geq 0$. Hence for any Borel set A we have that

$$\left| \int_A f_i \, d\mu - \int_A f_{m_0} \, d\mu \right| = \int_A (f_i - f_{m_0}) \, d\mu \leq \int (f_i - f_{m_0}) \, d\mu < \varepsilon.$$

By (5.8) and the last inequality we obtain (taking $A = \bigcup_{j \in L} E_j$)

$$\sum_{j \in L} \int_{E_j} f_i \, d\mu - 4\varepsilon = \int_{\bigcup_{j \in L} E_j} f_i \, d\mu - 4\varepsilon < \int_A f_{m_0} \, d\mu - 3\varepsilon$$

$$\leq \sum_{j \in L} \int_{E_j} f_{p(x_j)} d\mu - 3\varepsilon < \sum_{j \in L} f_i(x_j)\mu(E_j) < \sum_{j \in L} \int_{E_j} f_{p(x_j)} d\mu + 3\varepsilon$$

$$\leq \sum_{j \in L} \int_{E_j} f_i \, d\mu + 3\varepsilon.$$

Hence

$$\left| \sum_{p(x_j)<i} f_i(x_j)\mu(E_j) - \sum_{p(x_j)<i} \int_{E_j} f_i \, d\mu \right| < 4\varepsilon. \qquad (5.9)$$

From (5.4), (5.5) and (5.9) we conclude that for every $\varepsilon > 0$ there exists a gauge γ such that for every $\Pi \prec \gamma$ and every positive integer i

$$\left| \sum_{\Pi} f_i - \int f_i \, d\mu \right| < 5\varepsilon. \qquad (5.10)$$

Since $f_i(x) \to f(x)$ for every $x \in T$, also $\sum_{\Pi} f_i \to \sum_{\Pi} f$, and consequently there exists $i_0 \in \mathbb{N}$ such that

$$\left|\sum_{\Pi} f_i - \sum_{\Pi} f\right| < \varepsilon \qquad (5.11)$$

for every $i \geq i_0$. From (5.10) and (5.11) we obtain

$$\sum_{\Pi} f - 6\varepsilon < \int f_i d\mu < \sum_{\Pi} f + 6\varepsilon.$$

Since the last inequalities hold for all sufficiently large i, we have

$$\sum_{\Pi} f - 6\varepsilon \leq \lim_{i \to +\infty} \int f_i d\mu \leq \sum_{\Pi} f + 6\varepsilon.$$

Therefore there is $I = \lim_{i \to +\infty} \int f_i d\mu$ such that for every $\varepsilon > 0$ there exists a gauge γ such that

$$\left|I - \sum_{\Pi} f\right| \leq 6\varepsilon$$

for every partition $\Pi \prec \gamma$. Hence f is integrable and

$$\int f \, d\mu = I = \lim_{i \to +\infty} \int f \, d\mu.$$

This concludes the proof. ◆

6 Convergence Theorems

Abstract: We continue to investigate the topics of Chapter 5, following the approach there introduced, and prove some versions of the Henstock Lemma, the Beppo Levi and the Lebesgue dominated convergence theorem.

Note that the involved measures and the domain of our functions can be even unbounded.

6.1 Elementary Properties

In the book we have exposed the elementary theory of the Kurzweil-Henstock integral (Chapter 1), then basic facts about the Kurzweil-Henstock integral for functions from a compact interval to a Riesz space (Chapter 3). Of course, in Chapter 5 there have been considered functions defined on a compact topological space. In this chapter we deal with convergence theorems (monotone convergence theorem and Lebesgue dominated convergence theorem) for the Kurzweil-Henstock integral in an abstract setting, for functions defined in a compact topological space T, satisfying suitable properties, and with values in a Dedekind complete Riesz space, with respect to a positive Riesz space-valued measure μ, which can assume even the value $"+\infty"$. A particular case is $T = [A, B]$ an interval (possibly unbounded) of the extended real line, or the whole of $\tilde{\mathbb{R}}$, and $\mu =$ the Lebesgue measure (this case was investigated in [26]). We continue the investigation started in [24], in which μ is $\tilde{\mathbb{R}}$-valued, and in which some other kinds of convergence theorems were demonstrated. Our results given here are proved in [29] and extend the ones in [227], Chapter 5, which were proved in the case in which $\mu(T)$ is finite, and the ones of [170], which were proved in the case $T = [A, B]$, where $[A, B]$ is as above, and all the involved Riesz spaces coincide with the (eventually extended) real line. A similar Kurzweil-Henstock type integral was investigated in [36] for Banach space-valued maps.

From now on, in this chapter we shall always suppose that R is a Dedekind complete weakly σ-distributive Riesz space.

Assumption 6.1 Let $(T, \mathcal{T})$, $\mathcal{B}$ be as in Chapter 5, and let us consider a positive mapping $\mu : \mathcal{B} \to R$.

Definition 6.2 We say that $\mu : \mathcal{B} \to R$ is *additive*, if

$$\mu(E \cup F) + \mu(E \cap F) = \mu(E) + \mu(F)$$

whenever $E, F, E \cup F \in \mathcal{B}$.

Let $\mathcal{S}$ be the σ-algebra of all Borel subsets of T. We say that a positive set function $\mu : \mathcal{S} \to R$ is *regular*, if to any $E \in \mathcal{S}$ there exists a regulator $(a_{i,j})_{i,j}$ such that for every $\varphi \in \Phi$ there exist C compact, U open and $F, G \in \mathcal{S}$ such that $F \subset C \subset E \subset U \subset G$, $\mu(G) \leq \mu(F) + \bigvee_{i=1}^{\infty} a_{i,\varphi(i)}$.

The concepts of gauge, partition, decomposition, γ-fineness and separating family are given analogously as in Chapter 5.

Assumptions 6.3 Let R_1, R_2, R be three Dedekind complete Riesz spaces. We say that (R_1, R_2, R) is a *product triple* if there exists a map $\cdot : R_1 \times R_2 \to R$, which we will call *product*, satisfying the conditions given at the end of Chapter 2 and such that

$$\text{if } r_j \in R_j, \ j = 1, 2 \text{ and } \alpha \in \mathbb{R}, \text{ then } (\alpha r_1) \cdot r_2 = r_1 \cdot (\alpha r_2) = \alpha(r_1 \cdot r_2).$$

We will write often ab instead of $a \cdot b$. A Dedekind complete Riesz space R is called an *algebra* if (R, R, R) is a product triple.

We always assume that (R_1, R_2, R) is a product triple and that R is weakly σ-distributive. Furthermore, we add to R_2 two extra elements, $+\infty$ and $-\infty$, extending ordering and operations in a natural way, and denote $\overline{R_2} = R_2 \cup \{+\infty, -\infty\}$.

We now give our definition of Kurzweil-Henstock integrability. We suppose that $\mu : \mathcal{B} \to \overline{R_2}$ is an additive positive regular measure. If $\Pi = \{(E_i, t_i) : i = 1, \dots, q\}$ is a partition or a decomposition of a set $A \in \mathcal{B}$, and $f : T \to R_1$, then we define the *Riemann sum* as follows:

$$\sum_\Pi f = \sum_{i=1}^q f(t_i) \mu(E_i)$$

if the sum exists in R, with the conventions that $0 \cdot (+\infty) = 0 \cdot (-\infty) = 0$ and that only the E_i's with $\mu(E_i) \neq +\infty$ are involved. The points t_i, $i = 1, \dots, q$, are called *tags*.

Definition 6.4 A function $f : T \to R_1$ is (KH)-*integrable* (or, in short, *integrable*) if there exist $I \in R$ and a regulator $(a_{i,j})_{i,j}$ such that $\forall \varphi \in \Phi$ there exists a gauge γ such that

$$\left| \sum_\Pi f - I \right| \leq \bigvee_{i=1}^\infty a_{i,\varphi(i)} \tag{6.1}$$

whenever $\Pi = \{(E_i, t_i), i = 1, \dots, q\}$ is a γ-fine partition of T.

Evidently the number I is determined uniquely. It will be denoted by

$$(KH)\int_T f \, d\mu, \int_T f \, d\mu \text{ or } \int f \, d\mu.$$

It is easy to check that, even in this context, the involved integral is a linear positive functional.

Definition 6.5 A function $f : T \to R_1$ is (KH)-*integrable* (or, in short, *integrable*) on a set $E \in \mathcal{B}$ if there exist $I_E \in R$ and a regulator $(a_{i,j}^{(E)})_{i,j}$ such that $\forall \varphi \in \Phi$ there exists a gauge γ such that

$$\left| \sum_\Pi f - I \right| \leq \bigvee_{i=1}^\infty a_{i,\varphi(i)}^{(E)} \tag{6.2}$$

whenever $\Pi = \{(E_i, t_i), i = 1, \dots, q\}$ is a γ-fine partition of E.

The element I_E is denoted by $\int_E f\, d\mu$. When we will say simply "integrable", we will intend "integrable on T".

We now state the Bolzano-Cauchy condition.

Theorem 6.6 *A map* $f: T \to R_1$ *is* (KH)-*integrable on* $E \in \mathcal{B}$ *if and only if there exists a regulator* $(b_{i,j}^{(E)})_{i,j}$ *such that,* $\forall \varphi \in \Phi$, $\exists$ *a gauge* γ *such that for all* γ-*fine partitions* Π_1, Π_2 *of* E *we have*

$$\left| \sum_{\Pi_1} f - \sum_{\Pi_2} f \right| \leq \bigvee_{i=1}^{\infty} b_{i,\varphi(i)}^{(E)}.$$

The proof is similar to the one of [227], Proposition 5.2.9, pp. 77-79.

Proposition 6.7 *If* $E,F,G \in \mathcal{B}$, $E = F \cup G$, $\mu(F \cap G) = 0$ *and* f *is integrable on* E, *then* f *is integrable on* F *and on* G *too, and*

$$\int_E f\, d\mu = \int_F f\, d\mu + \int_G f\, d\mu.$$

The proof is similar to the one of [227], Proposition 5.2.10, pp. 79-80 (see also Chapter 5, Propositions 5.14 and 5.15).

By induction, it is possible to prove the following:

Proposition 6.8 *If* $E \in \mathcal{B}$, $E_i \in \mathcal{B}$, $i = 1,...,n$, *and* $E = \cup_{i=1}^{n} E_i$, $\mu(E_i \cap E_j) = 0$ *whenever* $i \neq j$ *and* f *is integrable on* E, *then* f *is integrable on* E_i *for every* i, *and*

$$\int_E f\, d\mu = \sum_{i=1}^{n} \int_{E_i} f\, d\mu.$$

(see also Chapter 5, Proposition 5.16).

The following result holds (see also [227], Proposition 5.2.11, pp. 80-81):

Theorem 6.9 *Let* $\mathcal{S}$ *be the class of all Borel sets of* T, $\mu: \mathcal{S} \to \overline{R_2}$ *be positive, additive and regular,* $E \in \mathcal{S}$, *with* $\mu(E) \in R_2$. *Let* $r \in R_1$, *and* $g = \chi_E r$ *(defined by the relation* $g(t) = r$, *if* $t \in E$ *and* $g(t) = 0$, *if* $t \notin E$*). Then* g *is integrable, and*

$$\int_T g\, d\mu = r\, \mu(E).$$

Proof: First of all, assume $r \geq 0$. By regularity of μ on $\mathcal{S}$ there exists a (D)-sequence $(a_{i,j})_{i,j}$ such that for every $\varphi \in \Phi$ there exist an open set U and a compact set C, $C \subset E \subset U$, such that

$$\mu(U \setminus C) \leq \bigvee_{i=1}^{\infty} a_{i,\varphi(i)}.$$

Since C is compact and U is open, there exists a gauge η such that $\eta(t) \subset U$ $\forall t \in C$, $\eta(t) \subset U \setminus C$ $\forall t \in U \setminus C$, $\eta(t) \cap C = \emptyset$ $\forall t \notin U$. Take any partition $\Pi \prec \eta$, $\Pi = \{(E_i, t_i) : i = 1,...,n\}$. Then

$$r\,\mu(C) \le r\,\mu(E) \le r\,\mu(U),$$

$$r\,\mu(U \setminus C) = r\,\mu(U) - r\,\mu(C),$$

and

$$r\,\mu(E) - r\bigvee_{i=1}^{\infty} a_{i,\varphi(i)} \le r\,\mu(U) - r\bigvee_{i=1}^{\infty} a_{i,\varphi(i)} \le r\,\mu(C)$$

$$\le r\mu\left(\bigcup_{t_i \in C} E_i\right) \le \sum_{t_i \in C} r\,\mu(E_i)$$

$$= \sum_{i=1}^{n} \chi_C(t_i) r\,\mu(E_i) \le \sum_{i=1}^{n} \chi_E(t_i) r\,\mu(E_i) = \sum_{\Pi} g$$

$$\le \sum_{t_i \in U} r\,\mu(E_i) \le r\,\mu(U) \le r\,\mu(E) + r\bigvee_{i=1}^{\infty} a_{i,\varphi(i)},$$

and hence

$$\left| \sum_{\Pi} g - r\,\mu(E) \right| \le r\bigvee_{i=1}^{\infty} a_{i,\varphi(i)}$$

for any partition $\Pi \prec \eta$. From the properties of the product it follows that the double sequence $(r \cdot a_{i,j})_{i,j}$ is a (D)-sequence. Thus, we get the assertion, at least when $r \ge 0$.

In the general case $(r \in R_1)$ we get:

$$\int_T \chi_E\, r\, d\mu = \int_T \chi_E\, (r^+ - r^-)\, d\mu$$
$$= \int_T \chi_E\, r^+\, d\mu - \int_T \chi_E\, r^-\, d\mu$$
$$= r^+\, \mu(E) - r^-\, \mu(E) = r\,\mu(E). \quad \blacksquare$$

6.2 General Convergence Theorems

We begin with proving a theorem, which will be used in the sequel, in order to demonstrate our versions of convergence theorems.

Theorem 6.10 *Let $(f_n : T \to R_1)_n$ be a sequence of integrable functions. Suppose that*

   6.10.1) *there is a regulator $(b_{i,j})_{i,j}$ such that to every $\varphi \in \Phi$ there exist a gauge ζ and $n_0 \in \mathbb{N}$ such that*

$$\left| \int_T f_n\, d\mu - \sum_{\Pi} f_n \right| \le \bigvee_{i=1}^{\infty} b_{i,\varphi(i)}$$

for every partition $\Pi \prec \zeta$ and $n \ge n_0$;

6.10.2) *there exist a function* $f : T \to R_1$, *a (KH)-integrable map* $h^* : T_0 \to \mathbb{R}^+$ *(with respect to* μ *) and a regulator* $(a^*_{i,j})_{i,j}$ *such that,* $\forall \varphi \in \Phi$, $\forall t \in T_0$, $\exists p(t) \in \mathbb{N} : \forall n \geq p(t)$,

$$|f_n(t) - f(t)| \leq h^*(t) \left(\bigvee_{i=1}^{\infty} a^*_{i,\varphi(i)} \right). \tag{6.3}$$

Then f *is integrable and*

$$(D) \lim_n \int_T f_n d\mu = \int_T f \, d\mu.$$

Proof: We shall use the Bolzano-Cauchy condition. Let $(b_{i,j})_{i,j}$, ζ and n_0 be as in 6.10.1). By 6.10.2) we get the existence of an element $0 \leq w \in R_2$ such that $\forall \varphi \in \Phi$ there exists a gauge $\eta \subset \zeta$ (without loss of generality) such that, for every η-fine partition Π of T, $\Pi = \{(E_i, t_i) : i = 1, ..., q\}$, we have:

$$\left| \sum_{\Pi} f - \sum_{\Pi} f_n \right| \leq \sum_{\Pi} [f(t_i) - f_n(t_i)]\mu(E_i) \tag{6.4}$$

$$\leq \sum_{\Pi} h^*(t_i) \left(\bigvee_{i=1}^{\infty} a^*_{i,\varphi(i)} \right) \mu(E_i) \tag{6.5}$$

$$\leq \left(\bigvee_{i=1}^{\infty} a^*_{i,\varphi(i)} \right) w$$

$\forall n \geq \max\{p(t_i) : i = 1, ..., q\}$. Put $a_{i,j} = a^*_{i,j} w$, $\forall i, j \in \mathbb{N}$.

Without loss of generality, we can suppose that $p(t_i) \geq n_0$ $\forall i = 1, ..., n$. Choose now a regulator $(c_{i,j})_{i,j}$ such that

$$2 \left(\bigvee_{i=1}^{\infty} a_{i,\varphi(i)} + \bigvee_{i=1}^{\infty} b_{i,\varphi(i)} \right) \leq \bigvee_{i=1}^{\infty} c_{i,\varphi(i)}.$$

Then for all partitions $\Pi_1, \Pi_2 \prec \eta$, we have (using sufficiently large n, depending on the involved partitions Π_1 and Π_2):

$$\left| \sum_{\Pi_1} f - \sum_{\Pi_2} f \right| \leq \left| \sum_{\Pi_1} f - \sum_{\Pi_1} f_n \right|$$

$$+ \left| \sum_{\Pi_1} f_n - \int_T f_n d\mu \right| + \left| \int_T f_n d\mu - \sum_{\Pi_2} f_n \right|$$

$$+ \left| \sum_{\Pi_2} f_n - \sum_{\Pi_2} f \right| \leq \bigvee_{i=1}^{\infty} c_{i,\varphi(i)}.$$

The integrability of f follows from this and the Bolzano-Cauchy condition.

By integrability of f we obtain

$$\left| \int_T f \, d\mu - \sum_{\Pi} f \right| \leq \bigvee_{i=1}^{\infty} \overline{a}_{i,\varphi(i+1)}$$

for every partition $\Pi \prec \eta_1$. By 6.10.1) there is a regulator $(b_{i,j})_{i,j}$ such that

$$\left| \sum_{\Pi} f_k - \int_T f_k d\mu \right| \leq \bigvee_{i=1}^{\infty} b_{i,\varphi(i+2)}$$

for every k greater than a suitable integer k_0 (depending on the involved φ) and for each partition $\Pi \prec \eta_2$. By 6.10.2), proceeding as in (6.4), we get the existence of a regulator $(c_{i,j})_{i,j}$ such that

$$\left| \sum_{\Pi} f - \sum_{\Pi} f_k \right| \leq \bigvee_{i=1}^{\infty} c_{i,\varphi(i+3)}$$

for every $k \geq k'$, where k' is a positive integer depending on the involved partition Π. Without loss of generality, we can assume $k' \geq k_0$. Choose a regulator $(d_{i,j})_{i,j}$ such that

$$\bigvee_{i=1}^{\infty} \overline{a}_{i,\varphi(i+1)} + \bigvee_{i=1}^{\infty} b_{i,\varphi(i+2)}$$
$$+ \bigvee_{i=1}^{\infty} c_{i,\varphi(i+3)} \leq \bigvee_{i=1}^{\infty} d_{i,\varphi(i)}.$$

Then (Π being a partition chosen arbitrarily, $\Pi \prec \eta_1 \cap \eta_2$)

$$\left| \int_T f \, d\mu - \int_T f_k d\mu \right| \leq \left| \int_T f \, d\mu - \sum_{\Pi} f \right|$$
$$+ \left| \sum_{\Pi} f - \sum_{\Pi} f_k \right| + \left| \sum_{\Pi} f_k - \int_T f_k d\mu \right| \leq \bigvee_{i=1}^{\infty} d_{i,\varphi(i)}$$

for every $k \geq k'$. We have proved that

$$(D) \lim_k \int_T f_k d\mu = \int_T f \, d\mu$$

with respect to the regulator $(d_{i,j})_{i,j}$. This concludes the proof.

6.3 Henstock Lemma

We now prove a version of the Henstock lemma.

Theorem 6.11 *Let $g : T \to R_1$ be an integrable function. Let $(a_{i,j})_{i,j}$ be a regulator such that to every $\varphi \in \Phi$ there exists a gauge η such that*

$$\left| \int_T g \, d\mu - \sum_{\Pi} g \right| \leq \bigvee_{i=1}^{\infty} a_{i,\varphi(i)}$$

for every partition $\Pi = \{(E_i, t_i) : i = 1, \ldots, n\} \prec \eta$. Then for every $L \neq \varnothing$, $L \subset \{1, \ldots, n\}$, we have:

$$\left| \sum_{i \in L} \int_{E_i} g \, d\mu - \sum_{i \in L} g(t_i)\mu(E_i) \right| \leq \bigvee_{i=1}^{\infty} a_{i,\varphi(i)}.$$

Proof: First of all, we note that integrability of g on all E_i's follows from Proposition 3. So there exists a (D)-sequence $(b_{i,j})_{i,j}$ such that to every $\psi \in \Phi$ there is a gauge η' such that

$$\left| \sum_{i \notin L} \int_{E_i} g \, d\mu - \sum_{i \notin L} \sum_{\Pi_i} g \right| \leq \bigvee_{i=1}^{\infty} b_{i,\psi(i)}$$

whenever $\Pi_i \prec \eta'_{|E_i}$, $i \notin L$. Put $\eta_i = (\eta_{|\overline{E}_i}) \cap \eta'$, take $\Pi'_i \prec \eta_i$, $i \notin L$, and set $\Pi' = \{(E_i, t_i) : i \in L\} \cup (\cup_{i \notin L} \Pi'_i)$. Then $\Pi' \prec \eta$, hence

$$\left| \int_T g \, d\mu - \sum_{\Pi'} g \right| \leq \bigvee_{i=1}^{\infty} a_{i,\varphi(i)},$$

and

$$\left| \sum_{i \notin L} \int_{E_i} g \, d\mu - \sum_{i \notin L} \sum_{\Pi'_i} g \right| \leq \bigvee_{k=1}^{\infty} b_{k,\psi(k)}.$$

Now

$$\left| \sum_{i \in L} \int_{E_i} g \, d\mu - \sum_{i \in L} g(t_i) \mu(E_i) \right|$$

$$= \left| \int_T g \, d\mu - \sum_{i \notin L} \int_{E_i} g \, d\mu - \sum_{\Pi'} g + \sum_{i \notin L} \sum_{\Pi'_i} g \right|$$

$$\leq \left| \int_T g \, d\mu - \sum_{\Pi'} g \right| + \left| \sum_{i \notin L} \sum_{\Pi'_i} g - \sum_{i \notin L} \int_{E_i} g \, d\mu \right|$$

$$\leq \bigvee_{i=1}^{\infty} a_{i,\varphi(i)} + \bigvee_{k=1}^{\infty} b_{k,\psi(k)}.$$

Since

$$\left| \sum_{i \in L} \int_{E_i} g \, d\mu - \sum_{i \in L} g(t_i) \mu(E_i) \right| - \bigvee_{i=1}^{\infty} a_{i,\varphi(i)} \leq \bigvee_{k=1}^{\infty} b_{k,\psi(k)}$$

for every $\psi \in \Phi$, by weak σ-distributivity of R we obtain:

$$\left| \sum_{i \in L} \int_{E_i} g \, d\mu - \sum_{i \in L} g(t_i) \mu(E_i) \right| - \bigvee_{i=1}^{\infty} a_{i,\varphi(i)} \leq 0.$$

6.4 Levi's Theorem

We now are ready to prove the monotone convergence theorem.

Theorem 6.12 *Let $(f_n : T \to R_1)_n$ be a sequence of integrable functions, $f_n \leq f_{n+1}$ $(n \in \mathbb{N})$, and let the sequence $\left(\int_T f_n \, d\mu \right)_n$ be bounded. Suppose that*

    6.12.1) there exist a function $f : T \to R_1$, a (KH)-integrable map

$h^*: T \to \mathbb{R}^+$ *(with respect to μ) and a (D)-sequence $(a_{i,j})_{i,j}$ such that, $\forall \varphi \in \Phi$, $\forall t \in T$,*
$\exists p(t) \in \mathbb{N}: \forall n \geq p(t)$,

$$|f(t) - f_n(t)| \leq h^*(t) \left(\bigvee_{i=1}^{\infty} a_{i,\varphi(i)} \right). \tag{6.6}$$

Furthermore, assume that

6.12.2) *there exist $a \in R$, $a \geq 0$, and a gauge $\hat{\gamma}$, such that, for every $\hat{\gamma}$-fine partition Π of T, we have*

$$\left| \sum_{\Pi} f_n - \int_T f_n \, d\mu \right| \leq a \quad \forall n \in \mathbb{N}.$$

Then f is integrable on T, and

$$\int_T f \, d\mu = (D) \lim_n \int_T f_n \, d\mu.$$

Remark 6.13 We note that regularity does imply σ-additivity of μ, at least for R_2-valued positive finitely additive measures.

Moreover we observe that, if needed, a distinguished point x_0 could be dropped from T, provided it is possible to define $f(x_0) = f_n(x_0) = 0$, $n \in \mathbb{N}$, without affecting the other hypotheses.

Furthermore, we observe that, when $R_1 = R_2 = R = \mathbb{R}$ and $T = [A, +\infty]$ or $T = [-\infty, B]$ is a halfline of the extended real line, condition 6.12.1) is equivalent to pointwise convergence of the sequence $(f_n)_n$ to f: indeed, it is sufficient to take the function h^* defined by setting

$$h^*(t) = \frac{1}{1+t^2}, \quad t \in [A, +\infty) \text{ or } t \in (-\infty, B].$$

Proof of Theorem 6.12: Since the sequence $\left(\int_T f_n d\mu \right)_n$ is bounded and increasing, it admits the (D)-limit in R. Thus, there exists a (D)-sequence $(c_{i,j})_{i,j}$ in R such that, for every $\varphi \in \Phi$, there exists $k_0 \in \mathbb{N}$ such that, $\forall k, l \geq k_0$,

$$\left| \int_T f_k d\mu - \int_T f_l d\mu \right| \leq \bigvee_{i=1}^{\infty} c_{i,\varphi(i)}. \tag{6.7}$$

Furthermore, from 6.12.1) we get the existence of an element $0 \leq w \in R_2$ such that $\forall \varphi \in \Phi$ there exists a gauge γ^* such that, for every γ^*-fine partition Π of T, $\Pi = \{(E_i, t_i) : i = 1, \ldots, q\}$, we have:

$$\sum_{\Pi} [f(t_i) - f_{p(t_i)}(t_i)] \mu(E_i) \leq \sum_{\Pi} h^*(t_i) \left(\bigvee_{i=1}^{\infty} a_{i,\varphi(i)} \right) \mu(E_i)$$

$$\leq \left(\bigvee_{i=1}^{\infty} a_{i,\varphi(i)} \right) w. \tag{6.8}$$

Note that in (6.8) the natural numbers $p(t_i)$ can be chosen greater than k_0. Since f_k is integrable $\forall k \in \mathbb{N}$, then for each $k \in \mathbb{N}$ there exists a (D)-sequence $(a_{i,j}^{(k)})_{i,j}$ such that, for every $\varphi \in \Phi$, there exists a gauge γ_k such that for every partition $\Pi \prec \gamma_k$ we have

$$\left| \sum_\Pi f_k - \int_T f_k d\mu \right| \leq \bigvee_{i=1}^{\infty} a_{i,\varphi(i+k+1)}^{(k)}. \qquad (6.9)$$

For each $i, j \in \mathbb{N}$, put $b_{i,j}^{(1)} = 2 a_{i,j} w$, and $b_{i,j}^{(m)} = a_{i,j}^{(m-1)}$ $(m = 2, 3, \ldots)$. Moreover, let a be as in 6.12.2). By virtue of the Fremlin lemma there exists a (D)-sequence $(b_{i,j})_{i,j}$ such that, $\forall \varphi \in \Phi$ and $\forall s \in \mathbb{N}$

$$a \wedge \left(\sum_{m=1}^{s} \left(\bigvee_{i=1}^{\infty} a_{i,\varphi(i+m)}^{(m)} \right) \right) \leq \bigvee_{i=1}^{\infty} b_{i,\varphi(i)}. \qquad (6.10)$$

Let $\varphi \in \Phi$ and $k_0 = k_0(\varphi)$ be as in (6.7). Put

$$\gamma_0(t) = \gamma^*(t) \cap \hat{\gamma}(t) \cap \gamma_1(t) \cap \gamma_2(t) \cap \ldots \cap \gamma_{p(t)}(t),$$

where the involved gauges are the ones associated with φ, as above. Choose a partition $\Pi \prec \gamma_0$, $\Pi = \{(E_i, t_i) : i = 1, \ldots, q\}$. Fix arbitrarily $k > k_0$, where k_0 is as in (6.7). We have:

$$\left| \sum_\Pi f_k - \int_T f_k d\mu \right| \leq \left| \sum_{p(t_i) \geq k} f_k(t_i) \mu(E_i) - \sum_{p(t_i) \geq k} \int_{E_i} f_k d\mu \right| \qquad (6.11)$$

$$+ \left| \sum_{p(t_i) < k} f_k(t_i) \mu(E_i) - \sum_{p(t_i) < k} \int_{E_i} f_k d\mu \right|. \qquad (6.12)$$

Construct Π similarly as in the proof of the Henstock lemma, i.e.

$$\Pi = \{(E_i, t_i) : k \leq p(t_i)\} \cup \left(\bigcup_{p(t_i) < k} \Pi_i \right),$$

where Π_i is a sufficiently fine partition of E_i, in such a way that $\Pi \prec \gamma_k$. Then

$$\left| \sum_\Pi f_k - \int_T f_k d\mu \right| \leq \bigvee_{i=1}^{\infty} a_{i,\varphi(i+k+1)}^{(k)}.$$

Hence, by the Henstock lemma ($L = \{i : p(t_i) \geq k\}$ and $L = \{i : p(t_i) = k\}$), we obtain

$$\left| \sum_{p(t_i) \geq k} f_k(t_i) \mu(E_i) - \sum_{p(t_i) \geq k} \int_{E_i} f_k d\mu \right| \leq \bigvee_{i=1}^{\infty} a_{i,\varphi(i+k+1)}^{(k)} \qquad (6.13)$$

and

$$\left| \sum_{p(t_i) = k} f_k(t_i) \mu(E_i) - \sum_{p(t_i) = k} \int_{E_i} f_k d\mu \right| \leq \bigvee_{i=1}^{\infty} a_{i,\varphi(i+k+1)}^{(k)} \qquad (6.14)$$

respectively. We now estimate (6.12). We have:

$$\left| \sum_{p(t_i)<k} f_k(t_i)\mu(E_i) - \sum_{p(t_i)<k} \int_{E_i} f_k \, d\mu \right| \qquad (6.15)$$

$$\leq \left| \sum_{m=k_0}^{k-1} \sum_{p(t_i)=m} f_k(t_i)\mu(E_i) - \sum_{m=k_0}^{k-1} \sum_{p(t_i)=m} f_{p(t_i)}(t_i)\mu(E_i) \right| + \qquad (6.16)$$

$$+ \left| \sum_{m=k_0}^{k-1} \sum_{p(t_i)=m} f_{p(t_i)}(t_i)\mu(E_i) - \sum_{m=k_0}^{k-1} \sum_{p(t_i)=m} \int_{E_i} f_{p(t_i)} \, d\mu \right|$$

$$+ \sum_{m=k_0}^{k-1} \sum_{p(t_i)=m} \int_{E_i} (f_k - f_m) \, d\mu$$

$$\leq \sum_{m=k_0}^{k-1} \sum_{p(t_i)=m} (f_k(t_i) - f_{p(t_i)}(t_i))\mu(E_i) \qquad (6.17)$$

$$+ \sum_{m=k_0}^{k-1} \left| \sum_{p(t_i)=m} f_m(t_i)\mu(E_i) - \sum_{p(t_i)=m} \int_{E_i} f_m \, d\mu \right| \qquad (6.18)$$

$$+ \sum_{m=k_0}^{k-1} \sum_{p(t_i)=m} \int_{E_i} (f_k - f_{k_0}) \, d\mu$$

$$\leq \bigvee_{i=1}^{\infty} b_{i,\varphi(i+1)}^{(1)} + \sum_{m=k_0}^{k-1} \bigvee_{i=1}^{\infty} a_{i,\varphi(i+m+1)}^{(m)} + \int_T (f_k - f_m) \, d\mu$$

$$\leq \bigvee_{i=1}^{\infty} b_{i,\varphi(i+1)}^{(1)} + \sum_{m=2}^{k} \bigvee_{i=1}^{\infty} b_{i,\varphi(i+m)}^{(m)} + \int_T (f_k - f_{k_0}) \, d\mu$$

$$= \sum_{m=1}^{k} \left(\bigvee_{i=1}^{\infty} b_{i,\varphi(i+m)}^{(m)} \right) + \int_T (f_k - f_{k_0}) \, d\mu. \qquad (6.19)$$

Thus, from (6.7)-(6.19) we get the existence of a (D)-sequence $(d_{i,j})_{i,j}$ such that, for every $\varphi \in \Phi$, there exist a gauge γ_0 and $k_0 \in \mathbb{N}$ such that, for each γ_0-fine partition Π and $\forall k > k_0$, we have:

$$\left| \sum_{\Pi} f_k - \int_T f_k \, d\mu \right| \leq \bigvee_{i=1}^{\infty} d_{i,\varphi(i)}. \qquad (6.20)$$

The assertion follows from Lemma 5. ♣

6.5 Lebesgue's Theorem

We now state and prove a version of the Lebesgue dominated convergence theorem.

Theorem 6.14   Let $(f_n : T \to R_1)_n$ be a sequence of integrable functions, and suppose that $\kappa : T \to R_1$ is an integrable map, such that $|f_n(x)| \leq \kappa(x)$ for all $x \in T$ and $n \in \mathbb{N}$. Suppose that

   6.14.1) there exist a function $f : T \to R_1$, a (KH)-integrable map

$h^*: T \to \mathbb{R}^+$ *(with respect to* μ*) and a (D)-sequence* $(a_{i,j})_{i,j}$ *such that,* $\forall \varphi \in \Phi$, $\forall t \in T$, $\exists p(t) \in \mathbb{N} : \forall n \geq p(t)$,

$$|f_n(t) - f(t)| \leq h^*(t) \left(\bigvee_{i=1}^{\infty} a_{i,\varphi(i)} \right). \qquad (6.21)$$

Then f *is integrable and*

$$\int_T f \, d\mu = (D) \lim_n \int_T f_n \, d\mu.$$

Proof: For all $s \in \mathbb{N}$ and $k \geq s$, put

$$g_{s,k} = \bigvee_{s \leq \min(n,m) \leq k} |f_n - f_m|;$$

moreover, for each $s \in \mathbb{N}$, set

$$g_s = \bigvee_{n,m \geq s} |f_n - f_m|.$$

We shall prove that, for each fixed $s \in \mathbb{N}$, the sequence $(g_{s,k})_{k \geq s}$ satisfies the hypothesis of Theorem 6.12.

First of all, it is easy to check that the sequence

$$\left(\int_T g_{s,k} \, d\mu \right)_k$$

is well-defined and bounded in R (Indeed, it is possible to check that the $g_{s,k}$'s are integrable, taking into account that κ is integrable and proceeding analogously as in [19], Theorem 4.33 and [168], Lemma 4.2).

Let now h^* and $(a_{i,j})_{i,j}$ be as in 6.14.1). We know that, $\forall \varphi \in \Phi$, $\forall t \in T$, $\forall s \in \mathbb{N}$, $\exists p \in \mathbb{N}$, with $p \geq s$, such that, $\forall k \geq p$,

$$\bigvee_{n,m \geq k} |f_n(t) - f_m(t)| \leq 2 h^*(t) \left(\bigvee_{i=1}^{\infty} a_{i,\varphi(i)} \right). \qquad (6.22)$$

Fix arbitrarily $s \in \mathbb{N}$. We have, for every $t \in T$:

$$\bigvee_{n,m \geq s} |f_n(t) - f_m(t)|$$

$$= \left(\bigvee_{s \leq \min(n,m) \leq k} |f_n(t) - f_m(t)| \right) \vee \left(\bigvee_{n,m \geq k} |f_n(t) - f_m(t)| \right)$$

$$\leq \left(\bigvee_{s \leq \min(n,m) \leq k} |f_n(t) - f_m(t)| \right) + \left(\bigvee_{n,m \geq k} |f_n(t) - f_m(t)| \right),$$

and hence

$$0 \leq g_s(t) - g_{s,k}(t) \leq \bigvee_{n,m \geq k} |f_n(t) - f_m(t)| \quad \forall k \geq s, \forall t \in T.$$

Thus, we get that the (D)-sequence $(a_{i,j})_{i,j}$ is such that, $\forall \varphi \in \Phi$, $\forall t \in T$, $\forall s \in \mathbb{N}$, $\exists p \in \mathbb{N}$, with $p \geq s$, such that, $\forall k \geq p$,

$$g_s(t) - g_{s,k}(t) \leq 2h^*(t)\left(\bigvee_{i=1}^{\infty} a_{i,\varphi(i)}\right).$$

So, 6.12.1) is satisfied.

We now turn to 6.12.2). As κ is integrable, there exist a gauge $\hat{\gamma}$ and a positive element $a^* \in R$ such that, for every $\hat{\gamma}$-fine partition $\Pi = \{(E_i, t_i) : i = 1, \ldots, q\}$, $\forall s \in \mathbb{N}$, $\forall k \geq s$, we get:

$$\sum_{i=1}^{q}\left[\bigvee_{s \leq \min(n,m) \leq k} \left|f_n(t_i) - f_m(t_i)\right| \mu(E_i)\right]$$

$$\leq 2\sum_{i=1}^{q} \kappa(t_i)\mu(E_i) \leq 2a^*, \qquad (6.23)$$

that is

$$\sum_{i=1}^{q} g_{s,k}(t_i)\mu(E_i) \leq 2a^*.$$

From this it follows that 6.12.2) is satisfied. Thus we get that, for every $s \in \mathbb{N}$, g_s is integrable and

$$\int_T g_s \, d\mu = \bigvee_{k \geq s} \int_T g_{s,k} \, d\mu.$$

We now prove that the sequence $(-g_s)_s$ satisfies the hypotheses of Theorem 6.12.

First of all, it is easy to check that the sequence $\left(\int_T g_s \, d\mu\right)_s$ is bounded. Furthermore, we know that $\forall \varphi \in \Phi$, $\forall t \in T$, $\exists p \in \mathbb{N}$ such that, $\forall s \geq p$,

$$\bigvee_{n,m \geq s} \left|f_n(t) - f_m(t)\right| \leq h^*(t)\left(\bigvee_{i=1}^{\infty} a_{i,\varphi(i)}\right),$$

that is

$$g_s(t) = \left|-g_s(t)\right| \leq h^*(t)\left(\bigvee_{i=1}^{\infty} a_{i,\varphi(i)}\right).$$

So, 6.12.1) is satisfied. Concerning 6.12.2), it is enough to check that the argument in (6.23) works even if we replace $\bigvee_{s \leq \min(n,m) \leq k} |f_n(t_i) - f_m(t_i)|$ with $\bigvee_{n,m \geq s} |f_n(t_i) - f_m(t_i)|$. Thus, we get that

$$(D)\lim_s \int_T g_s \, d\mu = \bigwedge_{s \in \mathbb{N}} \int_T g_s \, d\mu = 0. \qquad (6.24)$$

Proceeding analogously as in the proof of Theorem 6.12, it is possible to prove the existence of (D)-sequences $(e_{i,j}^{(m)})_{i,j}$, $m \in \mathbb{N}$, such that, $\forall \varphi \in \Phi$, there exist a gauge γ' and $k' \in \mathbb{N}$ such that, for each γ'-fine partition $\Pi = \{(E_i, t_i), i = 1, \ldots, q\}$, $\forall k > k'$, we have:

$$\left| \sum_{\Pi} f_k - \int_T f_k \, d\mu \right| \tag{6.25}$$

$$\leq \sum_{m=1}^{k} \left(\bigvee_{i=1}^{\infty} e_{i,\varphi(i+m)}^{(m)} \right) + \sum_{m=k'}^{k-1} \sum_{p(t_i)=m} \left| \int_{E_i} (f_k - f_m) \, d\mu \right| \tag{6.26}$$

$$\leq \sum_{m=1}^{k} \left(\bigvee_{i=1}^{\infty} e_{i,\varphi(i+m)}^{(m)} \right) + \int_T g_{k'} \, d\mu.$$

From (6.25) we get the existence of a (D)-sequence $(d'_{i,j})_{i,j}$ such that, $\forall \varphi \in \Phi$, there exist a gauge γ' and $k' \in \mathbb{N}$ such that, for each γ'-fine partition Π, $\forall k > k'$, we have:

$$\left| \sum_{\Pi} f_k - \int_T f_k \, d\mu \right| \leq \bigvee_{i=1}^{\infty} d'_{i,\varphi(i)}. \tag{6.27}$$

The assertion follows from (6.27) and Theorem 6.10. 🍎

7 Improper Integral

Abstract: In this chapter we deal with the Kurzweil-Henstock integral for functions defined in (possibly unbounded) subintervals of the extended real line.

We begin with real-valued maps and after we consider Riesz space-valued mappings.

All the basic properties are proved, together with Hake convergence-type theorems.

7.1 Real Valued Case

The aim of this chapter is to generalize the Kurzweil-Henstock integral to functions defined on an unbounded interval of the extended real line (and later, more generally, on a suitable locally compact topological space) and with values in R , in a Banach space and in a Riesz space, and to construct a type of integral containing the improper Riemann integral under suitable hypotheses. The case of real-valued functions was investigated in [170]. The cases of Banach- and Riesz-space-valued functions are topics of our research (see [36] and [26]). Moreover ([23, 24]), we considered also the case of real-valued or Riesz-space-valued functions, defined on abstract locally compact topological spaces, satisfying some suitable properties. The case of the (extended) real line is included in this general case; however, for the sake of clearness, we consider and investigate it separately at the beginning of this chapter, even because - in this case - we proved more detailed results.

We begin with the case of real-valued functions, defined on (possibly) unbounded subintervals of the extended real line. We will report in a more detailed way some proofs of [170]: the technique here used will be useful also in the case of Banach- and Riesz-space-valued functions.

We will construct a type of integral (with respect to the Lebesgue measure defined on subintervals of the extended real line, not necessarily bounded), containing the improper Riemann integral. From now on, we denote by $[A,B]$ a closed interval or halfline contained in $\tilde{R}$, or the whole of $\tilde{R}$, and by Δ the set of all positive real-valued functions, defined on $[A,B]$. Moreover, given a measurable set $E \subset \tilde{R}$, we denote by $|E|$ its Lebesgue measure (this quantity can be finite or $+\infty$). Throughout this paragraph, our integral deals with real-valued functions defined on $[A,B]$, but it can be investigated analogously if we take functions defined on R or on halflines of the type $[a,+\infty)$ or $(-\infty,a]$, with $a \in R$.

Definitions 7.1 A *decomposition* or *subpartition* Π of $[A,B]$ is a set of pairs (I_k,ξ_k) , $k=1,...,p$, such that $\xi_k \in I_k \quad \forall k$, and the I_k 's are non-overlapping closed intervals, contained in $[A,B]$. A *partition* $\Pi = \{(I_k,\xi_k):k=1,...,p\}$ of $[A,B]$ is a subpartition of $[A,B]$ with $\bigcup_{k=1}^{p} I_k = [A,B]$.

A *gauge* is a map γ defined in $[A,B]$ and taking values in the set of all open intervals in $\tilde{R}$, such that $\xi \in \gamma(\xi)$ for every $\xi \in [A,B]$ and $\gamma(\xi)$ is a bounded open interval for every $\xi \in R \cap [A,B]$. Given a gauge γ , we will say that a partition or decomposition $\Pi = \{(I_k,\xi_k):k=1,...,p\}$ of $[A,B]$ is γ -*fine* if $I_k \subset \gamma(\xi_k) \quad \forall k=1,...,p$. Given a bounded interval $[a,b] \subset R$ and a map $\delta:[a,b] \to R^+$, a partition or decomposition $\Pi = \{(I_k,\xi_k):k=1,...,p\}$ of $[a,b]$ is said to be δ -*fine* if $I_k \subset (\xi_k - \delta(\xi_k),\xi_k + \delta(\xi_k)) \quad \forall k=1,...,p$.

Antonio Boccuto / Beloslav Riečan / Marta Vrábelová

We note that, if I_k is an unbounded interval, then the element ξ_k associated with I_k is necessarily $+\infty$ or $-\infty$: otherwise $\gamma(\xi_k)$ should be a bounded interval and contain an unbounded interval: contradiction.

Given any partition or decomposition $\Pi = \{(I_k, \xi_k) : k = 1, \ldots, p\}$ of $[A, B]$ and a function $f : [A, B] \to R$, we call *Riemann sum* of f (and we write $\sum_\Pi f$) the quantity

$$\sum_{k=1}^{p} |I_k| f(\xi_k), \tag{7.1}$$

where in the sum in (7.1) only the terms for which I_k is a bounded interval are included. This can be required by simply postulating it or by defining the measure of an unbounded interval as $+\infty$, by requiring $f(+\infty) = f(-\infty) = 0$ and by means of the convention $0 \cdot (+\infty) = 0$ (see also [170], p. 65).

We now formulate our definition of Kurzweil-Henstock integral for functions defined on $[A, B]$.

Definition 7.2 We say that a function $f : [A, B] \to R$ is *Kurzweil-Henstock integrable* (in short (KH)-*integrable*) on $[A, B]$ if there exists an element $I \in R$ such that $\forall \varepsilon > 0$ there exist a function $\delta \in \Delta$ and a positive real number P such that

$$\left| \sum_\Pi f - I \right| \leq \varepsilon \tag{7.2}$$

whenever $\Pi = \{(I_k, \xi_k) : k = 1, \ldots, p\}$ is a δ-fine partition of any bounded interval $[a, b]$ with $[a, b] \supset [A, B] \cap [-P, P]$ and $[a, b] \subset [A, B]$. In this case we say that I is the (KH)-*integral of* f, and we denote the element I by the symbol $(KH) \int_A^B f$ or more simply $\int_A^B f$. Later we will prove that our integral is well-defined, that is such an I is uniquely determined.

We now prove the following characterization of (KH)-integrability:

Theorem 7.3 *A function* $f : [A, B] \to R$ *is* (KH)-*integrable if and only if there exists* $J \in R$ *such that* $\forall \varepsilon > 0$ *there exists a gauge* γ *such that*

$$\left| \sum_\Pi f - J \right| \leq \varepsilon \tag{7.3}$$

whenever $\Pi = \{(I_k, \xi_k) : k = 1, \ldots, p\}$ *is a* γ-*fine partition of* $[A, B]$, *and in this case we have* $\int_A^B f = J$.

Proof: We begin with the "only if" part. By hypothesis, $\forall \varepsilon > 0$ there exist a function $\delta \in \Delta$ and a positive real number P such that (7.2) holds. We now define on $[A, B]$ a gauge γ in the following way:

$$\gamma(\xi) = \begin{cases} (\xi - \delta(\xi), \xi + \delta(\xi)) & \text{if } \xi \in [A, B] \cap R, \\ [-\infty, -P) & \text{if } \xi = -\infty \text{ and } A = -\infty, \\ (P, +\infty] & \text{if } \xi = +\infty \text{ and } B = +\infty. \end{cases}$$

We observe that every γ-fine partition $\Pi = \{(I_k, \xi_k) : k = 1, \ldots, p\}$ of $[A, B]$ is such that $I_k \subset \gamma(\xi_k)$ $\forall k = 1, \ldots, p$. In the case $A = -\infty$, $B = +\infty$, the partition Π contains two unbounded intervals, which we call J and K: of course, if $\inf J = -\infty$ and $\sup K = +\infty$, then the ξ_k's associated with J

and K are $-\infty$ and $+\infty$ respectively. Then, since Π is γ-fine, we have $J \subset \gamma(-\infty)$ and $K \subset \gamma(+\infty)$. Then $J \subset [-\infty, -P)$ and $K \subset (P, +\infty]$. So, if $a = \sup J$ and $b = \inf K$, then $[a,b]$ is a bounded interval, containing $[-P,P]$. If Π' is the restriction of Π to $[a,b]$, then Π' is δ-fine, and by construction we get

$$\sum_{\Pi'} f = \sum_{\Pi} f. \qquad (7.4)$$

In this case, the assertion follows from (7.2) and (7.4).

In the case $A \in \mathbb{R}$, $B = +\infty$, the partition Π contains only an unbounded interval K, with $\sup K = +\infty$. Let P be associated with K as above, and $b = \inf K$: we have $P \leq b$. We note that, without loss of generality, P can be taken greater than $|A|$. Thus, $[A,b]$ is a bounded interval, containing $[-P,P]$, and the assertion follows by proceeding as in the previous case. The case $A = -\infty$, $B \in \mathbb{R}$ is analogous to the previous one. Finally, if $[A,B]$ is bounded, then the assertion is straightforward, because in this case the number P can be taken greater than $\max(|A|,|B|)$ and, of course, (7.2) holds even in the case $[a,b] = [A,B]$. This concludes the proof of the "only if" part.

We now turn to the "if" part. By hypothesis, we know that $\forall \varepsilon > 0$ there exists a gauge γ satisfying (7.3). By definition of gauge, there exist $\delta_1, \delta_2 \in \Delta$ such that

$$\gamma(\xi) = (\xi - \delta_1(\xi), \xi + \delta_2(\xi)) \quad \forall \xi \in [A,B] \cap \mathbb{R}.$$

For such ξ's, let $\delta(\xi) = \min\{\delta_1(\xi), \delta_2(\xi)\}$. Moreover, if $+\infty$ and $-\infty$ belong to $[A,B]$, and $\gamma(-\infty) = [-\infty, P_1^*)$, $\gamma(+\infty) = (P_2^*, +\infty]$, put $P_1 = \min\{P_1^*, -1\}$, $P_2 = \max\{P_2^*, 1\}$, $P = \max\{-P_1, P_2\}$: we note that, in the case $A \in \mathbb{R}$ (resp. $B \in \mathbb{R}$), P can be chosen greater than $|A|$ (resp. $|B|$); moreover, set $\delta(-\infty) = \delta(+\infty) = P$. Let now $[a,b] \subset [A,B]$ be any bounded interval, containing $[A,B] \cap [-P,P]$, and $\Pi = \{(I_k, \xi_k) : k = 1, \ldots, p\}$ be a δ-fine partition of $[a,b]$. Let Π' be that partition of $[A,B]$, whose elements are the ones of Π with the addition of $([A,a], A)$, if $A = -\infty$, and $([b,B], B)$, if $B = +\infty$: we note that Π' is γ-fine. This follows from the fact that, if (I_k, ξ_k) is any element of Π, then

$$I_k \subset (\xi_k - \delta(\xi_k), \xi_k + \delta(\xi_k)) \subset (\xi_k - \delta_1(\xi_k), \xi_k + \delta_2(\xi_k)) = \gamma(\xi_k),$$

and from the following inclusions:

$$(b, +\infty] \subset (P, +\infty] \subset (P_2, +\infty] \subset (P_2^*, +\infty] = \gamma(+\infty),$$

$$[-\infty, a) \subset [-\infty, P) \subset [-\infty, P_1) \subset [-\infty, P_1^*) = \gamma(-\infty).$$

Then, taking into account that the Riemann sum concerning the partition Π' is done without considering the unbounded intervals, we get $\sum_{\Pi'} f = \sum_{\Pi} f$. From this and (7.3) the assertion follows, by proceeding analogously as at the end of the proof of the converse implication. This concludes the proof of the theorem. $\square$

Remark 7.4 We note that the Kurzweil-Henstock integral is well-defined, that is there exists at most one element I, satisfying condition (7.3): indeed, if $\exists$ such two elements I, J, then $\forall \varepsilon > 0$ $\exists$ two gauges γ_1, γ_2 such that, for each γ_1-fine partition Π and for every γ_2-fine partition Π' of $[A,B]$ we have

$$\left| \sum_{\Pi} f - I \right| \le \varepsilon$$

and

$$\left| \sum_{\Pi'} f - J \right| \le \varepsilon$$

respectively. Let now $\gamma(\xi) = \gamma_1(\xi) \cap \gamma_2(\xi)$, $\forall \xi \in [A, B]$ and take any γ-fine partition Π'': then Π'' is both γ_1- and γ_2-fine, and thus we have

$$0 \le |I - J| \le 2\varepsilon.$$

By arbitrariness of $\varepsilon > 0$, it follows that $|I - J| = 0$, and thus $I = J$. So our (KH)-integral is well-defined.

We now state the main properties of the (KH)-integral.

Proposition 7.5 *If f_1, f_2 are (KH)-integrable on $[A, B]$ and $c_1, c_2 \in \mathbb{R}$, then $c_1 f_1 + c_2 f_2$ is (KH)-integrable on $[A, B]$ and*

$$\int_A^B (c_1 f_1 + c_2 f_2) = c_1 \int_A^B f_1 + c_2 \int_A^B f_2.$$

(See [170], Theorems 2.5.1 and 2.5.3.)

Proposition 7.6 *If f_1 and f_2 are (KH)-integrable on $[A, B]$ and $f_1 \le f_2$, then*

$$\int_A^B f_1 \le \int_A^B f_2.$$

(See [170], Theorem 2.5.8.)

Proposition 7.7 *Let $A, B \in \tilde{\mathbb{R}}$, and c be such that $A < c < B$. If $f : [A, B] \to \mathbb{R}$ is (KH)-integrable both on $[A, c]$ and on $[c, B]$, then f is (KH)-integrable on $[A, B]$ and*

$$\int_A^B f = \int_A^c f + \int_c^B f.$$

Proof: In correspondence with (KH)-integrability of f on $[A, c]$ and $[c, B]$, $\forall \varepsilon > 0$ there exist two mappings $\underline{\delta} : [A, c] \to \mathbb{R}^+$, $\overline{\delta} : [c, B] \to \mathbb{R}^+$, and two positive real numbers $\underline{P}$ and $\overline{P}$ (without loss of generality, $\underline{P} > |c|$, $\overline{P} > |c|$) such that, if $\underline{\Pi}$ is any $\underline{\delta}$-fine partition of any bounded interval $[a_1, b_1] \subset [A, c]$, $[a_1, b_1] \supset [A, c] \cap [-\underline{P}, \underline{P}]$ and $\overline{\Pi}$ is any $\overline{\delta}$-fine partition of any bounded interval $[a_2, b_2] \subset [c, B]$, $[a_2, b_2] \supset [c, B] \cap [-\overline{P}, \overline{P}]$, then

$$\left| \sum_{\underline{\Pi}} f - \int_{a_1}^{b_1} f \right| \le \frac{\varepsilon}{2}$$

and

$$\left| \sum_{\overline{\Pi}} f - \int_{a_2}^{b_2} f \right| \le \frac{\varepsilon}{2}.$$

If $A = -\infty$, let $\delta(-\infty) = \underline{\delta}(-\infty)$; if $B = +\infty$, let $\delta(+\infty) = \overline{\delta}(+\infty)$. Moreover, set

$$\delta(x) = \begin{cases} \min\left\{\underline{\delta}(x), \dfrac{1}{2}(c-x)\right\} & \text{if } x \in [A,c) \cap \mathrm{R}, \\[2mm] \min\left\{\overline{\delta}(x), \dfrac{1}{2}(x-c)\right\} & \text{if } x \in (c,B] \cap \mathrm{R}, \\[2mm] \min\{\underline{\delta}(c), \overline{\delta}(c)\} & \text{if } x = c, \end{cases}$$

and $P = \max\{\underline{P}, \overline{P}\}$. Take now any arbitrary bounded interval $[a,b] \subset [A,B]$, $[a,b] \supset [A,B] \cap [-P,P]$, and any δ-fine partition $\Pi = \{([u_k, v_k], \xi_k) : k = 1, \ldots, p\}$ of $[a,b]$. Then necessarily $c \in (a,b)$. We now claim that there exists $k \in \{1, \ldots, p\}$ such that $c = \xi_k$, or $c = u_k$, or $c = v_k$. Otherwise there would be an interval $[u_j, v_j]$ such that $u_j < c < v_j$ and either $c < \xi_j < v_j$ or $u_j < \xi_j < c$. Since Π is δ-fine, we should get $[u_j, v_j] \subset (\xi_j - \delta(\xi_j), \xi_j + \delta(\xi_j))$ and thus $v_j - u_j < 2\delta(\xi_j)$. So $v_j - u_j < \xi_j - c$ if $\xi_j > c$ or $v_j - u_j < c - \xi_j$ if $\xi_j < c$. This would imply that ξ_j is outside (u_j, v_j), contradiction. Thus we have:

$$\sum_\Pi f = \sum_{l=1}^{j-1} f(\xi_l)(v_l - u_l) + f(\xi_j)(v_j - u_j)$$

$$+ \sum_{l=j+1}^{p} f(\xi_l)(v_l - u_l) \qquad\qquad (7.5)$$

$$= \sum_{l=1}^{j-1} f(\xi_l)(v_l - u_l) + f(\xi_j)(\xi_j - u_j) + f(\xi_j)(v_j - \xi_j)$$

$$+ \sum_{l=j+1}^{p} f(\xi_l)(v_l - u_l).$$

The quantity $S_a^c = \sum_{l=1}^{j-1} f(\xi_l)(v_l - u_l) + f(\xi_j)(\xi_j - u_j)$ is a Riemann sum for a suitable $\underline{\delta}$-fine partition of $[a,c]$, which is a bounded interval contained in $[A,c]$ and containing $[A,c] \cap [-\underline{P}, \underline{P}]$, by construction.

Analogously, the quantity $S_c^b = f(\xi_j)(v_j - \xi_j) + \sum_{l=j+1}^{p} f(\xi_l)(v_l - u_l)$ is a Riemann sum for a suitable $\overline{\delta}$-fine partition of $[c,b]$, which is a bounded interval contained in $[c,B]$ and containing $[c,B] \cap [-\overline{P}, \overline{P}]$. Thus we have:

$$\left| S_a^c - \int_A^c f \right| \le \frac{\varepsilon}{2}, \qquad \left| S_c^b - \int_c^B f \right| \le \frac{\varepsilon}{2},$$

and hence

$$\left| \sum_\Pi f - \int_A^c f - \int_c^B f \right| \le \varepsilon.$$

Thus the assertion follows. $\square$

We now state two versions of the Bolzano-Cauchy condition.

Theorem 7.8   *A map* $f : [A, B] \to \mathrm{R}$ *is* (KH)*-integrable if and only if* $\forall \, \varepsilon > 0 \; \exists \;$ *a gauge* γ *such that for every* γ*-fine partition* Π_1, Π_2 *of* $[A, B]$ *we have*

$$\left| \sum_{\Pi_1} f - \sum_{\Pi_2} f \right| \leq \varepsilon. \tag{7.6}$$

Proof: (see also [207]) The necessary part is straightforward.

We now turn to the sufficient part. By hypothesis, condition (7.6) holds even for $\varepsilon = \dfrac{1}{n}$, with

$n \in \mathrm{N}$. Let γ_n be a corresponding gauge. Without loss of generality, we can suppose that

$$\gamma_{n+1}(x) \subset \gamma_n(x) \quad \forall \, x \in [A, B]. \tag{7.7}$$

Let $(\Pi_n)_n$ be a sequence of partitions of $[A, B]$ such that Π_n is γ_n-fine $\forall \, n \in \mathrm{N}$. From (7.7) it follows that, $\forall \, n, p \in \mathrm{N}$, every γ_{n+p}-fine partition is also γ_n-fine. Thus, in correspondence with $\varepsilon > 0$, let $\overline{n}$ be such that $\dfrac{1}{\overline{n}} \leq \varepsilon$: for $n \geq \overline{n}$ and $p \in \mathrm{N}$ we have:

$$\left| \sum_{\Pi_{n+p}} f - \sum_{\Pi_n} f \right| \leq \varepsilon.$$

Thus it follows that the sequence $\left(\sum_{\Pi_n} f \right)_n$ is Cauchy, and thus convergent, because of completeness of R. Let $I = \lim_n \sum_{\Pi_n} f$. Fix arbitrarily $\varepsilon > 0$. Then there exists an integer n^*,

$n^* > \dfrac{2}{\varepsilon}$, such that

$$\left| \sum_{\Pi_{n^*}} f - I \right| \leq \frac{\varepsilon}{2}.$$

Let $\gamma = \gamma_{n^*}$. If Π is any γ-fine partition of $[A, B]$, then

$$\left| \sum_{\Pi} f - I \right| \leq \left| \sum_{\Pi} f - \sum_{\Pi_{n^*}} f \right| + \left| \sum_{\Pi_{n^*}} f - I \right| \tag{7.8}$$

$$\leq \frac{1}{n^*} + \frac{\varepsilon}{2} < \frac{\varepsilon}{2} + \frac{\varepsilon}{2} = \varepsilon.$$

The assertion follows from (7.8). □

Theorem 7.9   *A map* $f : [A, B] \to \mathrm{R}$ *is* (KH)*-integrable if and only if* $\forall \, \varepsilon > 0 \; \exists \;$ *a map* $\delta \in \Delta$ *and a positive real number* P *such that*

$$\left| \sum_{\Pi_1} f - \sum_{\Pi_2} f \right| \leq \varepsilon$$

whenever Π_1, Π_2 *are* δ*-fine partitions of any bounded interval* $[a, b]$*, with* $[a, b] \subset [A, B]$ *and* $[a, b] \supset [A, B] \cap [-P, P]$.

Proof: The proof is similar to the one of Theorem 7.8. □

We now prove a result about (KH)-integrability on subintervals.

Theorem 7.10 *Let* $f:[A,B] \to \mathbf{R}$ *be* (KH)*-integrable, and* $A < c < B$. *Then* $f|_{[A,c]}$ *and* $f|_{[c,B]}$ *are* (KH)*-integrable too, and*

$$\int_A^B f = \int_A^c f + \int_c^B f. \tag{7.9}$$

Proof: By virtue of Theorem 7.8, $\forall \varepsilon > 0 \; \exists$ a gauge γ on $[A,B]$ such that for all γ-fine partitions Π_1 and Π_2 of $[A,B]$ we have

$$\left| \sum_{\Pi_1} f - \sum_{\Pi_2} f \right| \le \varepsilon. \tag{7.10}$$

Set $\gamma_0 = \gamma|_{[A,c]}$ and let Π, Π' be any two γ_0-fine partitions of $[A,c]$. By virtue of the Cousin Lemma there exists a γ-fine partition Π_0 of $[c,B]$. Put $\Pi_1 = \Pi \cup \Pi_0$, $\Pi_2 = \Pi' \cup \Pi_0$. Then Π_1 and Π_2 are γ-fine partitions of $[A,B]$. Moreover, we get

$$\sum_{\Pi_1} f = \sum_{\Pi} f + \sum_{\Pi_0} f, \quad \sum_{\Pi_2} f = \sum_{\Pi'} f + \sum_{\Pi_0} f. \tag{7.11}$$

From (7.10) and (7.11) we have

$$\left| \sum_{\Pi} f - \sum_{\Pi'} f \right| \le \varepsilon. \tag{7.12}$$

From (7.12) and Theorem 7.8 it follows that $f|_{[A,c]}$ is (KH)-integrable. The proof of (KH)-integrability of $f|_{[c,B]}$ is analogous. The equality (7.9) follows from this and Proposition 7.7. $\square$

We now prove the following.

Theorem 7.11 *Let* $f:[A,B] \to \mathbf{R}$ *be an* (KH)*-integrable function. Let* $A < c < B$. *Then the function* $g = f \, \chi_{[A,c]}$ *is* (KH)*-integrable on* $[A,B]$, *and* $\int_A^c f = \int_A^B g$.

Proof: First of all, we note that $c \in \mathbf{R}$, and g is (KH)-integrable on $[A,c]$, because g coincides with f in $[A,c]$ and, by virtue of Theorem 7.10, f is (KH)-integrable on $[A,c]$. Moreover, it is easy to see that g is (KH)-integrable on $[c,B]$ and $\int_c^B g = 0$. So, by virtue of Proposition 7.7, we get that g is (KH)-integrable on $[A,B]$ and

$$\int_A^B g = \int_A^c g + \int_c^B g = \int_A^c f. \tag{7.13}$$

This concludes the proof. $\square$

Remark 7.12 In an analogous way it is possible to prove that $h = f \, \chi_{[c,B]}$ is (KH)-integrable on $[A,B]$ and $\int_c^B f = \int_A^B h$.

Corollary 7.13 *Let* $f:[A,B] \to \mathrm{R}$ *be* (KH)*-integrable on* $[A,B]$*, and let* $A < c < c' < B$. *Then the map* $l = f \chi_{[c,c']}$ *is* (KH)*-integrable on* $[A,B]$*, and* $\int_c^{c'} f = \int_A^B l$.

Proof: First of all, we note that $c,c' \in \mathrm{R}$. Let $k = f|_{[A,\,c']}$: by virtue of Theorem 7.10, k is (KH)-integrable on $[A,c']$, and by Theorem 7.11, where the role of A, B, c is played by A, c', c respectively, the function

$$l' = k \chi_{[c,c']} = f|_{[A,c']} \, \chi_{[c,c']}$$

is (KH)-integrable on $[A,c']$, and $\int_c^{c'} f = \int_c^{c'} k = \int_A^{c'} l'$. Moreover, since l coincides with l' on $[A,c']$ and vanishes on $(c',B]$, then, thanks to Proposition 7.7, we get that l is (KH)-integrable on $[A,B]$ and $\int_A^B l = \int_A^{c'} l'$. From this the assertion follows. $\square$

Now, given an interval $[a,b] \subset \mathrm{R}$, a partition $\Pi = \{([x_{k-1},x_k],\xi_k): k = 1,2,\dots p\}$ and a point $c \in (a,b)$, if c coincides with some x_k, let $\Pi_1 (\Pi_2)$ be the partition of all elements of Π which are contained in $[a,c]$ $([c,b])$ respectively, and put

$$\sum_\Pi {}_a^c f = \sum_{\Pi_1} f, \quad \sum_\Pi {}_c^b f = \sum_{\Pi_2} f.$$

If $c \in (x_{k-1}, x_k)$ for some $k = 1,\dots, p$, then put

$$\sum_\Pi {}_a^c f = \sum_{l=1}^{k-1} f(\xi_l)(x_l - x_{l-1}) + f(c)(c - x_{k-1});$$

$$\sum_\Pi {}_c^b f = f(c)(x_k - c) + \sum_{l=k+1}^{p} f(\xi_l)(x_l - x_{l-1}).$$

In the sequel, when we will deal with the interval $[a,b]$ or $[A,B]$, sometimes we will write $\sum_\Pi {}_a^b f$ or $\sum_\Pi {}_A^B f$, instead of $\sum_\Pi f$, in order to avoid confusion. We now prove the following theorem (see also [170], Lemma 2.8.1., pp. 56-57):

Theorem 7.14 *Let* $[a,b] \subset \mathrm{R}$ *be a bounded interval,* $f:[a,b] \to \mathrm{R}$ *be a* (KH)*-integrable function,* $\varepsilon > 0$, *and* $\delta:[a,b] \to \mathrm{R}^+$ *such that, for every* δ*-fine partition* Π' *of* $[a,b]$,

$$\left| \sum_{\Pi'} {}_a^b f - \int_a^b f \right| \le \varepsilon. \qquad (7.14)$$

Then δ *is such that,* $\forall c \in (a,b)$ *and for every* δ*-fine partition* Π *of* $[a,b]$,

$$\left| \sum_\Pi {}_a^c f - \int_a^c f \right| \le 2\varepsilon, \quad \left| \sum_\Pi {}_c^b f - \int_c^b f \right| \le 2\varepsilon. \qquad (7.15)$$

Proof: Let Π be a δ-fine partition of $[a,b]$. By virtue of Theorem 7.10, f is (KH)-integrable in $[a,c]$, and thus there exists a function $\delta_c : [a,c] \to \mathrm{R}^+$ such that for every δ_c-fine partition Π'_c of $[a,c]$ we have:

$$\left| \sum_{\Pi'_c} {}_a^c f - \int_a^c f \right| \le \varepsilon. \qquad (7.16)$$

Let now Π_c be a δ- and δ_c-fine partition of $[a,c]$. Moreover, let Π_0 be that partition of $[c,b]$ consisting of those elements $([x_{l-1},x_l],\xi_l)$ of Π such that the intervals $[x_{l-1},x_l]$ are contained in $[c,b]$ and eventually of (J,c), where J is the intersection of $[c,b]$ with that (eventual) interval

$[x_{k-1}, x_k]$ for which $x_{k-1} < c < x_k$. Let Π' be that partition consisting of the "union" of Π_c and Π_0: Π' is δ-fine, and we have:

$$\sum_{\Pi} {}^b_c f - \int_c^b f = \sum_{\Pi_0} {}^b_c f - \int_c^b f$$

$$= \sum_{\Pi'} {}^b_c f - \int_c^b f = \sum_{\Pi'} {}^b_a f - \int_a^b f$$

$$-\left(\sum_{\Pi'} {}^c_a f - \int_a^c f \right) = \sum_{\Pi'} {}^b_a f - \int_a^b f$$

$$-\left(\sum_{\Pi_c} {}^c_a f - \int_a^c f \right).$$

By virtue of (7.14) and (7.16) we get:

$$\left| \sum_{\Pi} {}^b_c f - \int_c^b f \right| \le \left| \sum_{\Pi'} {}^b_a f - \int_a^b f \right| + \left| \sum_{\Pi_c} {}^c_a f - \int_a^c f \right| \le 2\varepsilon.$$

This proves the second inequality of (7.15). The proof of the first inequality of (7.15) is analogous. □

We now prove that, in the case of real-valued functions, the (KH)-integral contains the improper Riemann integral (see also [170], Theorem 2.9.3., pp. 61-63).

Theorem 7.15 *Let $a \in \mathbb{R}$, and $f : [a, +\infty] \to \mathbb{R}$ be (KH)-integrable on $[a, +\infty]$. Then f is (KH)-integrable on every interval $[a, b]$ with $a < b < +\infty$, and*

$$\lim_{b \to +\infty} \int_a^b f = \int_a^{+\infty} f.$$

Conversely, if $f : [a, +\infty] \to \mathbb{R}$ is (KH)-integrable on every interval $[a, b]$ with $a < b < +\infty$ and there exists in $\mathbb{R}$ the limit $l = \lim_{b \to +\infty} \int_a^b f$, then f is (KH)-integrable on $[a, +\infty]$ and $\int_a^{+\infty} f = l$.

Proof: We begin with the first part of the theorem. Since $f : [a, +\infty] \to \mathbb{R}$ is (KH)-integrable, then $\forall \varepsilon > 0 \ \exists \delta : [a, +\infty] \to \mathbb{R}^+$ and $\exists P > |a|$, such that for each bounded interval $[d_1, d_2]$ with $[d_1, d_2] \subset [a, +\infty]$, $[d_1, d_2] \supset [a, +\infty] \cap [-P, P]$, and for every δ-fine partition Π of $[d_1, d_2]$ we have:

$$\left| \sum_{\Pi} f - \int_a^{+\infty} f \right| \le \frac{\varepsilon}{2}. \tag{7.17}$$

Now, by virtue of Theorem 7.10, f is (KH)-integrable on $[a, b]$ for every $b \in (a, +\infty]$, and hence we get that $\forall \varepsilon > 0$, $\forall b \in (a, +\infty]$, $\exists \delta_1 : [a, b] \to \mathbb{R}^+$ such that for each δ_1-fine partition Π' of $[a, b]$ we get:

$$\left| \sum_{\Pi'} f - \int_a^b f \right| \le \frac{\varepsilon}{2}. \tag{7.18}$$

Let us define $\delta_2 : [a, b] \to \mathbb{R}^+$ by setting $\delta_2(x) = \min\{\delta(x), \delta_1(x)\}$, and let Π be a δ_2-fine partition of $[a, b]$, $b > P$. Then, thanks to (7.17) and (7.18), $\forall \varepsilon > 0 \ \exists P > 0 : \forall b > P$,

$$\left| \int_a^b f - \int_a^{+\infty} f \right| \leq \left| \sum_{\Pi} f - \int_a^b f \right| + \left| \sum_{\Pi} f - \int_a^{+\infty} f \right| \leq \varepsilon.$$

Thus the first part is completely proved.

We now turn to the second part. By hypothesis, $\forall \varepsilon > 0$, $\exists P > 0 : \forall b > P$ we get

$$\left| \int_a^b f - l \right| \leq \frac{\varepsilon}{2}. \tag{7.19}$$

Let now $(b_n)_n$ be a strictly increasing sequence of real numbers, such that $\lim_n b_n = +\infty$ and $b_1 = a$. We observe that, by virtue of Theorem 7.10, f is (KH)-integrable in $[b_n, b_{n+1}]$ for each n. So, $\forall \varepsilon > 0$ and $\forall n \in \mathrm{N}$, $\exists$ a function $\delta_n : [b_n, b_{n+1}] \to \mathrm{R}^+$ such that

$$\left| \sum_{\Pi_n} f - \int_{b_n}^{b_{n+1}} f \right| \leq \frac{\varepsilon}{2^{n+1}} \tag{7.20}$$

whenever Π_n is any δ-fine partition of $[b_n, b_{n+1}]$.

Let now $\delta : [a, +\infty] \to \mathrm{R}^+$ be such that, $\forall n \in \mathrm{N}$,

$$\begin{cases} \delta(\xi) \leq \delta_n(\xi) & \text{if } \xi \in [b_n, b_{n+1}], \\ [\xi - \delta(\xi), \xi + \delta(\xi)] \subset (b_n, b_{n+1}) & \text{if } \xi \in (b_n, b_{n+1}), \\ (b_n - \delta(b_n), b_n + \delta(b_n)) \subset (b_{n-1}, b_{n+1}). \end{cases} \tag{7.21}$$

Choose now arbitrarily $b > P$.

If $b_N < b \leq b_{N+1}$ and $\Pi = \{([x_{k-1}, x_k], \xi_k) : k = 1, 2, \dots p\}$ is a partition of $[a, b]$, then each b_n, with $n \leq N$, must belong to some interval $[x_{k-1}, x_k]$. So, either b_n coincides with some x_k's, or $b_n \in (x_{k-1}, x_k)$. In this last case, from (7.21) and the fact that Π is δ-fine it follows that $\xi_k \notin (b_n, b_{n+1})$, otherwise

$$[x_{k-1}, x_k] \subset (\xi_k - \delta(\xi_k), \xi_k + \delta(\xi_k)) \subset (b_n, b_{n+1}):$$

this is a contradiction. Analogously, $\xi_k \notin (b_{n-1}, b_n)$, and in general, if $j \in \mathrm{N}$ is such that $b_j \in (x_{k-1}, x_k)$, we have necessarily $\xi_k \notin (b_{j-1}, b_j)$, $\xi_k \notin (b_j, b_{j+1})$: otherwise $[x_{k-1}, x_k] \subset (b_{j-1}, b_j)$ or $[x_{k-1}, x_k] \subset (b_j, b_{j+1})$: this is absurd. Thus ξ_k does coincide with some b_{j_0}. From the third condition in (7.21) and the fact that Π is δ-fine it follows that

$$[x_{k-1}, x_k] \subset (\xi_k - \delta(\xi_k), \xi_k + \delta(\xi_k)) \tag{7.22}$$
$$= (b_{j_0} - \delta(b_{j_0}), b_{j_0} + \delta(b_{j_0})) \subset (b_{j_0-1}, b_{j_0+1}).$$

But we know that, by hypothesis, $b_n \in (x_{k-1}, x_k)$, and from (7.22) it follows that $j_0 = n$ and that no b_j but b_n belongs to (x_{k-1}, x_k). So, all the b_n's do coincide either with some x_k or with some ξ_k. So, Π is the partition of $[a, b]$ "determined" by the x_k's and the b_n's. We have:

$$\sum_{\Pi} \int_a^b f = \sum_{n=1}^{N-1} \left(\sum_{\Pi} \int_{b_n}^{b_{n+1}} f \right) + \sum_{\Pi} \int_{b_N}^b f. \tag{7.23}$$

Since the restriction of Π to $[b_n, b_{n+1}]$ is δ_n-fine, from (7.20) it follows that

$$\sum_{n=1}^{N-1} \left| \sum_{\Pi} \int_{b_n}^{b_{n+1}} f - \int_{b_n}^{b_{n+1}} f \right| \leq \frac{\varepsilon}{2}. \tag{7.24}$$

From (7.19), (7.23) and (7.24) we have

$$\left|\sum_{\Pi} {}_{a}^{b} f - l\right| \le \frac{\varepsilon}{2} + \frac{\varepsilon}{2} + \left|\sum_{\Pi} {}_{b_N}^{b} f - \int_{b_N}^{b} f\right|.$$

Since the restriction of Π to $[b_N, b]$ is δ_N-fine, then Π can be "extended" to a δ_N-fine partition Π' of $[b_N, b_{N+1}]$. By Theorem 7.14, where the roles of $[a,b]$ and c are played by $[b_N, b_{N+1}]$ and b respectively, we get

$$\left|\sum_{\Pi} {}_{b_N}^{b} f - \int_{b_N}^{b} f\right| \le \frac{\varepsilon}{2^N} < \varepsilon.$$

From this the assertion follows. □

Remark 7.16 We observe that theorems similar to Theorem 7.15 hold even if we consider open, semi-open and/or left halflines, R or $\tilde{R}$, instead of $[a, +\infty]$.

We now prove that every simple measurable function defined on R, and assuming values different from zero only on a set of finite Lebesgue measure, is (KH)-integrable according to our definition, and in this case our integral coincides with the usual one. To do this, thanks to Proposition 7.5, it is sufficient to prove the following:

Theorem 7.17 *Let $E \subset R$ be a Lebesgue measurable set with $|E| < +\infty$, $r \in R$, and χ_E be the characteristic function associated with E. Then the function $\chi_E r$ is (KH)-integrable, and*

$$\int_{-\infty}^{+\infty} \chi_E r = |E| r.$$

Proof: Without loss of generality, we can suppose that $r \ge 0$. First of all, it is easy to check that the assertion holds when E is a bounded interval of R. Now, let us prove the assertion when E is an open set with $|E| < +\infty$. To this aim, we prove that $\forall \varepsilon > 0$ there exists a gauge γ, defined on R, such that for all γ-fine partitions Π of R we get

$$\left|\sum_{\Pi} \chi_E r - |E| r\right| \le \varepsilon. \tag{7.25}$$

Since E is open, then there exists a sequence of disjoint open bounded intervals (a_n, b_n), $n \in N$, such that

$$E = \bigcup_{n=1}^{\infty} (a_n, b_n).$$

Without loss of generality, we can suppose that $b_j \ne a_l$ whenever $l \ne j + 1$.
By virtue of the previous step, we know that

$$\int_{-\infty}^{+\infty} \chi_{(a_n,b_n)} r = \int_{-\infty}^{+\infty} \chi_{[a_n,b_n]} r = (b_n - a_n) r \quad \forall n \in N.$$

For every $\varepsilon > 0$ and $n \in N$, let γ_n be a gauge, such that

$$\left|\sum_{\Pi^{(n)}} \chi_{(a_n,b_n)} r - (b_n - a_n) r\right| \le \frac{\varepsilon}{2^n} \tag{7.26}$$

whenever $\Pi^{(n)} = \{(I_k^{(n)}, \xi_k^{(n)}) : k = 1, \ldots, p^{(n)}\}$ is a γ_n-fine partition of $(-\infty, +\infty)$. Let now γ be a gauge, such that $\gamma(x) \subset \gamma_n(x)$ whenever $x \in (a_n, b_n)$, $n \in \mathbb{N}$, and such that

$$\gamma(x) \subset \gamma_n(x) \cap \gamma_{n+1}(x)$$

whenever $x \in [a_n, b_n] \cap [a_{n+1}, b_{n+1}]$, $n \in \mathbb{N}$, and choose arbitrarily any γ-fine partition Π. We get:

$$\left| \sum_\Pi \chi_E r - |E| r \right| \tag{7.27}$$

$$\leq \sum_{n=1}^\infty \left| \sum_{\Pi^{(n)}} \chi_{[a_n, b_n]} r - (b_n - a_n) r \right| \leq \varepsilon.$$

Thus the assertion follows, at least in the case in which E is open. Let now $E \subset \mathbb{R}$ be a compact set. Then there exists a bounded interval $[c, d] \subset \mathbb{R}$ such that $E \subset [c, d]$. Since $\chi_E r = \chi_{(c,d)} r - \chi_{(c,d) \setminus E} r$ and $(c, d) \setminus E$ is an open set with finite Lebesgue measure, then, by virtue of the previous steps, χ_E is (KH)-integrable on $(-\infty, +\infty)$, and we have:

$$\int_{-\infty}^{+\infty} \chi_E r = \int_{-\infty}^{+\infty} \chi_{(c,d)} r - \int_{-\infty}^{+\infty} \chi_{(c,d) \setminus E} r \tag{7.28}$$

$$= (d - c) r - [(d - c) - |E|] r = |E| r,$$

that is the assertion when $E \subset \mathbb{R}$ is compact. In the general case, when E is a measurable set with finite Lebesgue measure, then for every $\varepsilon > 0$ there exist an open set U and a compact set K such that $K \subset E \subset U$ and $|U| \leq |E| + \varepsilon \leq |K| + 2\varepsilon$. By the previous steps, we know that, in correspondence with ε, there exists a gauge γ such that for all γ-fine partitions Π of $(-\infty, +\infty)$ we have:

$$\left| \sum_\Pi \chi_U r - |U| r \right| \leq \varepsilon, \qquad \left| \sum_\Pi \chi_K r - |K| r \right| \leq \varepsilon. \tag{7.29}$$

Moreover, since $r \geq 0$, for every partition Π of $(-\infty, +\infty)$ we get:

$$\sum_\Pi \chi_K r \leq \sum_\Pi \chi_E r \leq \sum_\Pi \chi_U r. \tag{7.30}$$

From (7.29) and (7.30) it follows that, $\forall \varepsilon > 0$, there exists a gauge γ such that for all γ-fine partitions Π of $(-\infty, +\infty)$ we have:

$$\sum_\Pi \chi_E r - |E| r \leq \sum_\Pi \chi_U r - |U| r + \varepsilon r \leq \varepsilon (1 + r);$$

$$|E| r - \sum_\Pi \chi_E r \leq |K| r - \sum_\Pi \chi_K r + \varepsilon r \leq \varepsilon (1 + r);$$

and hence

$$\left| \sum_\Pi \chi_E r - |E| r \right| \leq \varepsilon (1 + r)$$

(for the technique, see also [219]). This concludes the proof. $\square$

In [23] we generalized some of the above results to the Kurzweil-Henstock integral for real-valued functions, defined on a suitable (locally) compact topological spaces, satisfying certain axioms.

Let X be a Hausdorff compact topological space. If $A \subset X$, then the interior of the set A is denoted by $\text{int } A$.

We shall work with a family F of compact subsets of X such that $X \in \mathrm{F}$ and closed under the intersection and the finite union, and a monotone, additive mapping $\lambda : \mathrm{F} \to [0, +\infty]$. The additivity means that

$$\lambda(A \cup B) + \lambda(A \cap B) = \lambda(A) + \lambda(B) \qquad (7.31)$$

whenever $A, B, A \cup B \in F$.

By a *partition* (detaily, (F, λ)-*partition*) of a set $A \in F$ we mean a finite collection $\{(U_1, t_1), \ldots, (U_k, t_k)\}$ such that

(i) $U_1, \ldots, U_k \in F$,

(ii) $\displaystyle\bigcup_{i=1}^{k} U_i = A$,

(iii) $\lambda(U_i \cap U_j) = 0$ whenever $i \neq j$,

(iv) $t_i \in U_i \ (i = 1, \ldots, k)$.

A finite collection $\{(U_1, t_1), \ldots, (U_k, t_k)\}$ of subsets of $A \in F$, satisfying conditions (i), (iii) and (iv), but not necessarily (ii), is said to be *decomposition* of A. We shall assume that F *separates points* in the following way: to any $A \in F$ there exists a sequence $(A_n)_n$ of partitions of A such that

(i) A_{n+1} is a refinement of A_n,

(ii) to any $x, y \in A$, $x \neq y$, there exist $n \in N$ and $B \in A_n$ such that $x \in B$ and $y \notin B$.

We note that this assumption is fulfilled if the topological space X is metrizable or it satisfies the second axiom of countability (see [217]) and F is the family of all compact subsets of X.

A *gauge* on a set $A \subset X$ is a mapping γ assigning to every point $x \in A$ a neighborhood $\gamma(x)$ of x. If $\Pi = \{(U_1, t_1), \ldots, (U_k, t_k)\}$ is a decomposition of A and γ is a gauge on A, then we say that Π is γ-*fine* if $U_i \subset \gamma(t_i)$ for any $i \in \{1, 2, \ldots, k\}$.

We obtain a simple example putting $X = [a, b] \subset R$ with the usual topology, F = the family of all finite unions of closed subintervals of X, $\lambda([\alpha, \beta]) = \beta - \alpha$, $a \leq \alpha < \beta \leq b$. Any gauge can be represented by a real function $\delta : [a, b] \to R^+$, if we put $\gamma(x) = (x - \delta(x), x + \delta(x))$.

Another example is the unbounded interval $[a, +\infty] = [a, +\infty) \cup \{+\infty\}$ considered as the one-point compactification of the locally compact space $[a, +\infty)$. The basis of open sets consists on open subsets of $[a, +\infty)$ and the sets of the type $(b, +\infty) \cup \{+\infty\}$, $a \leq b < +\infty$. Any gauge in $[a, +\infty]$ has the form

$\gamma(x) = (x - \delta(x), x + \delta(x))$, if $x \in [a, +\infty] \cap R$,

and $\gamma(+\infty) = (b, +\infty] = (b, +\infty) \cup \{+\infty\}$,

where δ denotes a positive real-valued function defined on $[a, +\infty)$, and b denotes a real number.

Let us return to the definition of Kurzweil-Henstock integral $((KH)$-integral) on X.

If $\Pi = \{(U_1, t_1), \ldots, (U_k, t_k)\}$ is a decomposition of a set A, and $f : X \to R$, then we define the Riemann sum as follows

$$\sum_{\Pi} f = \sum_{i=1}^{k} f(t_i) \lambda(U_i),$$

if the sum exists in R, with the convention $0 \cdot (+\infty) = 0$.

We note that the fact that F separates points guarantees the existence of at least one γ-fine partition Π such that $\sum_{\Pi} f$ is well-defined for any gauge γ (see [217,232]).

Definition 7.18   A function $f : X \to \mathrm{R}$ is (KH)-*integrable* (in short, *integrable*) on a set A if there exists $I \in \mathrm{R}$ such that $\forall \, \varepsilon > 0$ there exists a gauge γ on A such that

$$\left| \sum_{\Pi} f - I \right| \leq \varepsilon \qquad (7.32)$$

whenever Π is a γ-fine partition of A such that $\sum_{\Pi} f$ exists in R. We denote

$$I = \int_{A} f$$

(see also [217], Definition 1.8., p. 154).

We now prove the following convergence theorem:

Theorem 7.19   *Let* $X = X_0 \cup \{x_0\}$ *be the one-point compactification of a locally compact space* X_0. *Let* $f : X \to \mathrm{R}$ *be a function such that* $f(x_0) = 0$. *Let* $(A_n)_n$ *be a sequence of sets, such that* $A_n \in \mathrm{F}$, $A_n \subset \mathrm{int}\, A_{n+1}$, $A_{n+1} \ \mathrm{int}\, A_n \in \mathrm{F}$, $\lambda(A_n \ \mathrm{int}\, A_n) = 0$ $(n \in \mathrm{N})$, $\bigcup_{n=1}^{\infty} A_n = X_0$. *Let* f *be integrable on every* $A \in \mathrm{F}$, *with* $A \subset X_0$, *and let there exist in* R *an element* I *such that,* $\forall \, \varepsilon > 0$, *there exists an integer* n_0 *such that*

$$\left| \int_{A} f - I \right| \leq \varepsilon \quad \forall \, A \in \mathrm{F}, \quad X_0 \supset A \supset A_{n_0}.$$

Then f *is integrable on* X *and* $\int_{X} f = I$.

Proof: Let ε be an arbitrary positive real number, and $n_0 \in \mathrm{N}$ be as in the hypotheses of the theorem. Put $A_0 = \varnothing$, $B_0 = A_1$, $B_n = A_{n+1} \ \mathrm{int}\, A_n$ $(n \in \mathrm{N})$. For all $(n \in \mathrm{N})$ there exists a gauge γ_n on B_n such that

$$\left| \int_{B_n} f - \sum_{\Pi} f_n \right| \leq \frac{\varepsilon}{2^{n+3}} \qquad (7.33)$$

for any γ_n-fine partition Π_n of B_n. From (7.33) and Henstock's Lemma (see also [217], Lemma 2.1., pp. 158-159; [170], Theorem 3.2.1., pp. 81-83), it follows that

$$\left| \int_{\bigcup_{i=1}^{h} V_i} f - \sum_{\Pi_n} f \right| \leq \frac{\varepsilon}{2^{n+2}} \qquad (7.34)$$

for each γ_n-fine decomposition $\Pi_n = \{(V_1, t_1), \ldots, (V_h, t_h)\}$ of B_n. Evidently

$$B_n \cap B_{n-1} = A_n \ \mathrm{int}\, A_n \quad \forall \, n \in \mathrm{N}.$$

Therefore

$$B_n = (B_n \cap B_{n-1}) \cup (\mathrm{int}\, B_n) \cup (B_n \cap B_{n+1}) \quad \forall \, n.$$

Moreover, it is easy to check that

$$B_j \cap B_l = \varnothing \ \text{whenever} \ |j - l| \geq 2 \qquad (7.35)$$

and that

$$(\mathrm{int}\, B_n) \cap (\mathrm{int}\, B_{n+1}) = \varnothing \quad \forall \, n \in \mathrm{N}. \qquad (7.36)$$

Now define a gauge γ on X by the following formula:

$$\gamma(x) = \begin{cases} \gamma_n(x) \cap (\mathrm{int}\, B_n) & \text{if } x \in \mathrm{int}\, B_n, \\ \gamma_n(x) \cap \gamma_{n+1}(x) \cap (\mathrm{int}\, A_{n+1}) & \text{if } x \in B_n \cap B_{n+1}, \quad (n \in \mathrm{N}) \\ (X_0 \ A_{n_0}) \cup \{x_0\} & \text{if } x = x_0. \end{cases} \qquad (7.37)$$

Let $\Pi = \{(U_1, t_1), \ldots, (U_k, t_k)\}$ be a γ-fine partition of X. There exists $(U_{i_0}, t_{i_0}) \in \Pi$, with $i_0 \in \{1, 2, \ldots, k\}$, such that $x_0 \in U_{i_0}$. We shall prove that $t_{i_0} = x_0$. Namely, in the opposite case,

$$x_0 \in U_{i_0} \subset \gamma(t_{i_0}) \subset \gamma_n(t_{i_0})$$

for some n. But $\gamma_n(t) \subset X_0$ for $t \neq x_0$. We have obtained $x_0 \in X_0$, that is a contradiction.

Since $f(x_0) = 0$, the Riemann sum $\sum_{\Pi} f$ has the form

$$\sum_{i=1,\ldots,k, i \neq i_0} f(t_i) \lambda(U_i),$$

and $t_i \in X_0$ $(i = 1, \ldots, k, i \neq i_0)$. Let

$$A = \bigcup_{i=1,\ldots,k, i \neq i_0} U_i,$$

and

$$T = \{n \in \mathbb{N} : \exists i \in \{1, \ldots, k\}, i \neq i_0 : B_n \cap U_i \neq \varnothing\}. \qquad (7.38)$$

By (7.37), and since Π is a γ-fine *partition* of X, we get that

$$X_0 \supset A \supset A_{n_0}. \qquad (7.39)$$

By hypothesis we have

$$\left| \int_A f - I \right| \leq \varepsilon. \qquad (7.40)$$

We claim that, if U_i, $i \neq i_0$, has nonempty intersection with at least two of the $\operatorname{int} B_n$'s, then necessarily there exists $n \in \mathbb{N}$ such that the point t_i corresponding to U_i belongs to $B_n \cap B_{n+1}$. Indeed, if $t_i \in \operatorname{int} B_n$ for some n, then, from (7.37) and the fact that Π is a γ-fine partition of X, we'd have

$$U_i \subset \gamma(t_i) \subset \operatorname{int} B_n :$$

this is impossible, by virtue of (7.35) and (7.36). From this and since

$$(B_{n-1} \cap B_n) \cap (B_n \cap B_{n+1}) = \varnothing \quad \forall n,$$

it follows that, for every $i = 1, 2, \ldots, k$, $i \neq i_0$, the B_n's having nonempty intersection with U_i are at most two, while the B_n's which have nonempty intersection with U_{i_0} can be infinitely many (even all the B_n's). Thus we proved that the set T in (7.38) is finite.

For $n \in T$ define a decomposition Π_n of B_n in the following way:

$$\Pi_n = \{(U_i, t_i) : t_i \in \operatorname{int} B_n\}$$
$$\cup \{(U_i \cap B_n, t_i) : t_i \in B_n \cap B_{n-1}\}$$
$$\cup \{(U_i \cap B_n, t_i) : t_i \in B_n \cap B_{n+1}\}.$$

Then, by construction, we have:

$$\sum_{\Pi} f = \sum_{n \in T} \sum_{\Pi_n} f, \qquad (7.41)$$

by additivity of λ and since $A_n \operatorname{int} A_n = B_n \cap B_{n+1} \subset \operatorname{int} A_{n+1}$ and $\lambda(A_n \operatorname{int} A_n) = 0$ $\forall n \in \mathbb{N}$. Similarly,

$$\sum_{n \in T} \int_{\bigcup_{U_i \subset \operatorname{int} B_n, i \neq i_0} U_i} f = \int_A f. \qquad (7.42)$$

Since Π_n is γ_n-fine, we have (7.33). From (7.33), (7.40), (7.41) (7.42) and (7.39) we obtain:

$$\left| \sum_{\Pi} f - I \right| = \left| \sum_{n \in T} \sum_{\Pi_n} f - I \right|$$

$$= \left| \sum_{n \in T} \left(\sum_{\Pi_n} f - \int_{\bigcup_{U_i \subset \mathrm{int}\, B_n, i \neq i_0} U_i} f \right) + \sum_{n \in T} \int_{\bigcup_{U_i \subset \mathrm{int}\, B_n, i \neq i_0} U_i} f - I \right|$$

$$\leq \sum_{n \in T} \left| \sum_{\Pi_n} f - \int_{\bigcup_{U_i \subset \mathrm{int}\, B_n, i \neq i_0} U} f \right| + \left| \int_A f - I \right|$$

$$\leq \sum_{n \in T} \frac{\varepsilon}{2^{n+2}} + \varepsilon < 2\varepsilon.$$

From this the assertion follows. □

7.2 Vector Valued Case

In [36], the Authors extended all results from 1.3 to 1.16 (except, of course, Proposition 6), and extended also Theorem 7.17 to Banach-valued functions, by using its classical version (that is Theorem 7.17 itself) and uniform continuity of the map "norm". We report only the definition of (KH)-integrability and the extension of Theorem 7.17 to this context.

Definition 7.20 Let S be a Banach space. We say that a function $f:[A,B] \to S$ is *Kurzweil-Henstock integrable* (in short (KH)-*integrable*) on $[A,B]$ if there exists an element $I \in S$ such that $\forall \varepsilon > 0$ there exist a function $\delta:[A,B] \to R^+$ and a positive real number P such that

$$\left\| \sum_{\Pi} f - I \right\| \leq \varepsilon \tag{7.43}$$

whenever $\Pi = \{(I_k, \xi_k): k = 1,\ldots, p\}$ is a δ-fine partition of any bounded interval $[a,b]$ with $[a,b] \supset [A,B] \cap [-P,P]$ and $[a,b] \subset [A,B]$. In this case we say that I is the (KH)-*integral of* f, and we denote the element I by the symbol $(KH)\int_A^B f$ or more simply $\int_A^B f$.

Theorem 7.21 *Let* S *be a Banach space,* $E \subset R$ *be a Lebesgue measurable set with* $|E| < +\infty$, $r \in S$, *and* χ_E *be the characteristic function associated with* E. *Then the function* $\chi_E r$ *is* (KH)-*integrable, and* $\int_{-\infty}^{+\infty} \chi_E r = |E| r$, *where* $|E|$, *as usual, denotes the Lebesgue measure of* E.

Proof: By virtue of Theorem 7.17, we know that the theorem is true in the particular case $S = R$ and $r = 1$. Thus for every $\varepsilon > 0$ there exists a gauge γ, defined on R, such that for each γ-fine partition Π of R we get

$$\left| \sum_{\Pi} \chi_E - |E| \right| \leq \varepsilon. \tag{7.44}$$

Moreover, it is easy to see that for each partition Π of R we have

$$\sum_{\Pi} \chi_E r = \left(\sum_{\Pi} \chi_E \right) r. \tag{7.45}$$

The assertion follows from (7.44), (7.45) and (uniform) continuity of the "norm" map in Banach spaces. □

We now turn to Riesz spaces. We always suppose that our involved space R is Dedekind complete and weakly σ-distributive (see Chapter 2). We now formulate our definition of Kurzweil-Henstock integral.

Definition 7.22 We say that a function $f:[A,B] \to R$ is *Kurzweil-Henstock integrable* (in short (KH)-*integrable*) on $[A,B]$ if there exist an element $I \in R$ and a (D)-sequence $(a_{i,j})_{i,j}$ in R (see Chapter 2) such that $\forall \varphi \in \Phi$ there exist a function $\delta:[A,B] \to R^+$ and a positive real number P such that

$$\left| \sum_{\Pi} f - I \right| \leq \bigvee_{i=1}^{\infty} a_{i,\varphi(i)} \qquad (7.46)$$

whenever $\Pi = \{(I_k, \xi_k): k=1,\ldots,p\}$ is a δ-fine partition of any bounded interval $[a,b]$ with $[a,b] \supset [A,B] \cap [-P,P]$ and $[a,b] \subset [A,B]$. In this case we say that I is the (KH)-*integral of* f, and we denote the element I by the symbol $(KH)\int_A^B f$ or shortly $\int_A^B f$. It is possible to check that our integral is well-defined, that is such an I is uniquely determined.

The following characterization of (KH)-integrability holds (see [26]):

Theorem 7.23 *A function $f:[A,B] \to R$ is (KH)-integrable if and only if there exist $J \in R$ and a (D)-sequence $(a_{i,j})_{i,j}$ such that $\forall \varphi \in N^N$ there exists a gauge γ such that*

$$\left| \sum_{\Pi} f - J \right| \leq \bigvee_{i=1}^{\infty} a_{i,\varphi(i)} \qquad (7.47)$$

whenever $\Pi = \{(I_k, \xi_k): k=1,\ldots,p\}$ is a γ-fine partition of $[A,B]$, and in this case we have $\int_A^B f = J$.

We now observe that the results from 1.3 to 1.16 hold even in the case of Riesz spaces (see [26]), with similar techniques, except Proposition 7.6, which holds, even the technique of the proof is slightly different from the classical one and uses weak σ-distributivity. For the sake of clearness, we report it.

Proposition 7.24 *If f and g are (KH)-integrable on $[A,B]$ and $f \leq g$, then*

$$\int_A^B f \leq \int_A^B g.$$

Proof: By hypothesis, there exist two (D)-sequences $(a_{i,j})_{i,j}$ and $(b_{i,j})_{i,j}$ such that, $\forall \varphi \in \Phi$, there exist two gauges γ_1, γ_2 such that, whenever Π is a γ_1-fine partition of $[A,B]$ and Π' is a γ_2-fine partition of $[A,B]$, we have

$$\int_A^B f - \bigvee_{i=1}^{\infty} a_{i,\varphi(i)} \leq \sum_{\Pi} f \leq \int_A^B f + \bigvee_{i=1}^{\infty} a_{i,\varphi(i)}$$

and

$$\int_A^B g - \bigvee_{i=1}^{\infty} b_{i,\varphi(i)} \leq \sum_{\Pi'} g \leq \int_A^B g + \bigvee_{i=1}^{\infty} b_{i,\varphi(i)}$$

respectively. For every $\xi \in [A,B]$, let $\gamma(\xi) = \gamma_1(\xi) \cap \gamma_2(\xi)$, and take any γ-fine partition Π'' of $[A,B]$: then Π'' is both γ_1- and γ_2-fine. Thus we get

$$\int_A^B f - \bigvee_{i=1}^{\infty} a_{i,\varphi(i)} \leq \sum_{\Pi''} f \leq \sum_{\Pi''} g \leq \int_A^B g + \bigvee_{i=1}^{\infty} b_{i,\varphi(i)}$$

and hence, $\forall \varphi \in \Phi$,

$$\int_A^B f - \int_A^B g \leq \bigvee_{i=1}^{\infty} a_{i,\varphi(i)} + \bigvee_{i=1}^{\infty} b_{i,\varphi(i)} \leq \bigvee_{i=1}^{\infty} c_{i,\varphi(i)},$$

where $c_{i,j} = 2(a_{i,j} + b_{i,j}) \forall i,j \in \mathbb{N}$. By arbitrariness of $\varphi \in \Phi$, since $(c_{i,j})_{i,j}$ is a (D)-sequence and taking into account of weak σ-distributivity of R, we get

$$\int_A^B f - \int_A^B g \leq \bigwedge_{\varphi \in \Phi} \left(\bigvee_{i=1}^{\infty} c_{i,\varphi(i)} \right) = 0,$$

that is $\int_A^B f \leq \int_A^B g$. This concludes the proof. $\square$

Corollary 7.25 *If both f and $|f|$ are (KH)-integrable in $[A,B]$, then*

$$\left| \int_A^B f \right| \leq \int_A^B |f|.$$

The following theorem, whose proof is similar to the one of the real-valued case, will be useful in the sequel:

Theorem 7.26 *Let $[a,b] \subset \mathbb{R}$ be a bounded interval, $f:[a,b] \to R$ be a (KH)-integrable function, and suppose that there exists a (D)-sequence $(a_{i,j})_{i,j}$ such that $\forall \varphi \in \Phi$ there exists $\delta:[a,b] \to \mathbb{R}^+$ such that, for every δ-fine partition Π' of $[a,b]$,*

$$\left| \sum_{\Pi'} {}_a^b f - \int_a^b f \right| \leq \bigvee_{i=1}^{\infty} a_{i,\varphi(i)}. \tag{7.48}$$

Then δ is such that, $\forall c \in (a,b)$ and for every δ-fine partition Π of $[a,b]$,

$$\left| \sum_{\Pi} {}_a^c f - \int_a^c f \right| \leq 2\bigvee_{i=1}^{\infty} a_{i,\varphi(i)}, \quad \left| \sum_{\Pi} {}_c^b f - \int_c^b f \right| \leq 2\bigvee_{i=1}^{\infty} a_{i,\varphi(i)}. \tag{7.49}$$

We now prove that, under suitable hypotheses, the (KH)-integral contains the improper Riemann integral even in the case of Riesz spaces with respect to (D)-convergence.

Theorem 7.27 *Let $a \in \mathbb{R}$, and $f:[a,+\infty] \to R$ be (KH)-integrable on $[a,+\infty]$. Then f is (KH)-integrable on every interval $[a,b]$ with $a < b < +\infty$, and*

$$(D) \lim_{b \to +\infty} \int_a^b f = \int_a^{+\infty} f.$$

Conversely, let $f:[a,+\infty] \to R$ be (KH)-integrable on every interval $[a,b]$ with $a < b < +\infty$ and let there exist in R the limit $l = (D) \lim_{b \to +\infty} \int_a^b f$. Moreover, suppose that

7.27.1) there exist $u \in R$, $u \geq 0$, and a map $\delta_0:[a,+\infty] \to \mathbb{R}^+$, such that, for every b with $a < b < +\infty$ and for every δ_0-fine partition Π of $[a,b]$, we have:

$$\left| \sum_{\Pi} {}_a^b f - \int_a^b f \right| \leq u.$$

Then f is (KH)-integrable on $[a,+\infty]$ and $\int_a^{+\infty} f = l$.

Proof: We begin with the first part of the theorem. Since $f:[a,+\infty] \to R$ is (KH)-integrable, there exists a (D)-sequence $(a_{i,j})_{i,j}$ such that, $\forall \varphi \in \Phi$, $\exists \delta:[a,+\infty] \to R^+$ and $\exists P > |a|$, such that for each bounded interval $[d_1,d_2]$ with $[d_1,d_2] \subset [a,+\infty]$, $[d_1,d_2] \supset [a,+\infty] \cap [-P,P]$, and for every δ-fine partition Π of $[d_1,d_2]$ we have:

$$\left| \sum_{\Pi} f - \int_a^{+\infty} f \right| \leq \bigvee_{i=1}^{\infty} a_{i,\varphi(i)}. \qquad (7.50)$$

Now, we get that f is (KH)-integrable on $[a,b]$ for every $b \in (a,+\infty]$ with respect to the *same* regulator $(a_{i,j})_{i,j}$, and hence we get that $\forall \varphi \in \Phi$, $\forall b \in (a,+\infty]$, $\exists \delta_1 : [a,b] \to R^+$ such that for each δ_1-fine partition Π' of $[a,b]$ we get:

$$\left| \sum_{\Pi'} f - \int_a^b f \right| \leq \bigvee_{i=1}^{\infty} a_{i,\varphi(i)}. \qquad (7.51)$$

Let us define $\delta_2 : [a,b] \to R^+$ by setting $\delta_2(x) = \min\{\delta(x), \delta_1(x)\}$, and let Π be a δ_2-fine partition of $[a,b]$, $b > P$. Then, thanks to (7.50) and (7.51), $\forall \varphi \in \Phi$ $\exists P > 0 : \forall b > P$,

$$\left| \int_a^b f - \int_a^{+\infty} f \right| \leq \left| \sum_{\Pi} f - \int_a^b f \right| + \left| \sum_{\Pi} f - \int_a^{+\infty} f \right| \leq 2 \bigvee_{i=1}^{\infty} a_{i,\varphi(i)}.$$

Thus the first part is completely proved.

We now turn to the second part. By hypothesis, there exists a (D)-sequence $(a_{i,j})_{i,j}$ such that, $\forall \varphi \in \Phi$, $\exists P > 0 : \forall b > P$ we get

$$\left| \int_a^b f - l \right| \leq \bigvee_{i=1}^{\infty} a_{i,\varphi(i)}. \qquad (7.52)$$

Let now $(b_n)_n$ be a strictly increasing sequence of real numbers, such that $\lim_n b_n = +\infty$ and $b_1 = a$. We observe that f is (KH)-integrable in $[b_n, b_{n+1}]$ for each n (with respect to the same regulator $(a_{i,j}^{(n)})_{i,j}$, which is the one "associated" to the interval $[a, b_{n+1}]$). So, $\forall \varphi \in \Phi$ and $\forall n \in N$, $\exists$ a function $\delta_n : [b_n, b_{n+1}] \to R^+$ such that

$$\left| \sum_{\Pi_n} f - \int_{b_n}^{b_{n+1}} f \right| \leq \bigvee_{i=1}^{\infty} a_{i,\varphi(i+n)}^{(n)} \qquad (7.53)$$

whenever Π_n is any δ-fine partition of $[b_n, b_{n+1}]$. Let now $(b_{i,j})_{i,j}$ be a (D)-sequence such that

$$u \wedge \left(\sum_{n=1}^{\infty} \left(\bigvee_{i=1}^{\infty} a_{i,\varphi(i+n)}^{(n)} \right) \right) \leq \bigvee_{i=1}^{\infty} b_{i,\varphi(i)}, \quad \forall \varphi \in \Phi, \qquad (7.54)$$

where u is as in 7.27.1): such a sequence does exist, by virtue of the Fremlin Lemma (see Chapter 2).

Let now $\delta : [a,+\infty] \to R^+$ be as in (7.21), and such that $\delta \leq \delta_0$, where δ_0 is as in 7.27.1).

Choose now arbitrarily $b > P$. If $b_N < b \leq b_{N+1}$ and $\Pi = \{([x_{k-1}, x_k], \xi_k) : k = 1, 2, \ldots p\}$ is a partition of $[a,b]$, then Π is the partition of $[a,b]$ determined by the x_k's and the b_n's. We have:

$$\sum_{\Pi} {}_{a}^{b} f = \sum_{n=1}^{N-1} \left(\sum_{\Pi} {}_{b_n}^{b_{n+1}} f \right) + \sum_{\Pi} {}_{b_N}^{b} f. \tag{7.55}$$

Since the restriction of Π to $[b_N, b]$ is δ_N-fine, then Π can be "extended" to a δ_N-fine partition Π' of $[b_N, b_{N+1}]$. By (7.53) and Theorem 7.26, where the roles of $[a,b]$ and c are played by $[b_N, b_{N+1}]$ and b respectively, we get

$$\left| \sum_{\Pi} {}_{b_N}^{b} f - \int_{b_N}^{b} f \right| \le 2 \bigvee_{i=1}^{\infty} a_{i, \varphi(i+N)}^{(N)}. \tag{7.56}$$

Since the restriction of Π to $[b_n, b_{n+1}]$ is δ_n-fine, from (7.53) and (7.56) it follows that

$$\sum_{n=1}^{N-1} \left| \sum_{\Pi} {}_{b_n}^{b_{n+1}} f - \int_{b_n}^{b_{n+1}} f \right| + \left| \sum_{\Pi} {}_{b_N}^{b} f - \int_{b_N}^{b} f \right| \le 2 \left(\sum_{n=1}^{\infty} \left(\bigvee_{i=1}^{\infty} a_{i,\varphi(i+n)}^{(n)} \right) \right). \tag{7.57}$$

From 7.27.1), (7.52), (7.54), (7.55), (7.56) and (7.57) we get:

$$\left| \sum_{\Pi} {}_{a}^{b} f - l \right| \le \left| \sum_{\Pi} {}_{a}^{b} f - \int_{a}^{b} f \right| + \left| \int_{a}^{b} f - l \right| \le 2 \bigvee_{i=1}^{\infty} b_{i,\varphi(i)} + \bigvee_{i=1}^{\infty} a_{i,\varphi(i)}.$$

Thus the assertion follows. $\square$

Remark 7.28 We observe that theorems similar to Theorem 7.27 hold even if we consider open, semi-open and/or left halflines, R or $\tilde{R}$, instead of $[a, +\infty]$.

Remark 7.29 First of all we observe that, in the classical cases, 7.27.1) is readily fulfilled. There are also several other situations in which 7.27.1) is satisfied: we now prove it when $R = L^0(X, B, \mu)$, where (X, B, μ) is a measure space, with μ positive, σ-additive and σ-finite. We note that, by using the following technique, it is possible also to prove directly Theorem 7.27 for such an R.

We now introduce an important property in the context of Riesz spaces.

Definition 7.30 We say that a Riesz space R has *property* σ, if for every sequence $(u_n)_n$ in R, with $u_n \ge 0$ $\forall n$, there exists a sequence $(\lambda_n)_n$ in R^+ and $\exists\ v \in R$, $v \ge 0$, such that $\lambda_n u_n \le v$ $\forall n \in \mathbb{N}$.

By proceeding similarly as in the proof in the classical case, let $(b_n)_n$ be a strictly increasing sequence of real numbers, such that $\lim_n b_n = +\infty$ and $b_1 = a$. We observe that f is (KH)-integrable in $[b_n, b_{n+1}]$ for each n. So, since in such a space R (D)-convergence, order convergence (see Chapter 2) and (r)-convergence (see Chapter 9) do coincide (see also [175]), we get that $\forall n \in \mathbb{N}$, there exists $u_n \in R$, $u_n \ge 0$, such that, $\forall \varepsilon > 0$, $\exists$ a function $\delta_n : [b_n, b_{n+1}] \to R^+$ such that

$$\left| \sum_{\Pi_n} f - \int_{b_n}^{b_{n+1}} f \right| \le \frac{\varepsilon}{2^{n+1}} u_n \tag{7.58}$$

whenever Π_n is any δ-fine partition of $[b_n, b_{n+1}]$. Since R satisfies property σ (see [175]), then, in correspondence with the sequence $(u_n)_n$, there exist a sequence $(\lambda_n)_n$ of positive real-valued numbers and a positive element $u \in R$, such that

$$\lambda_n u_n \le u \qquad \forall n \in \mathbb{N}.$$

So, we note that, $\forall n \in \mathbb{N}$, $\forall \varepsilon > 0$, $\exists$ a function $\delta_n : [b_n, b_{n+1}] \to R^+$ such that

$$\left| \sum_{\Pi_n} f - \int_{b_n}^{b_{n+1}} f \right| \le \frac{\varepsilon \lambda_n}{2^{n+1}} u_n \le \frac{\varepsilon}{2^{n+1}} u. \qquad (7.59)$$

Let now $\delta : [a, +\infty] \to R^+$ be as in (7.21). Choose now arbitrarily $b > P$. If $b_N < b \le b_{N+1}$ and $\Pi = \{([x_{k-1}, x_k], \xi_k) : k = 1, 2, \dots p\}$ is a δ-fine partition of $[a, b]$, then Π is the partition of $[a, b]$ "determined" by the x_k's and the b_n's. We have:

$$\sum_{\Pi} {}_a^b f = \sum_{n=1}^{N-1} \left(\sum_{\Pi} {}_{b_n}^{b_{n+1}} f \right) + \sum_{\Pi} {}_{b_N}^b f. \qquad (7.60)$$

Since the restriction of Π to $[b_n, b_{n+1}]$ is δ_n-fine, from (7.59) it follows that

$$\sum_{n=1}^{N-1} \left| \sum_{\Pi} {}_{b_n}^{b_{n+1}} f - \int_{b_n}^{b_{n+1}} f \right| \le \frac{\varepsilon}{2} u. \qquad (7.61)$$

From (7.60) and (7.61) we have:

$$\left| \sum_{\Pi} {}_a^b f - \int_a^b f \right| \le \frac{\varepsilon}{2} u + \left| \sum_{\Pi} {}_{b_N}^b f - \int_{b_N}^b f \right|.$$

Since the restriction of Π to $[b_N, b]$ is δ_N-fine, then Π can be "extended" to a δ_N-fine partition Π' of $[b_N, b_{N+1}]$. By Theorem 7.26, where the roles of $[a, b]$ and c are played by $[b_N, b_{N+1}]$ and b respectively, we get

$$\left| \sum_{\Pi} {}_{b_N}^b f - \int_{b_N}^b f \right| \le \frac{\varepsilon}{2^N} u \le \varepsilon u.$$

From this, taking $\varepsilon = 1$, 7.27.1) follows.

We now prove that every simple measurable function defined on R, and assuming values different from zero only on a set of finite Lebesgue measure, is (KH)-integrable according to our definition, and in this case our integral coincides with the usual one. To do this, it is sufficient to prove the following:

Theorem 7.31 *Let $E \subset R$ be a Lebesgue measurable set with $|E| < +\infty$, $r \in R$, and χ_E be the characteristic function associated with E. Then the function $\chi_E r$ is (KH)-integrable, and*

$$\int_{-\infty}^{\infty} \chi_E r = |E| r.$$

Proof: Without loss of generality, we can suppose that $r \ge 0$: indeed every element r of a Riesz space R is the difference between r^+ and r^-, which are two positive elements of R. In order to demonstrate the theorem, we prove that $\forall \varepsilon > 0$ there exists a gauge γ, defined on R, such that for all γ-fine partitions Π of R we get

$$\left| \sum_{\Pi} \chi_E r - |E| r \right| \le \varepsilon r; \qquad (7.62)$$

from (7.62) it will follow that there exists a (D)-sequence $(d_{i,j})_{i,j}$ such that $\forall \varphi \in N^N$ there exists a gauge γ, defined on R, such that for each γ-fine partition Π of R we have

$$\left| \sum_{\Pi} \chi_E r - |E| r \right| \le \bigvee_{i=1}^{\infty} d_{i,\varphi(i)}. \tag{7.63}$$

Indeed, for every $i, j \in \mathbb{N}$, put $d_{i,j} = \dfrac{1}{j} r$. It is easy to check that the double sequence $(d_{i,j})_{i,j}$ is a

(D)-sequence. Fix arbitrarily a map $\varphi \in \Phi$, and set $i_0 = \min\{\varphi(i) : i \in \mathbb{N}\}$, $\varepsilon = \dfrac{1}{i_0}$. Then we get:

$$\bigvee_{i=1}^{\infty} d_{i,\varphi(i)} = \bigvee_{i=1}^{\infty} \frac{1}{\varphi(i)} r = \left(\bigvee_{i=1}^{\infty} \frac{1}{\varphi(i)} \right) r = \frac{1}{i_0} r = \varepsilon r. \tag{7.64}$$

So the assertion of the theorem, that is (7.63), will follow from (7.62) and (7.64). Thus, for our purposes, it will be enough to prove (7.62). By virtue of [170], p. 136, we know that the theorem is true in the particular case $R = \mathbb{R}$ and $r = 1$. Thus for every $\varepsilon > 0$ there exists a gauge γ, defined on $\mathbb{R}$, such that for each γ-fine partition Π of $\mathbb{R}$ we get

$$\left| \sum_{\Pi} \chi_E - |E| \right| \le \varepsilon. \tag{7.65}$$

Moreover, it is easy to see that for each partition Π of $\mathbb{R}$ we have

$$\sum_{\Pi} \chi_E r = \left(\sum_{\Pi} \chi_E \right) r. \tag{7.66}$$

Thus (7.62) follows from (7.65) and (7.66). This concludes the proof. $\square$

Property σ plays a fundamental role in the theory of the Kurzweil-Henstock integral in Riesz spaces. We prove the following:

Proposition 7.32   *Let R be a Dedekind complete Riesz space, satisfying property σ, $Q \subset [A,B]$ be a countable set, and $f : [A,B] \to R$ be a function, such that $f(x) = 0$ for all $x \in [A,B] \setminus Q$. Then f is (KH)-integrable on $[A,B]$ and $\int_A^B f = 0$.*

Proof   Let $Q = \{x_n : n \in \mathbb{N}\}$ and $f(x_n) = u_n$ for all $n \in \mathbb{N}$; without loss of generality, we can suppose that $u_n \ge 0 \ \forall n \in \mathbb{N}$. We note that, in order to prove the Proposition, it is enough to show that there exists an element $z \in R$, $z \ge 0$, $z \ne 0$, such that $\forall \varepsilon > 0$ there exist a map $\delta : [A,B] \to \mathbb{R}^+$ and a positive real number P such that

$$\left| \sum_{\Pi} f \right| \le \varepsilon z \tag{7.67}$$

for each bounded interval $[a,b]$ with $[A,B] \cap [-P,P] \subset [a,b] \subset [A,B]$ and for every δ-fine partition Π of $[a,b]$. By property σ, in correspondence with the sequence $(u_n)_n$, there exist a sequence $(\lambda_n)_n$ of positive real numbers and an element $z \in R$ such that $0 \le \lambda_n u_n \le z$ for all $n \in \mathbb{N}$. Fix now an arbitrary $\varepsilon > 0$ and set $\delta(x_n) = \varepsilon \, 2^{-n-2} \lambda_n \ \forall n \in \mathbb{N}$, and $\delta(x) = 1$ if $x \notin Q$. For all $P > 0$, for every bounded interval $[a,b]$ with $[A,B] \cap [-P,P] \subset [a,b] \subset [A,B]$ and for each δ-fine partition $\Pi = \{(J_i, \xi_i) : i = 1, \dots, q\}$ of $[a,b]$ we have

$$0 \le \sum_{\Pi} f = \sum_{i=1}^{q} |J_i| f(\xi_i) \le \sum_{n=1}^{\infty} \left(\sum_{\xi_i = x_n} |J_i| u_i \right)$$

$$\le \sum_{n=1}^{\infty} \left(\sum_{\xi_i = x_n} 2\delta(\xi_i) u_i \right) \le 4\varepsilon \sum_{n=1}^{\infty} 2^{-n-2} \lambda_n u_n$$

$$\le 4\varepsilon \left(\sum_{n=1}^{\infty} 2^{-n-2} \right) z = \varepsilon z,$$

proving (7.67) and thus the assertion. $\square$

We note that Proposition 7.32 does not hold, if R does not have property σ:

Example 7.33 Let R be a Dedekind complete Riesz space without property σ. In R there exists a sequence $(u_n)_n$ such that for all sequences $(\lambda_n)_n$ of positive real numbers, the sequence $(\lambda_n u_n)_n$ is not bounded in R. We now check that the function $f : [0,1] \to R$, defined by setting

$$f(x) = \begin{cases} u_n & \text{if } x = 1/n \\ 0 & \text{otherwise} \end{cases} \tag{7.68}$$

is not (KH)-integrable on $[0,1]$.

Indeed, fix arbitrarily $\delta : [0,1] \to R^+$ and $n \in N$, $n \ge 2$. For every $i = 1,\ldots,n-1$, let $\xi_i = \dfrac{1}{n+1-i}$ and choose an interval $]y_i, x_i[$ such that $\xi_i \in]y_i, x_i[$, $x_i - y_i < \delta(\xi_i)$, $[y_i, x_i] \cap [y_j, x_j] = \varnothing$ $\forall i \ne j$, $0 < y_1$ and $x_{n-1} < 1$. We have:

$$0 < y_1 < x_1 < y_2 < x_2 < \ldots < y_{n-1} < x_{n-1} < 1.$$

Let $x_0 = 0$, $y_n = 1$, and let us divide each of the intervals $[x_{i-1}, y_i]$, $i = 1,\ldots,n$, in subintervals, in such a way to have δ-fine partitions: this is possible, by virtue of the Cousin Lemma. These subintervals and the elements $([y_i, x_i], \xi_i)$, $i = 1,\ldots,n-1$, form a δ-fine partition $\{([t_{j-1}, t_j], \eta_j) : j = 1,\ldots,p\}$. Since $f = 0$ on each of the intervals $[x_{i-1}, y_i]$, $i = 1,\ldots,n$, we have:

$$\sum_{j=1}^{p} (t_j - t_{j-1}) f(\eta_j) = \sum_{i=1}^{n-1} (x_i - y_i) f(\xi_i).$$

Let $\lambda_i^{(n)} = x_{n+1-i} - y_{n+1-i}$, $i = 2,\ldots,n$: then we get

$$\sum_{j=1}^{p} (t_j - t_{j-1}) f(\eta_j) = \sum_{i=1}^{n-1} \lambda_{n+1-i}^{(n)} f(\xi_i)$$

$$= \sum_{i=1}^{n-1} \lambda_{n+1-i}^{(n)} u_{n+1-i} \ge \lambda_n^{(n)} u_n. \tag{7.69}$$

We note that the $\lambda_j^{(n)}$'s can be chosen in such a way that $\lambda_j^{(n)} = \lambda_j^{(m)}$ for every $j \in N$, $j \ge 2$, and for each $n, m \in N$, with $m, n \ge j$. Put $\lambda_n = \lambda_n^{(n)}$ $\forall n \in N$: in conclusion we have that, for every $\delta : [0,1] \to R^+$, there exists a sequence $(\lambda_n)_n$ in R^+ such that, $\forall n \in N$, there is a δ-fine partition $\Pi(n)$ of $[0,1]$ such that

$$\sum_{\Pi(n)} f \ge \lambda_n u_n.$$

Thus, the sequence $\left(\sum_{\Pi(n)} f\right)_n$ is unbounded in R.

If f was (KH)-integrable on $[0,1]$, then there would exist a map $\delta_0 : [0,1] \to R^+$ such that

$$\sup\left\{\sum_{\Pi} f : \Pi \text{ is a } \delta_0 - \text{fine partition of } [0,1]\right\} \in R :$$

this is a contradiction. Hence, f is not (KH)-integrable on $[0,1]$. □

Analogously, it is possible to show that the function $f^* : [0,+\infty) \to R$, defined by setting

$$f^*(x) = \begin{cases} u_n & \text{if } x = n \\ 0 & \text{otherwise} \end{cases} \tag{7.70}$$

is not (KH)-integrable on $[0,+\infty)$. The function f^* defined in (7.70) is also an example to show that Theorem 7.27 in general does not hold, if R does not have property σ. Obviously, f^* does not satisfy condition 7.27.1).

We now turn to the case in which our involved functions are defined in suitable (locally) compact topological spaces and take values in (Dedekind complete weakly σ-distributive) Riesz spaces (see [24]).

Let X be a Hausdorff compact topological space. If $A \subset X$, we denote by $\text{int } A$ the interior of A. Let F, λ be as in the previous paragraph.

Let us turn to the definition of Kurzweil-Henstock integral ((KH)-integral) on X. If $\Pi = \{(U_1, t_1), \ldots, (U_k, t_k)\}$ is a decomposition of a set A, and $f : X \to R$, then we define the Riemann sum as follows:

$$\sum_{\Pi} f = \sum_{i=1}^{k} f(t_i)\lambda(U_i),$$

if the sum exists in R, with the convention $0 \cdot (+\infty) = 0 \cdot (-\infty) = 0$.

Definition 7.34 A function $f : X \to R$ is said to be (KH)-*integrable* (in short, *integrable*) on a set A if there exist $I \in R$ and a (D)-sequence $(b_{i,j})_{i,j}$ such that $\forall \varphi \in \Phi$ there exists a gauge γ on A such that $\sum_{\Pi} f$ exists, and

$$\left|\sum_{\Pi} f - I\right| \leq \bigvee_{i=1}^{\infty} b_{i,\varphi(i)} \tag{7.71}$$

whenever Π is a γ-fine partition of A such that $\sum_{\Pi} f$ exists in R. We denote

$$I = \int_A f.$$

We now state the Bolzano-Cauchy condition for (KH)-integrability.

Theorem 7.35 *A map $f : X \to R$ is (KH)-integrable on A if and only if there exists a (D)-sequence $(b_{i,j})_{i,j}$ such that, $\forall \varphi \in \Phi$, $\exists$ a gauge γ such that for every γ-fine partition Π_1, Π_2 of A we have*

$$\left|\sum_{\Pi_1} f - \sum_{\Pi_2} f\right| \leq \bigvee_{i=1}^{\infty} b_{i,\varphi(i)}.$$

Proof: The proof is similar to the one of Theorem 5.2.9, p. 77, of [227]. □

We now state a result about (KH)-integrability on subsets.

Theorem 7.36 *If* $f : X \to R$ *is integrable on* $A \in F$ *and* $B \in F$, $B \subset A$, *then* f *is integrable on* B *too. Moreover, if* $(b_{i,j})_{i,j}$ *is a regulator, satisfying condition (2) of integrability relatively to* A, *then* $(b_{i,j})_{i,j}$ *satisfies condition (2) of integrability with respect to* B *too.*

Proof: The proof is similar to the one of Proposition 5.2.10, p. 79, of [227]. □

Similarly as in [227], Theorem 5.3.1, pp. 82-83, it is possible to prove the following version of Henstock's Lemma:

Theorem 7.37 *Let* $f : X \to R$ *be integrable on* A, *and* $(b_{i,j})_{i,j}$ *be a regulator, such that* $\forall \varphi \in \Phi$ *there exists a gauge* γ *such that*

$$\left| \sum_\Pi f - \int_A f \right| \leq \bigvee_{i=1}^\infty b_{i,\varphi(i)}$$

for all γ*-fine partitions* Π *of* A.
Then, if γ *is a gauge and* $\Pi = \{(J_i, \xi_i) : i = 1, \dots, k\}$ *is a* γ*-fine decomposition of* A, *we have*

$$\sum_{i=1}^k \left| f(\xi_i)\lambda(J_i) - \int_{J_i} f \right| \leq 2 \bigvee_{i=1}^\infty b_{i,\varphi(i)}.$$

We now prove the following convergence theorem:

Theorem 7.38 *Let* $X = X_0 \cup \{x_0\}$ *be the one-point compactification of a locally compact space* X_0. *Let* $f : X \to R$ *be a function such that* $f(x_0) = 0$. *Let* $(A_n)_n$ *be a sequence of sets, such that* $A_n \in F$, $A_n \subset \operatorname{int} A_{n+1}$, $A_{n+1} \setminus \operatorname{int} A_n \in F$, $\lambda(A_n \setminus \operatorname{int} A_n) = 0$ $(n \in N)$, $\bigcup_{n=1}^\infty A_n = X_0$. *Let* f *be integrable on every* $A \in F$ *with* $A \subset X_0$, *and let there exist a regulator* $(a_{i,j})_{i,j}$ *and an element* $I \in R$ *such that,* $\forall \varphi \in \Phi$, *there exists an integer* n_0 *such that*

$$\left| \int_A f - I \right| \leq \bigvee_{i=1}^\infty a_{i,\varphi(i)} \quad \forall A \in F, X_0 \supset A \supset A_{n_0}. \tag{7.72}$$

Moreover, suppose that

7.38.1) *there exist* $u \in R$, $u \geq 0$, *and a gauge* γ_0, *such that for every* γ_0*-fine partition* Π *of* X, $\Pi = \{(U_1, t_1), \dots, (U_k, t_k)\}$, *we have:*

$$\left| \sum_\Pi f - \int_{\bigcup_{i=1,..,k, t_i \neq x_0} U_i} f \right| \leq u.$$

Then f *is integrable on* X *and* $\int_X f = I$.

Proof: Let $(a_{i,j})_{i,j}$ be as in the hypotheses of the theorem, choose arbitrarily an element $\varphi \in \Phi$ and, in correspondence with φ, let n_0 be as in (7.72). Put $A_0 = \varnothing$, $B_0 = A_1$, $B_n = A_{n+1}$, $\operatorname{int} A_n$ $(n \in N)$. For any $n \in N$ there is a (D)-sequence $(b_{i,j}^{(n)})_{i,j}$ such that $\forall \varphi \in \Phi$ there exists a gauge γ_n on B_n such that

$$\left| \int_{B_n} f - \sum_\Pi f_n \right| \leq \bigvee_{i=1}^\infty b_{i,\varphi(i+n)}^{(n)} \tag{7.73}$$

for any γ_n-fine partition Π_n of B_n. From (7.73) and the Henstock Lemma it follows that

$$\left| \int_{\cup_{i=1}^h V_i} f - \sum_{\Pi_n} f \right| \leq 2 \bigvee_{i=1}^\infty b_{i,\varphi(i+n)}^{(n)} \tag{7.74}$$

for each γ_n-fine decomposition $\Pi_n = \{(V_1,t_1),\ldots,(V_h,t_h)\}$ of B_n. Let now $(c_{i,j})_{i,j}$ be a (D)-sequence such that

$$u \wedge \left(\sum_{n=1}^\infty \left(\bigvee_{i=1}^\infty b_{i,\varphi(i+n)}^{(n)} \right) \right) \leq \bigvee_{i=1}^\infty c_{i,\varphi(i)}, \quad \forall \varphi \in \Phi,$$

where u is as in 7.38.1): such a sequence does exist, by virtue of the Fremlin Lemma (see [227]). Now define a gauge γ on X as in (7.37). Let $\Pi = \{(U_1,t_1),\ldots,(U_k,t_k)\}$ be a γ-fine partition of X. There exists $(U_{i_0},t_{i_0}) \in \Pi$, with $i_0 \in \{1,2,\ldots,k\}$, such that $x_0 \in U_{i_0}$. In this case we have $t_{i_0} = x_0$. Since $f(x_0) = 0$, the Riemann sum $\sum_\Pi f$ has the form

$$\sum_{i=1,\ldots,k, i \neq i_0} f(t_i) \lambda(U_i),$$

and $t_i \in X_0$ $(i=1,\ldots,k, i \neq i_0)$. Let

$$A = \bigcup_{i=1,\ldots,k, i \neq i_0} U_i$$

and T be as in (7.38). We have that the set T is finite.

By (7.37), and since Π is a γ-fine *partition* of X, we get

$$X_0 \supset A \supset A_{n_0}. \tag{7.75}$$

By hypothesis we have

$$\left| \int_A f - I \right| \leq \bigvee_{i=1}^\infty a_{i,\varphi(i)}. \tag{7.76}$$

For $n \in T$ define a decomposition Π_n of B_n in the following way:

$$\Pi_n = \{(U_i,t_i) : t_i \in \text{int } B_n\}$$
$$\cup \{(U_i \cap B_n, t_i) : t_i \in B_n \cap B_{n-1}\}$$
$$\cup \{(U_i \cap B_n, t_i) : t_i \in B_n \cap B_{n+1}\}.$$

Then, by construction, we have:

$$\sum_\Pi f = \sum_{n \in T} \sum_{\Pi_n} f. \tag{7.77}$$

Moreover, we get:

$$\sum_{n \in T} \int_{\cup_{U_i \subset \text{int } B_n, i \neq i_0} U_i} f = \int_A f. \tag{7.78}$$

Since Π_n is γ_n-fine, we have (7.73). From 7.38.1), (7.73), (7.76), (7.77), (7.78) and (7.75) we obtain

$$\left| \sum_\Pi f - I \right| = \left| \sum_{n \in T} \sum_{\Pi_n} f - I \right|$$

$$= \left| \sum_{n \in T} \left(\sum_{\Pi_n} f - \int_{\cup_{U_i \subset \text{int } B_n, i \neq i_0} U_i} f \right) + \sum_{n \in T} \int_{\cup_{U_i \subset \text{int } B_n, i \neq i_0} U_i} f - I \right|$$

$$\leq \sum_{n \in T} \left| \sum_{\Pi_n} f - \int_{\cup_{U_i \subset \text{int } B_n, i \neq i_0} U_i} f \right| + \left| \int_A f - I \right|$$

$$\leq 2 \bigvee_{i=1}^\infty c_{i,\varphi(i)} + \bigvee_{i=1}^\infty a_{i,\varphi(i)} \leq \bigvee_{i=1}^\infty d_{i,\varphi(i)},$$

where $d_{i,j} = 2(c_{i,j} + a_{i,j})$ $\forall i, j \in \mathrm{N}$. From this the assertion follows, since the double sequence $(d_{i,j})_{i,j}$ is a (D)-sequence. $\square$

Remark 7.39 First of all we observe that, in the classical cases, 7.38.1) is readily fulfilled. There are also several other situations in which 7.38.1) is satisfied: we now prove it when $R = L^0(X, \mathrm{B}, \mu)$, where (X, B, μ) is a measure space, with μ positive, σ-additive and σ-finite. By proceeding similarly as in the proof of the classical case and by using the same notations as above, since in such a space R (D)-convergence, order convergence and (r)-convergence do coincide (see also [175]), we get that $\forall n \in \mathrm{N}$, there exists $u_n \in R$, $u_n \geq 0$, such that, $\forall \varepsilon > 0$, $\exists$ a gauge γ_n on B_n such that

$$\left| \int_{\cup_{i=1}^{h} V_i} f - \sum_{\Pi_n} f \right| \leq \frac{\varepsilon}{2^{n+2}} u_n \qquad (7.79)$$

for each γ_n-fine decomposition $\Pi_n = \{(V_1, t_1), \ldots, (V_h, t_h)\}$ of B_n. Since R satisfies property σ (see [175]), then, in correspondence with the sequence $(u_n)_n$, there exist a sequence $(\lambda_n)_n$ of positive real-valued numbers and a positive element $u \in R$, such that

$$\lambda_n u_n \leq u \qquad \forall n \in \mathrm{N}.$$

So, we note that, $\forall n \in \mathrm{N}$, $\forall \varepsilon > 0$, $\exists$ a gauge γ_n on B_n such that

$$\left| \int_{\cup_{i=1}^{h} V_i} f - \sum_{\Pi_n} f \right| \leq \frac{\varepsilon \lambda_n}{2^{n+2}} u_n \leq \frac{\varepsilon}{2^{n+2}} u \qquad (7.80)$$

for every γ_n-fine decomposition $\Pi_n = \{(V_1, t_1), \ldots, (V_h, t_h)\}$ of B_n. Let now γ be a gauge as in (7.37): proceeding analogously as in the proof of Theorem 7.19, we get that, if $\Pi = \{(U_1, t_1), \ldots, (U_h, t_h)\}$ is any γ-fine partition of X, then

$$\left| \sum_{\Pi} f - \int_{\cup_{i=1, \ldots, k, t_i \neq x_0} U_i} f \right| \leq \frac{\varepsilon}{4} u. \qquad (7.81)$$

From this, taking arbitrarily $\varepsilon \in (0, 4]$, 7.38.1) follows. $\square$

Remark 7.40 We note that, under the hypotheses as in 7.39, it is possible to prove Theorem 7.38 directly, by using (r)-convergence and its properties.

Remark 7.41 We note that Theorem 7.27 is a consequence of Theorem 7.38, where $X = [a, +\infty]$, $x_0 = +\infty$. Indeed, with the same notations of these theorems, we have $A_N = [a, b_N]$, $N \in \mathrm{N}$. So, if $b_N < b \leq b_{N+1}$, to any partition of the interval $[a, b]$ corresponds a partition Π of $[0, +\infty]$, $\Pi = \{(U_1, t_1), \ldots, (U_h, t_h)\}$, in which the U_i's can be taken as closed intervals of the *real* line, except U_{i_0}, $U_{i_0} = [b, +\infty]$, and $\max[\cup_{i \neq i_0} U_i] = b$.

8 Choquet and Šipoš Integrals

Abstract: In this chapter we introduce some integrals for real-valued maps with respect to Riesz space-valued set functions, which are not necessarily finitely additive, but in general can be simply only increasing. First we deal with the Šipoš (symmetric) integral and prove the monotone and Lebesgue dominated convergence theorems, Fatou's lemma and the submodular theorem.

Moreover we introduce the Choquet (asymmetric) integral, giving in particular some applications to the weak and strong laws of large numbers in the context of Riesz spaces.

8.1 Symmetric Integral

In this chapter we deal with some recent research and results about Choquet-type integrals for real-valued (or extended real-valued) functions, with respect to set functions P with values in a Dedekind complete Riesz space R, and which can be σ-additive, finitely additive or simply only increasing. In the literature, in the case $R = \mathbb{R}$, there are substantially two kinds of integrals in this context: the symmetric and the asymmetric integral (see also [73]). Here we investigate in detail only the symmetric integral, because our main recent researches in this topic deal with the case in which P is finitely additive (here the two integrals will coincide) and, when P is only monotone, just (mainly) with the symmetric integral.

In [32] we introduced a "monotone-type" (that is, Choquet-type) integral for real-valued maps, with respect to finitely additive positive set functions, with values in a Dedekind complete Riesz space. In [98], a Choquet-type integral for real-valued functions with respect to Riesz-space-valued "capacities", that is monotone set functions (not necessarily finitely additive), is investigated.

We introduce a Šipoš-type integral, that is "symmetric" integral, for real-valued functions with respect to Riesz-space-valued capacities, we investigate the fundamental properties and prove some convergence theorems (for real-valued capacities see also [249] and [204], pp. 152-176).

Throughout this paragraph, we always suppose that R is a Dedekind complete Riesz space. In some suitable cases, we add to R two extra elements, which we call $+\infty$ and $-\infty$, extending ordering and operations, having the same role as the usual $+\infty$ and $-\infty$ with the real numbers (see also [15, 136]). We denote with the symbol $\overline{R}$ the set $R \cup \{+\infty\} \cup \{-\infty\}$.

We now extend to general directed nets the concept of (O)-convergence.

A directed net $(r_\alpha)_{\alpha \in \Xi}$ is called (O)-*net* if $r_\alpha \downarrow 0$, that is if it is decreasing and $\inf_{\alpha \in \Xi} r_\alpha = 0$. We say that the directed net $(r_\alpha)_{\alpha \in \Xi}$ is (O)-*convergent* to r, if

$$(O)\limsup_\alpha r_\alpha := \inf_\alpha [\sup_{\beta \geq \alpha} r_\beta] = (O)\liminf_\alpha r_\alpha := \sup_\alpha [\inf_{\beta \geq \alpha} r_\beta] = r,$$

and in this case we will write $(O)\lim_{\alpha \in \Xi} r_\alpha = r$.

Definition 8.1 Let X be an arbitrary nonempty set, and $f \in \tilde{\mathbb{R}}^X$. The class

$$Q_f := \{\{x \in X : f(x) > t\} : t \in \mathbb{R}\} \cup \{\{x \in X : f(x) \geq t\} : t \in \mathbb{R}\}$$

is called the *upper set system* of f.

Antonio Boccuto / Beloslav Riečan / Marta Vrábelová

Definition 8.2 We say that a class $\mathcal{C}$ of elements of $\tilde{\mathbb{R}}^X$ is *comonotonic* if $\bigcup_{f\in\mathcal{C}} Q_f$ is a chain, or equivalently, if, for each pair of $f, g \in \mathcal{C}$, there is no pair of elements $x_1, x_2 \in X$ such that $f(x_1) < f(x_2)$ and $g(x_1) > g(x_2)$ (see [73, 204]).

We begin with recalling the Choquet integral, introduced in [98], and we introduce and investigate the Šipoš (that is the symmetric Choquet) integral for (extended) real-valued functions with respect to Riesz space-valued capacities.

Definition 8.3 Let X be any nonempty set, and $\mathcal{A} \subset \mathcal{P}(X)$ be a σ-algebra (We suppose this for the sake of simplicity, though several results remain true if we consider more general structures). We say that a set function $P : \mathcal{A} \to R$ is a *capacity* if $P(\varnothing) = 0$, and $P(A) \le P(B)$ whenever $A, B \in A$, $A \subset B$; P is said to be *submodular* if

$$A, B \in A \Rightarrow P(A \cup B) + P(A \cap B) \le P(A) + P(B);$$

supermodular, if

$$A, B \in A \Rightarrow P(A \cup B) + P(A \cap B) \ge P(A) + P(B);$$

subadditive, if

$$A, B \in A \Rightarrow P(A \cup B) \le P(A) + P(B);$$

superadditive, if

$$A, B \in A \Rightarrow P(A \cup B) \ge P(A) + P(B).$$

An R-valued capacity P is said to be *continuous from below* if for every increasing sequence $(E_n)_n$ of elements of A we have

$$P\left(\bigcup_{n=1}^{\infty} E_n\right) = (O)\lim_n P(E_n) = \sup_n P(E_n);$$

continuous from above, if for every decreasing sequence $(E_n)_n$ of elements of A we have

$$P\left(\bigcap_{n=1}^{\infty} E_n\right) = (O)\lim_n P(E_n) = \inf_n P(E_n);$$

continuous, if it is continuous both from below and from above.

A map $P : \mathcal{A} \to R$ called *mean* (or *finitely additive set function*) if $P(A) \ge 0$ $\forall A \in \mathcal{A}$, and $P(A \cup B) = P(A) + P(B)$, whenever $A \cap B = \varnothing$. It is easy to check that every mean is a capacity, but the converse is in general not true. We say that a set function P is a *measure* or that P is σ-*additive* if it is a continuous mean. We say that a map $f : X \to \tilde{\mathbb{R}}$ is *measurable* if $\Sigma_t^f \in \mathcal{A}$, $\forall t \in \mathbb{R}$. A real-valued measurable map is called *random variable* too. Similarly as in [32], given a measurable mapping $f : X \to \tilde{\mathbb{R}}$ and a capacity P, for all $t \in \mathbb{R}$, set: $\Sigma_t^f = \Sigma_t := \{x \in X : f(x) \ge t\}$; and, for every $t \in \mathbb{R}$, let $u_f(t) = u(t) := P(\Sigma_t)$.

We now introduce the Choquet integral for non-negative functions with respect to Riesz space-valued capacities (see also [33, 98]).

Definition 8.4 A measurable non-negative map $f \in \tilde{\mathbb{R}}^X$ is said to be *Choquet integrable* if there exists in R the quantity

$$\int_0^{+\infty} u(t)\,dt := \sup_{a>0}(Ri)\int_0^a u(t)\,dt = (O)\lim_{a\to+\infty}(Ri)\int_0^a u(t)\,dt,$$

where $u(t) = P(\Sigma_t)$, $t \geq 0$, and $(Ri)\int$ is the Riemann integral for Riesz-space-valued maps, defined on an interval of the real line, as in Chapter 7. If f is Choquet integrable, we denote its integral by the symbol $(C)\int_X f\,dP$.

We now introduce the Šipoš integral, that is the symmetric Choquet integral, for extended real-valued functions with respect to Riesz space-valued capacities. We begin with the following:

Definitions 8.5 A measurable function $f : X \to \mathbb{R}$ is said to be *simple* if its range is finite.

Let $\mathcal{F}$ be the family of all finite subsets of $\mathbb{R}$ which contain zero. Given $F \in \mathcal{F}$ and $a \in \mathbb{R}$, set

$$F + a := \{d \in R : d = b + a, \text{ with } b \in F\}$$

and

$$aF := \{d \in R : d = ab, \text{ with } b \in F\}.$$

Let now $F \in \mathcal{F}$, $F = \{b_k, b_{k-1}, \ldots, b_1, b_0, a_0, a_1, \ldots, a_n\}$, where

$b_k < b_{k-1} < \ldots < b_1 < b_0 = 0 = a_0 < a_1 < \ldots < a_n$, and let f be a measurable function. As in [204], p. 153, set

$$f_F = \sum_{i=1}^n (a_i - a_{i-1})\chi_{A_i} + \sum_{j=1}^k (b_j - b_{j-1})\chi_{B_j},$$

where

$$A_i = \{x \in X : f(x) \geq a_i\}, i = 0, 1, \ldots, n; \tag{8.1}$$
$$B_j = \{x \in X : f(x) \leq b_j\}, j = 0, 1, \ldots, k. \tag{8.2}$$

If P is an R-valued capacity, we define the *integral sum* (with respect to P) associated to f and F as follows:

$$S_F(f) = \sum_{i=1}^n (a_i - a_{i-1})P(A_i) + \sum_{j=1}^k (b_j - b_{j-1})P(B_j),$$

(where the A_i's and the B_j's are as in (8.1) and (8.2) respectively) if the right-hand side expression contains no expression of the type $+\infty - \infty$; moreover, we put *by convention* $S_{\{0\}}(f) = 0$. We note that the set $\mathcal{F}$ is directed. We say that $f : X \to \tilde{\mathbb{R}}$ (not necessarily positive) is *Šipoš integrable*

((S)-*integrable*) if there exists in R the limit

$$(O)\lim_{F \in \mathcal{F}} S_F(f), \tag{8.3}$$

and in this case we denote the limit in (8.3) by the symbol $(S)\int_X f\,dP$. If the limit in (8.3) is $+\infty$ or $-\infty$, we say that

$$(S)\int_X f \, dP = +\infty$$

or

$$(S)\int_X f \, dP = -\infty$$

respectively, though f is, of course, not (S)-integrable. Furthermore, given a set $A \subset X$, $A \in A$, we say that f is (S)-*integrable on* A if $f\chi_A$ is (S)-integrable, and in this case we put, *by definition,*

$$(S)\int_A f \, dP := (S)\int_X f \chi_A \, dP. \qquad (8.4)$$

We now state the following properties.

Proposition 8.6 *Let $f : X \to \tilde{\mathbb{R}}$ be a measurable function. The following results hold:*

a) *If $f \geq 0$ and $F \ni F_1 \subset F_2 \in \mathcal{F}$, then $S_{F_1}(f) \leq S_{F_2}(f)$.*

b) *If $f \geq 0$, then $(S)\int_X f \, dP$ exists in $R \cup \{+\infty\}$ and $(S)\int_X f \, dP \geq 0$. Moreover in this case we get:*

$$(S)\int_X f \, dP = \sup_{F \in F} S_F(f).$$

c) *The Šipoš integral is a monotone functional.*

d) *If $(S)\int_X f \, dP$ exists in R, then for every $c \in \mathbb{R}$ we have:*

$$(S)\int_X (c f) \, dP = c \cdot (S)\int_X f \, dP.$$

Proof: The proof is similar to the one of Lemma 7.3 (ii), p. 156, and Theorem 7.10 (i)-(iii), p. 155, of [204].

We now prove that, for measurable non negative extended real-valued maps, the Šipoš and Choquet integrals do coincide.

Theorem 8.7 *Let $f : X \to \tilde{\mathbb{R}}$ be a non-negative measurable function. Then f is Šipoš integrable if and only if it is Choquet integrable, and in this case*

$$(S)\int_X f \, dP = (C)\int_X f \, dP.$$

Proof: First of all we prove that every non-negative bounded measurable function f is Šipoš integrable.

Let $K \in \mathbb{N}$ be such that $f(x) \leq K$ for all $x \in X$. Then the function $u(t) := P(\{x \in X : f(x) \geq t\})$ vanishes on $[K, +\infty)$, and therefore u is (Ri)-integrable in $[0, K]$, and we get:

$$R \ni (Ri)\int_0^K u(t)\,dt = \int_0^{+\infty} u(t)\,dt = (C)\int_X f\,dP. \qquad (8.5)$$

Thus f is Choquet integrable. Moreover, thanks also to Proposition 8.6 a), it is easy to check that $0 \le S_F(f) \le K\,P(X)$ for every $F \in \mathcal{F}$. From this and Proposition 8.6 b) it follows that f is Šipoš integrable too and

$$(S)\int_X f\,dP \le K\,P(X).$$

Now, given $F \in \mathcal{F}$, let $\Sigma(u,F,K)$ be the Riemann sum of u associated to that division of $[0,K]$ whose elements, which we denote by α_j, $j=1,...,s$, are the points of F belonging to $[0,K]$ and the points 0 and K, with respect to the (right end-)points α_j's themselves, $j=1,...,s$. We have:

$$0 \le \left| (S)\int_X f\,dP - (C)\int_X f\,dP \right|$$

$$= \left| (S)\int_X f\,dP - (Ri)\int_0^K u(t)\,dt \right|$$

$$= (O)\limsup_{F \in \mathcal{F}} \left| (S)\int_X f\,dP - S_F(f) + S_F(f) - (Ri)\int_0^K u(t)\,dt \right|$$

$$\le (O)\limsup_{F \in \mathcal{F}} \left| (S)\int_X f\,dP - S_F(f) \right|$$

$$+ (O)\limsup_{F \in \mathcal{F}} \left| \Sigma(u,F,K) - (Ri)\int_0^K u(t)\,dt \right|$$

$$= (O)\limsup_{F \in \mathcal{F}} \left| \Sigma(u,F,K) - (Ri)\int_0^K u(t)\,dt \right| = 0.$$

From this the assertion follows, at least when f is bounded and measurable.

If f is not bounded, then we have (finite or $+\infty$):

$$(S)\int_X f\,dP = \sup_{F \in \mathcal{F}} S_F(f)$$

$$= \sup_{K \in \mathbb{N}} \left[\sup_{F \in \mathcal{F}} S_F(f \wedge K) \right]$$

$$= \sup_{K \in \mathbb{N}} (S)\int_X (f \wedge K)\,dP = \sup_{K \in \mathbb{N}} (C)\int_X (f \wedge K)\,dP \qquad (8.6)$$

$$= \sup_{K \in \mathbb{N}} (Ri)\int_0^K P(\{x \in X : f(x) \ge t\})\,dt$$

$$= \int_0^{+\infty} P(\{x \in X : f(x) \ge t\})\,dt = (C)\int_X f\,dP.$$

This concludes the proof. ⌘

Proceeding in an analogous way as in (8.6), it is possible to prove the following:

Proposition 8.8 *A non-negative measurable extended real-valued function f is (S)-integrable if and only if there exists in R one of the following three elements:*

$$(O) \lim_{K \to +\infty} (S)\int_X (f \wedge K)\,dP,$$

$$\sup_{K \in \mathbb{R}, K \ge 0} (S)\int_X (f \wedge K)\,dP,$$

$$\sup_{K \in \mathbb{N}} (S) \int_X (f \wedge K) \, dP,$$

and these quantities coincide with $(S) \int_X f \, dP$, not only if they belong to R, but also if they are equal to $+\infty$.

We now prove the following:

Theorem 8.9 *If f is a measurable extended real-valued function (not necessarily non-negative) and $a \in \mathbb{R}$, $a \geq 0$, then*

$$(S) \int_X f \, dP = (S) \int_X (f \wedge a) \, dP + (S) \int_X (f - f \wedge a) \, dP \qquad (8.7)$$

(finite or $+\infty$), if one of the right-hand side expression belongs to R.

Proof: We prove the theorem in the case in which $(S) \int_X (f \wedge a) \, dP \in R$: the proof in the other case is analogous.

By Proposition 8.6 b), the quantity $(S) \int_X (f - f \wedge a) \, dP$ exists in $R \cup \{+\infty\}$. We consider firstly the case $(S) \int_X (f - f \wedge a) \, dP \in R$. By definition of Šipoš integral and (O)-convergence of nets, there exist two (O)-nets $(p_F)_{F \in \mathcal{F}}$ and $(q_F)_{F \in \mathcal{F}}$ such that

$$\left| S_F(f \wedge a) - (S) \int_X (f \wedge a) \, dP \right| \leq p_F \quad \forall F \in \mathcal{F} \qquad (8.8)$$

and

$$\left| S_F(f - f \wedge a) - (S) \int_X (f - f \wedge a) \, dP \right| \leq q_F \quad \forall F \in \mathcal{F}. \qquad (8.9)$$

Fix arbitrarily $F_1, F_2 \in \mathcal{F}$, let $F_0 := F_1 \cup (F_2 + a)$ and pick $F \supset F_0$. Since $F - a \supset F_2$, we get:

$$\begin{aligned}
&\left| S_F(f) - (S) \int_X (f - f \wedge a) \, dP - (S) \int_X (f \wedge a) \, dP \right| \\
&\leq \left| S_{F-a}(f - f \wedge a) - (S) \int_X (f - f \wedge a) \, dP \right| \qquad (8.10) \\
&\quad + \left| S_F(f \wedge a) - (S) \int_X (f \wedge a) \, dP \right| \leq q_{F_1} + p_{F_2}.
\end{aligned}$$

From (8.10), thanks also to Proposition 8.6 a), it follows that

$$(O) \limsup_{F \in \mathcal{F}} \left| S_F(f) - (S) \int_X (f - f \wedge a) \, dP - (S) \int_X (f \wedge a) \, dP \right| = 0. \qquad (8.11)$$

From (8.11) it follows that $(S) \int_X f \, dP$ exists in R and formula (8.7) holds true. In the case

$$(S)\int_X (f - f \wedge a)\,dP = +\infty,$$

proceeding with the analogous notations as above, we get:

$$S_F(f) = S_F(f \wedge a) + S_{F-a}(f - f \wedge a) \tag{8.12}$$
$$\geq S_F(f \wedge a) + \int_X (f \wedge a)\,dP - p_{\{0\}},$$

and taking the supremum in (8.12) it follows that

$$(S)\int_X f\,dP = +\infty \text{ if } (S)\int_X (f - f \wedge a)\,dP = +\infty,$$

and hence (8.7) holds, with the value $+\infty$. This concludes the proof. $\blacksquare$

Theorem 8.10 *Let $f : X \to \tilde{\mathbb{R}}$ be a measurable map. If $(S)\int_X f^+\,dP$ or $(S)\int_X f^-\,dP$ belongs to R, then $(S)\int_X f\,dP$ belongs to R too and*

$$(S)\int_X f\,dP = (S)\int_X f^+\,dP - (S)\int_X f^-\,dP. \tag{8.13}$$

Moreover, if f is (S)-integrable, then (8.13) holds true, and f^+, f^- are (S)-integrable too.

Proof: The first part is an easy consequence of Theorem 8.9 (see also [204], p. 159).

We now turn to the second part. In order to do this, it will be sufficient to prove that $(S)\int_X f^+\,dP$ and $(S)\int_X f^-\,dP$ belong to R. We now report in detail only the proof of the first property. Since f is (S)-integrable, there exists an (O)-net $(p_F)_{F \in \mathcal{F}}$ such that

$$\left| S_F(f) - (S)\int_X f\,dP \right| \leq p_F \quad \forall F \in \mathcal{F}. \tag{8.14}$$

Fix now arbitrarily $F_0 \in \mathcal{F}$ and choose $F \in \mathcal{F}$ with $F \supset F_0$ and $F \cap \mathbb{R}^- = F_0 \cap \mathbb{R}^-$. Proceeding analogously as in [204], pp.159-160, we get:

$$0 \leq S_{F \cap \mathcal{F}^+}(f^+) = S_F(f^+) = S_{-F}(f^-) + S_F(f) \tag{8.15}$$
$$= S_{-F_0}(f^-) + S_F(f) \leq S_{-F_0}(f^-) + (S)\int_X f\,dP + p_{\{0\}}.$$

We have

$$(S)\int_X f^+\,dP = \sup_{F \in \mathcal{F}} S_F(f^+) = \sup_{F \in \mathcal{F}} S_{F \cap \mathbb{R}^+}(f^+). \tag{8.16}$$

By virtue of Proposition 8.6 a), the involved supremum in (8.16) is equal to supremum with respect to those elements F of $\mathcal{F}$ which contain F_0 and such that $F \cap \mathbb{R}^- = F_0 \cap \mathbb{R}^-$. Since F_0 was *fixed*, from (8.15) and (8.16) it follows that $(S)\int_X f^+\,dP \in R$. $\blacksquare$

Proposition 8.11 *Let $X \supset A_1 \supset A_2 \supset ... \supset A_n \in \mathcal{A}$. Let c_i be positive real numbers and $f_i := c_i \chi_{A_i}$, $i = 1, 2, ..., n$. Then*

$$(S)\int_X \left(\sum_{i=1}^n f_i \right) dP = \sum_{i=1}^n c_i P(A_i).$$

Proposition 8.12 *If f is measurable and $|f|$ is (S)-integrable, then f is (S)-integrable too. Moreover, if f is measurable, g is (S)-integrable and $|f| \leq g$, then f is (S)-integrable too.*

From now on, given a non-negative measurable function $f : X \to \tilde{\mathbb{R}}$, let S_f be the set of all simple functions g, such that $0 \leq g(x) \leq f(x) \ \forall x \in X$.

Proposition 8.13 *If $f \geq 0$ is (S)-integrable, then*

$$(S)\int_X f \, dP = \sup_{g \in S_f} (S) \int_X g \, dP.$$

Conversely, if $f \geq 0$ is measurable and such that the quantity $\sup_{g \in S_f} (S)\int_X g \, dP$ exists in R, then f is (S)-integrable and

$$(S)\int_X f \, dP = \sup_{g \in S_f} (S) \int_X g \, dP.$$

Furthermore, if f is non-negative and (S)-integrable, then there exists a sequence of simple functions $(g_n)_n$ such that

$$(S)\int_X f \, dP = \sup_n (S) \int_X g_n \, dP.$$

Proposition 8.14 *If f is (S)-integrable, then*

$$(O) \lim_{t \to +\infty} P(\{x \in X : |f(x)| \geq t\}) = 0 = P(\{x \in X : |f(x)| = +\infty\}).$$

We now show absolute continuity of the Šipoš integral. In order to do this, firstly we state a preliminary lemma (for the case $R = \mathbb{R}$, see [204], Lemma 7.5. (i), p. 163).

Lemma 8.15 *If f is a non-negative (S)-integrable function, then*

$$(O) \lim_{A \to +\infty} (S)\int_X (f - f \wedge A) \, dP = 0.$$

Proof: Fix arbitrarily $F \in \mathcal{F}$, $F = \{b_k, b_{k-1}, \ldots, b_1, b_0 = 0 = a_0, a_1, \ldots, a_n\}$, where the elements of F are ordered in the increasing order, and let

$$f_F = \sum_{i=1}^n (a_i - a_{i-1}) \chi_{A_i}.$$

For $A \in \mathbb{R}^+$ large enough, we get:

$$f_F \leq f \wedge A \leq f. \tag{8.17}$$

Now, given $F \in \mathcal{F}$, let A satisfy property (8.17). From (8.17) and monotonicity of the Šipoš integral we have:

$$S_F(f) = (S)\int_X f_F \, dP \leq (S)\int_X (f \wedge A) \, dP \leq (S)\int_X f \, dP. \tag{8.18}$$

Moreover, by virtue of Theorem 8.9 and (8.18), we get:

$$(S)\int_X (f - f \wedge A) dP = (S)\int_X f \, dP - (S)\int_X (f \wedge A) \, dP \qquad (8.19)$$

$$\leq (S)\int_X f \, dP - S_F(f).$$

From (8.19) and Šipoš integrability of f it follows that

$$0 \leq (O)\limsup_{A \in R^+} \left[(S)\int_X (f - f \wedge A) dP \right] \qquad (8.20)$$

$$\leq (O)\limsup_{F \in \mathcal{F}} \left[(S)\int_X f \, dP - S_F(f) \right] = 0.$$

Thus the assertion follows. 🍎

A consequence of Lemma 8.15 is the following:

Theorem 8.16 *If $f : X \to \tilde{\mathbb{R}}$ is (S)-integrable, then the integral $(S)\int f \, dP$ is absolutely continuous, that is*

$$(O)\lim_n \int_{A_n} f \, dP = 0$$

whenever $(A_n)_n$ is a sequence in $\mathcal{A}$ such that $(O)\lim_n P(A_n) = 0$.

Remark 8.17 Let $P : \mathcal{A} \to R$ be a capacity, and Ω be as in the Maeda-Ogasawara-Vulikh Theorem (see Chapter 2). It is clear that there exists a nowhere dense set $N \subset \Omega$, such that, $\forall \omega \notin N$, the map $P_\omega : \mathcal{A} \to \tilde{\mathbb{R}}$, defined by setting

$$P_\omega(A) := P(A)(\omega),$$

is a real-valued capacity. Moreover, if P is finitely additive, then the involved P_ω's are finitely additive too.

Furthermore, for each integrable function f, there exists a meager set $M \supset N$, depending only on f, such that, $\forall \omega \in \Omega \setminus M$ and $\forall A \in \mathcal{A}$,

$$(S)\int_A f \, dP_\omega = \left((S)\int_A f \, dP \right)(\omega)$$

(for a proof, see also [32], and take into account of Theorem 8.10).

We now prove some convergence theorems for the Šipoš integral with respect to Riesz space-valued capacities, not necessarily finitely additive.

We always assume that X is any nonempty set, $\mathcal{A} \subset \mathcal{P}(X)$ is a σ-algebra, R is a Dedekind complete Riesz space, and $P : \mathcal{A} \to R$ is a *continuous* capacity.

We begin with the following theorem (for the real case, see [204], Theorem 7.13, pp. 162-163):

Theorem 8.18 *Let $c \in R$, $c \geq 0$, $(f_n : X \to \tilde{\mathbb{R}})_n$ be an increasing sequence of non-negative (S)-integrable functions with $(S)\int_X f_n dP \leq c$ for every $n \in \mathbb{N}$, and $f := \sup_n f_n$ be the pointwise supremum.*

Then f is (S)-integrable, $(S)\int_X f \, dP \leq c$ and

$$(S)\int_X f\, dP = \sup_n (S)\int_X f_n\, dP = (O)\lim_n (S)\int_X f_n\, dP.$$

Proof: Fix arbitrarily $\varepsilon > 0$ and $F \in \mathcal{F}$, $F = \{a_0, a_1, \ldots, a_n\}$, where $0 = a_0 < a_1 < \ldots < a_n$. We choose δ such that

$$0 < 2\delta < \min\{(a_j - a_{j-1}) : j = 1, 2, \ldots, k\}$$

and

$$\frac{\delta}{a_1 - \delta} < \varepsilon:$$

such a δ does exist. Proceeding analogously as in the proof of Theorem 7.13 of [204], we get:

$$S_F(f) \le (O)\lim_n (S)\int_X f_n\, dP + \frac{\delta}{a_1 - \delta}(O)\lim_n (S)\int_X f_n\, dP$$

$$\le (O)\lim_n (S)\int_X f_n\, dP + \varepsilon\, c,$$

and hence

$$(S)\int_X f\, dP = \sup_{F \in F} S_F(f) \le (O)\lim_n (S)\int_X f_n\, dP + \varepsilon\, c. \qquad (8.21)$$

By arbitrariness of $\varepsilon \in \mathbb{R}^+$, from (8.21) we get:

$$(S)\int_X f\, dP \le (O)\lim_n (S)\int_X f_n\, dP.$$

The converse inequality follows easily from the monotonicity of the integral $(S)\int_X f\, dP$. 🍎

Consequences of Theorem 8.18 are the following:

Corollary 8.19 *If $(\alpha_n)_n$ is any decreasing sequence of positive real numbers with $\inf_n \alpha_n = 0$, then*

$$(O)\lim_{n \to +\infty} (S)\int_X (f \wedge \alpha_n)\, dP = 0.$$

Proof: The proof is similar to the one of Lemma 7.5 (ii) of [204], p. 163. 🍎

Corollary 8.20 *(Fatou's Lemma) Let $c \in R$, $c \ge 0$, $(f_n : X \to \tilde{\mathbb{R}})_n$ be any sequence of non-negative (S)-integrable functions with $(S)\int_X f_n dP \le c$ for every $n \in \mathbb{N}$, and $f := \liminf_n f_n$.*

Then

$$(S)\int_X f\, dP \le (O)\liminf_n (S)\int_X f_n\, dP.$$

Proof: First of all, we note that f is (S)-integrable, thanks to Theorem 8.18.

For each $n \in \mathbb{N}$, let $h_n = \inf_{i \ge n} f_i$. Then $0 \le h_n \uparrow f$ and

$$(S)\int_X h_n \, dP \le (S)\int_X f \, dP \quad \forall n \in \mathbb{N}.$$

Again by Theorem 8.18, we get:

$$(S)\int_X f \, dP = (O)\lim_n (S)\int_X h_n \, dP$$

$$= (O)\liminf_n (S)\int_X h_n \, dP \le (O)\liminf_n (S)\int_X f_n \, dP.$$

This concludes the proof. $\blacksquare$

We now turn to another version of the monotone convergence theorem. In order to prove it, we first demonstrate the following:

Lemma 8.21 *Let* $[a,b] \subset \mathbb{R}$, $u_n : [a,b] \to R$ $(n \in \mathbb{N} \cup \{0\})$ *be monotone decreasing functions, such that*

$$u_n(t) = \inf_{s<t} u_n(s) \quad \forall t \in (a,b], \forall n \in \mathbb{N} \cup \{0\}; \qquad (8.22)$$

$$u_n(t) \ge u_{n+1}(t) \quad \forall t \in [a,b], \forall n \ge 1; \qquad (8.23)$$

$$\inf_n u_n(t) = u_0(t) \quad \forall t \in [a,b]. \qquad (8.24)$$

Then

$$(Ri)\int_a^b u_0(t) \, dt = \inf_{n \ge 1} (Ri)\int_a^b u_n(t) \, dt = (O)\lim_{n \to +\infty} (Ri)\int_a^b u_n(t) \, dt.$$

Proof: First of all, we observe that for every $n \in \mathbb{N} \cup \{0\}$ the integral $\int_a^b u_n(t) \, dt$ is the limit, as $l \to +\infty$, of the Riemann sums of the type

$$\sum_{i=1}^{2^l} (a_i^{(l)} - a_{i-1}^{(l)}) u_n(a_i^{(l)}), \qquad (8.25)$$

where the $a_i^{(l)}$'s, $l \in \mathbb{N}$, $i = 1, 2, ..., 2^l$, are taken in such a way that $a_0^{(l)} = a$, $a_{2^l}^{(l)} = b$, and the division generated by the $a_i^{(l)}$'s divides the interval $[a,b]$ in 2^l equal parts. Denote by $\mathcal{Y}$ the set of points of all these divisions and let Q be the union between $\mathcal{Y}$ and the rational numbers contained in $[a,b]$: we note that Q is a countable dense subset of $[a,b]$.

Let now Ω be as in the Maeda-Ogasawara-Vulikh theorem. We note that there exists a meager set $N^* \subset \Omega$ such that, for all $\omega \notin N^*$, we have:

$$\left[\inf_n \left[\int_a^b u_n(t) \, dt\right](\omega)\right] = \left[\inf_n \left[\int_a^b u_n(t) \, dt\right]\right](\omega), \qquad (8.26)$$

and for each $\omega \notin N^*$ and $s \in Q$ we get:

$$u_n(s)(\omega) \ge u_{n+1}(s)(\omega) \quad \forall n \ge 1;$$

$$\lim_{n \to +\infty} u_n(s)(\omega) = \inf_{n \ge 1} u_n(s)(\omega) = u_0(s)(\omega),$$

and all involved quantities are *real* numbers. Now, for all $s \in [a,b] \cap Q$, $\forall \omega \notin N^*$ and $\forall n \in \mathbb{N} \cup \{0\}$, set

$$w_{n,\omega}(s) = u_n(s)(\omega).$$

For each $t \in [a,b]$, $\omega \notin N^*$ and $n \in \mathbb{N} \cup \{0\}$, put

$$w_{n,\omega}(t) = \inf_{s \le t, s \in Q} w_{n,\omega}(s). \tag{8.27}$$

By (8.27) and since the $w_{n,\omega}$'s are decreasing, their integrals can be evaluated analogously as in (8.25), and thus we get, $\forall n \in \mathbb{N} \cup \{0\}$ and $\forall \omega \notin N^*$,

$$\int_a^b w_{n,\omega}(t)\,dt = \left[\int_a^b u_n(t)\,dt \right](\omega). \tag{8.28}$$

We note that

$$w_{n,\omega}(s) \downarrow w_{0,\omega}(s) \quad \forall \omega \notin N^*, \quad \forall s \in [a,b] \cap Q. \tag{8.29}$$

Furthermore, $\forall \omega \notin N^*$ and $t \in [a,b]$, by "interchanging the involved infima" we get:

$$\inf_n w_{n,\omega}(t) = \inf_n \left[\inf_{s \le t, s \in Q} w_{n,\omega}(s) \right] \tag{8.30}$$

$$= \inf_{s \le t, s \in Q} [\inf_n w_{n,\omega}(s)] = \inf_{s \le t, s \in Q} [w_{0,\omega}(s)] = w_{0,\omega}(t),$$

and thus

$$w_{n,\omega}(t) \downarrow w_{0,\omega}(t) \quad \forall \omega \notin N^*, \quad \forall t \in [a,b]. \tag{8.31}$$

From (8.26), (8.28) and (8.31), and by applying the classical (dominated) convergence theorem for real-valued functions, we get, $\forall \omega \notin N^*$:

$$\left[\int_a^b u_0(t)\,dt \right](\omega) = \int_a^b w_{0,\omega}(t)\,dt = \inf_n \left[\int_a^b w_{n,\omega}(t)\,dt \right]$$

$$= \inf_n \left[\left[\int_a^b u_n(t)\,dt \right](\omega) \right] = \left[\inf_n \int_a^b u_n(t)\,dt \right](\omega). \tag{8.32}$$

From this, since N^* is meager and the complement of every meager subset of Ω is dense in Ω, it follows that

$$\int_a^b u_0(t)\,dt = \inf_n \int_a^b u_n(t)\,dt.$$

Thus we get the assertion. ✪

We now are in position to prove the following:

Theorem 8.22 *Let* $(f_n : X \to \tilde{\mathbb{R}})_n$ *be a decreasing sequence of non-negative* (S)*-integrable functions, and let* $f = \inf_n f_n$ *be the pointwise infimum. Then* f *is* (S)*-integrable and*

$$(S)\int_X f \, dP = \inf_n (S)\int_X f_n \, dP = (O)\lim_n (S)\int_X f_n \, dP.$$

Proof: First of all, since $0 \le f \le f_1$, from Proposition 8.12 it follows that f is integrable. Moreover, we observe that, proceeding similarly as in the first half of p. 164 of [204] and thanks to Lemma 8.15, we can suppose, without loss of generality, that the functions f_n and f are equibounded by a positive number A.

For each $t \ge 0$ and $n \in \mathbb{N}$, $n \ge 1$, let $u_n(t) = P(\{x \in X : f_n(x) \ge t\})$, and $\forall t \ge 0$ let $u_0(t) = P(\{x \in X : f(x) \ge t\})$.

Proceeding analogously as in the proof of Theorem 8.7, we get

$$(S)\int_X f \, dP = (Ri)\int_0^A u_0(t) \, dt \tag{8.33}$$

and

$$(S)\int_X f_n \, dP = (Ri)\int_0^A u_n(t) \, dt \quad \forall n \ge 1. \tag{8.34}$$

Since P is a continuous capacity and $f_n \downarrow f$, the functions u_n, $n \in \mathbb{N} \cup \{0\}$, satisfy conditions (8.22), (8.23) and (8.24). By applying Lemma 8.21 with $[a,b] = [0, A]$ and from (8.33) and (8.34), it follows that

$$(S)\int_X f \, dP = (Ri)\int_0^A u_0(t) \, dt = \inf_{n \ge 1} (Ri)\int_a^b u_n(t) \, dt$$

$$= (O)\lim_{n \to +\infty} (Ri)\int_a^b u_n(t) \, dt = \inf_n (S)\int_X f_n \, dP = (O)\lim_n (S)\int_X f_n \, dP,$$

that is the assertion. 🍎

We now prove the following:

Theorem 8.23 *Let $c \in R$, $(f_n)_n$ be a sequence of (S)-integrable functions, and f be a measurable map, such that $f_n \downarrow f$ and*

$$\int_X f_n \, dP \ge c \quad \forall n \in \mathbb{N}.$$

Then f is (S)-integrable and

$$(S)\int_X f \, dP = (O)\lim_n (S)\int_X f_n \, dP = \inf_n (S)\int_X f_n \, dP.$$

Proof: Since $f_n \downarrow f$, then $f_n^+ \downarrow f^+$ and $f_n^- \uparrow f^-$. We have:

$$0 \le (S)\int_X f_n^- \, dP = (S)\int_X f_n^+ \, dP - (S)\int_X f_n \, dP$$

$$\le (S)\int_X f_1^+ \, dP - (S)\int_X f_n \, dP \le (S)\int_X f_1^+ \, dP - c,$$

and thus we get that the integrals $(S)\int_X f_n^- \, dP$, $n \in \mathbb{N}$, are bounded from above by an element of R. By Theorem 8.18, f is Šipoš-integrable and

$$(S)\int_X f^- \, dP = (O)\lim_n (S)\int_X f_n^- \, dP = \sup_n (S)\int_X f_n^- \, dP. \tag{8.35}$$

Moreover, by Theorem 8.22, we get integrability of f^+ and

$$(S)\int_X f^+\, dP = (O)\lim_n (S)\int_X f_n^+\, dP = \inf_n (S)\int_X f_n^+\, dP. \qquad (8.36)$$

Thus, from (8.35), (8.36) and Theorem 8.10 we obtain

$$(O)\lim_n (S)\int_X f_n\, dP = (O)\lim_n (S)\int_X f_n^+\, dP - (O)\lim_n (S)\int_X f_n^-\, dP$$

$$= (S)\int_X f^+\, dP - (S)\int_X f^-\, dP = (S)\int_X f\, dP,$$

that is the assertion. $\blacksquare$

The proof of the following theorem is similar to the ones of Theorem 8.23 and of Theorem 7.15, p. 166, of [204], taking into account that $f_n \uparrow f$ implies $f_n^+ \uparrow f^+$ and $f_n^- \downarrow f^-$.

Theorem 8.24 Let $c \in R$, $c \geq 0$, $(f_n)_n$ be a sequence of (S)-integrable functions, and f be a measurable mapping, such that $f_n \uparrow f$ and

$$\int_X f_n\, dP \leq c \quad \forall n \in \mathbb{N}.$$

Then f is (S)-integrable and

$$(S)\int_X f\, dP = (O)\lim_n (S)\int_X f_n\, dP = \sup_n (S)\int_X f_n\, dP.$$

The following version of the Lebesgue convergence dominated theorem is a consequence of Theorems 8.23 and 8.24.

Theorem 8.25 If $(f_n)_n$ is a sequence of measurable functions which converges pointwise to a measurable function f, and if g is an (S)-integrable function with $|f_n| \leq g \ \forall n \in \mathbb{N}$, then f is (S)-integrable and

$$(S)\int_X f\, dP = (O)\lim_n (S)\int_X f_n\, dP.$$

We now prove some theorems for the Šipoš integral, in the case in which the involved capacities are submodular.

Theorem 8.26 Let $P : \mathcal{A} \to R$ be a submodular capacity, and $f, g : X \to \tilde{\mathbb{R}}$ be two non-negative measurable maps. Then

$$(S)\int_X (f \wedge g)\, dP + (S)\int_X (f \vee g)\, dP \leq (S)\int_X f\, dP + (S)\int_X g\, dP$$

(finite or $+\infty$). Moreover, if f and g are integrable, then $f \wedge g$ and $f \vee g$ are integrable too.

Proof: If $(S)\int_X f\, dP = +\infty$ or $(S)\int_X g\, dP = +\infty$ the assertion is trivial. Let f and g be both (S)-integrable. (S)-integrability of $f \wedge g$ follows immediately from Proposition 8.12.

We now prove that $f \vee g$ is (S)-integrable. To this aim, pick arbitrarily $F \in \mathcal{F}$ with $F = \{a_0, a_1, \ldots, a_n\}$, where $a_0 = 0 < a_1 < \ldots < a_n$. Set

$$A_i = \{x : f(x) \geq a_i\}, \quad B_i = \{x : g(x) \geq a_i\}, \qquad i = 0, 1, \ldots, n.$$

Proceeding analogously as in the proof of Theorem 7.17 of [204], thanks to submodularity of P, we get:

$$S_F(f \wedge g) + S_F(f \vee g) \leq S_F(f) + S_F(g) \leq \int_X f \, dP + \int_X g \, dP. \qquad (8.37)$$

From (8.37), taking into account of Dedekind completeness of R, we have:

$$(O) \lim_{F \in \mathcal{F}} [S_F(f \wedge g) + S_F(f \vee g)]$$

$$= (O) \lim_{F \in \mathcal{F}} [S_F(f \wedge g)] + (O) \lim_{F \in \mathcal{F}} [S_F(f \vee g)] \qquad (8.38)$$

$$= \sup_{F \in \mathcal{F}} [S_F(f \wedge g)] + \sup_{F \in \mathcal{F}} [S_F(f \vee g)] \in R.$$

From (S)-integrability of the function $f \wedge g$ and (8.38), we get:

$$(O) \lim_{F \in \mathcal{F}} [S_F(f \vee g)] = \sup_{F \in \mathcal{F}} [S_F(f \vee g)] \in R,$$

that is (S)-integrability of $f \vee g$. Taking the order limits as $F \in \mathcal{F}$, from (8.37) and (8.38) we obtain:

$$(S) \int_X f \, dP + (S) \int_X g \, dP \geq (O) \lim_{F \in \mathcal{F}} [S_F(f \wedge g) + S_F(f \vee g)]$$

$$= (S) \int_X (f \wedge g) \, dP + (S) \int_X (f \vee g) \, dP, \qquad (8.39)$$

that is the assertion. $\blacksquare$

Proceeding analogously as in Theorem 8.26, it is possible to prove the following:

Proposition 8.27 *If f and g are non-negative measurable functions and P is a R-valued subadditive capacity, then*

$$(S) \int_X (f \vee g) \, dP \leq (S) \int_X f \, dP + (S) \int_X g \, dP$$

(For the real case, see [204], Corollary 7.5, p. 168).

We now state the submodular theorem (see also [33]).

Proposition 8.28 *Let $P : \mathcal{A} \to R$ be a submodular capacity, and $f, g \in \tilde{\mathbb{R}}^X$ two non-negative (S)-integrable maps. Then*

$$(S) \int_X (f + g) \, dP \leq (S) \int_X f \, dP + (S) \int_X g \, dP.$$

Moreover, if f and g are (S)-integrable, then $f + g$ is (S)-integrable too.

Proof: If either f or g are not (S)-integrable, then the assertion is trivial. If f and g are (S)-integrable, then, by virtue of the inequality $0 \leq f + g \leq 2(f \vee g)$ and Proposition 8.12, we get that $f + g$ is (S)-integrable. For the remaining part, see [33]. $\blacksquare$

Proceeding analogously to Corollary 7.6 of [204], p. 173, it is possible to prove the following:

Theorem 8.29 *Let f be a measurable function and P be a R-valued submodular capacity. Then f is (S)-integrable if and only if $|f|$ is (S)-integrable.*

Remark 8.30 We observe that, in general, the hypothesis of submodularity of P cannot be dropped, even in the case $R = \mathbb{R}$: indeed, if P is a real-valued not submodular capacity, there exist some (S)-integrable functions f (with respect to P), such that $|f|$ is not (S)-integrable (see [204], Example 3.16, p. 161).

Similarly as in [204], Corollary 7.7, p. 174 and Corollary 7.8, p. 175, it is easy to prove the following two theorems:

Theorem 8.31 *If $P : \mathcal{A} \to R$ is a mean, and f, g are (S)-integrable, then*

$$(S)\int_X (f+g)\,dP = (S)\int_X f\,dP + (S)\int_X g\,dP.$$

Theorem 8.32 *If $P : \mathcal{A} \to R$ is a capacity, and f, g are (S)-integrable and comonotonic, then*

$$(S)\int_X (f+g)\,dP = (S)\int_X f\,dP + (S)\int_X g\,dP.$$

8.2 Asymmetric Integral

We now introduce the asymmetric integral.

We note that in general, even in the case $R = \mathbb{R}$, the symmetric and the asymmetric integrals are different, but, even when R is *any* general Dedekind complete Riesz space, the two integrals will coincide when the Choquet integral if the involved function f is non-negative (in this case, even if P is simply a monotone set function with $P(\varnothing) = 0$), and they coincide each other also when f may change sign, provided that P is finitely additive (we recall that a finitely additive set function with $P(\varnothing) = 0$ is positive if and only if it is increasing): indeed, in both cases, we will have:

$$(S)\int_X f\,dP = (A)\int_X f\,dP = (C)\int_X f^+\,dP - (C)\int_X f^-\,dP$$

(see also [30,32]), where (A) indicates the asymmetric integral which will be defined in a moment, X is the set on which f is defined, $f^+ := f \vee 0$, $f^- := (-f) \vee 0$ (that is $f^+(x) := f(x) \vee 0$, $f^-(x) := (-f(x)) \vee 0$ for all $x \in X$). Among the applications existing in the literature, that sometimes are more "natural" for the asymmetric than for the symmetric integral, see also [73] and its bibliography.

We now give here the definition of asymmetric integral in the context of Riesz spaces.

Definition 8.33 Under the same notations as above, a measurable map (not necessarily non-negative) $f \in \tilde{\mathbb{R}}^X$ is (A)-*integrable* with respect to a capacity $P : \mathcal{A} \to R$ if there exist in R the following quantities:

$$I_1 := \int_0^{+\infty} u(t)\, dt = \sup_{a>0} \int_0^a u(t)\, dt = (O) \lim_{a \to +\infty} \int_0^a u(t)\, dt$$

and

$$I_2 := (O) \lim_{b \to -\infty} \int_b^0 [u(t) - P(X)]\, dt,$$

where $u(t) := P(\{x \in X : f(x) \geq t\})$, $t \in \mathbb{R}$, is the *distribution function,* and the involved integrals are intended in the Riemann sense. In this case, we set

$$(A)\int_X f\, dP := I_1 + I_2.$$

We call the quantity $(A)\int_X f\, dP$ *the asymmetric integral* of f with respect to P, and, when it coincides with the symmetric integral, we denote it by $E(f)$ (expectation of f) or simply by the symbol $\int_X f\, dP$, when no confusion can arise.

8.3 Applications

In the literature, in certain types of studies (for example, in stochastic processes), it would be "natural" to investigate some kinds of "probabilities", which to every event associate not simply a real number, but a real-valued function or something else.

Indeed, one can give different valuations of the uncertainty of some event E, depending, for example, on the "informations", which one can receive, during his study about E or about some other events, "related" in some way to E.

For example, given a measurable space (X, Σ), we can consider applications $P : \Sigma \to [0,1]^T$ in order to stress that the "probability" of each event A depends on the "time": $P(A)$ is a function of $t \in T$.

As a second example, given a sub-σ-algebra $\mathcal{Z}$ of Σ and a probability P on (X, Σ), we can define the "conditional probability" as follows: $\widetilde{P}(A) = P(A \mid \mathcal{Z}) = E(1_A \mid \mathcal{Z})$ for every $A \in \Sigma$. So $\widetilde{P} : \Sigma \to L^0$.

More generally, it would be advisable to associate to each event an element of a Riesz space R: indeed, we note that, thanks to the Maeda-Ogasawara-Vulikh representation Theorem, every Archimedean Riesz space can be viewed as a suitable space of continuous extended real-valued functions.

On the other hand, in the literature there exist several contributions to the foundations of "qualitative probabilities " and their "realizations", which can be represented not only by additive functions but also by submodular capacities (see [65, 124, 125]). So, it will be natural, in certain problems, to consider "probabilities", as just monotonic functions, with values in Riesz spaces.

As a further example, we can consider stochastic integration, when we define the integral of a scalar function with respect to a stochastic measure I_X, where $X : \Omega \times \mathbb{R}^+ \to L^p$ is a process.

Another motivation for the study of the integral with respect to capacities is that, in the theory of decisions, the "preference" relations between "measurable" functions, which are the "acts" of the considered individual, can be represented by means of the Choquet integral of some suitable utility functions. In particular, if X denotes the space of all "choices" (i.e., states of nature) and Γ is the space of all possible "consequences", an *act* is a "measurable" mapping $f : X \to \Gamma$. Roughly speaking, to state that "f is preferable to g" is equivalent to say that it is possible to determine a capacity P and a utility function $u : \Gamma \to \mathbb{R}$, in such a way that

$$\int_X u \circ f \, dP \geq \int_X u \circ g \, dP.$$

In particular, if we operate under risk condition, P must be a measure (see [105]).

Here, among the various applications of these kinds of integral (see [33]), we give one to the probability, and more precisely to the strong and weak law of large numbers: this is one of the research results of [33]. From now on, we assume that P is a (σ-additive) measure, and thus the symmetric and the asymmetric integral will coincide (as said before, we denote the common value by the symbol $\int_X f \, dP$). Moreover, we assume that the symbol of series has the same meaning as introduced in Chapter 2. We introduce the following condition:

H1) For all $i \in \mathbb{N}$, for each A_i belonging to the σ-algebra $\sigma(f_1, ..., f_i)$ generated by $f_1, ..., f_i$, $\forall j > i \geq h \in \mathbb{N}$, it holds

$$\int_{A_i} f_h \, f_j \, dP = 0.$$

We observe that, when $R = \mathbb{R}$, condition H1) is equivalent to the following hypothesis:

$$E(f_{n+j} \mid \sigma(f_1, ..., f_{n-1})) = 0, \quad \forall n, j \in \mathbb{N}.$$

Remark 8.34 We note that it is not advisable to proceed in terms of "conditional expectation". Indeed, let $R := \mathbb{R}^2$, $P : A \to R$ be a (σ-additive) measure, $P = (P_1, P_2)$, where $P_1(X) = P_2(X) = 1$, and assume that

$$\int_X f \, dP_1 \neq \int_X f \, dP_2.$$

Put $B := \{\varnothing, X\}$: then, it is easy to check that a mapping $g \in \mathbb{R}^X$ is B-measurable if and only if it is constant. Then,

$$c = c \, P_j(X) = \int_X f \, dP_j \quad (j = 1, 2)$$

and so

$$\int_X f \, dP_1 = \int_X f \, dP_2 :$$

contradiction.

Thus, in this case, we cannot define a real-valued map g, playing the same role, as the conditional expectation in the case $R = \mathbb{R}$.

We now turn to the strong law of large numbers.

Theorem 8.35 *Let $(f_n)_n$ be a sequence of random variables, such that f_n^2 is integrable, $\forall\, n$, and satisfying condition H1). Moreover, suppose that the series $\sum_{j=1}^{\infty} \dfrac{E(f_j^2)}{j^2}$ converges. Set*

$$f_n := \frac{1}{n}\sum_{i=1}^{n} f_i, \ \forall n \in \mathbb{N}. \ \textit{Then, the sequence } (f_n)_n \textit{ converges to } 0 \textit{ almost everywhere.}$$

In order to prove this theorem, we introduce two lemmas.

Lemma 8.36 *Let $f_1,\ldots,f_n,\ldots$ be random variables, satisfying H1), and suppose that f_n^2 is integrable, $\forall\, n \in \mathbb{N}$. If $S_j := \sum_{i=1}^{j} f_i$, and $u_1 \leq \ldots \leq u_n$ are positive real numbers, then*

$$P(\{x \in X : |S_j(x)| < u_j, \forall\, j\}) \geq P(X) - \sum_{j=1}^{n} \frac{E(f_j^2)}{u_j^2}.$$

Proof: First of all, we prove that

(*) $\qquad E((f_h + f_j)^2) = E(f_h^2) + E(f_j^2), \forall\, h, j \in N, h \neq j.$

Without loss of generality, we can suppose that $h < j$. In H1), choose $i = h$, and $A_i = X$; then, $E(f_h\, f_j) = 0$, and so (*) follows. From (*) we obtain

$$E(S_n^2) = \sum_{i=1}^{n} E(f_i^2).$$

Set now $\alpha_i := \dfrac{1}{u_i^2}, \forall\, i = 1,\ldots,n; \alpha_{n+1} := 0,$ and

$$T(x) := \sum_{j=1}^{n} (\alpha_j - \alpha_{j+1})[S_j(x)]^2, \forall\, x \in X.$$

We have:

$$E(T) = \sum_{j=1}^{n} (\alpha_j - \alpha_{j+1})\, E(S_j^2).$$

For every j, put

$$B_j := \{x \in X : |S_i(x)| < u_i, \forall\, i < j; |S_j(x)| \geq u_j\},$$

and set

$$A := \{x \in X : |S_j(x)| < u_j, \forall\, j = 1,\ldots,n\}.$$

It is easy to check that $B_i \cap B_l = \emptyset, \forall\, i \neq l,$ and $\bigcup_{j=1}^{n} B_j = A^c.$

For every i and j, with $i < j$, we have:

$$\int_{B_i} S_j^2\, dP \geq \int_{B_i} S_i^2\, dP$$

by virtue of H1), and hence

$$E(S_j^2) \geq \sum_{i=1}^{j} u_i^2 \, P(B_i).$$

Thus,

$$\sum_{j=1}^{n} \frac{1}{u_j^2} E(f_j^2) = \sum_{j=1}^{n} (\alpha_j - \alpha_{j+1}) \, E(S_j^2) \geq \sum_{j=1}^{n} (\alpha_j - \alpha_{j+1}) \sum_{i=1}^{j} u_i^2 \, P(B_i) =$$

$$= \sum_{i=1}^{n} \alpha_i \, u_i^2 \, P(B_i) = \sum_{i=1}^{n} P(B_i) = P(X) - P(A),$$

that is the assertion. 🍎

Lemma 8.37 *Let R be a Dedekind complete Riesz space, $R \ni \alpha_1, ..., \alpha_n \geq 0$ be such that the series $\sum_{j=1}^{\infty} \frac{\alpha_j}{j^2}$ converges, and set $\sigma_n = \sum_{j=1}^{n} \alpha_j$. Then, $(O)\lim_n \frac{\sigma_n}{n^2} = 0$.*

Proof: Fix $n \in \mathbb{N}$, and set $k := [\sqrt{n}]$. It holds:

$$0 \leq \frac{\sigma_n}{n^2} = \frac{1}{n^2} \sum_{j=1}^{k} \alpha_j + \frac{1}{n^2} \sum_{j=k+1}^{n} \alpha_j \leq \frac{1}{n^2} \sum_{j=1}^{k} \frac{k^2}{j^2} \alpha_j +$$

$$+ \sum_{j=k+1}^{n} \frac{\alpha_j}{j^2} = \frac{k^2}{n^2} \sum_{j=1}^{k} \frac{\alpha_j}{j^2} + \sum_{j=k+1}^{n} \frac{\alpha_j}{j^2} \leq$$

$$\leq \frac{1}{n} \sum_{j=1}^{\infty} \frac{\alpha_j}{j^2} + \sum_{j=k}^{\infty} \frac{\alpha_j}{j^2} \longrightarrow 0.$$

Thus, the assertion follows. 🍎

We are now in position to prove our version of the strong law of large numbers (Theorem 8.35).

Proof: Fix $n, m \in \mathbb{N}$ and $\varepsilon > 0$, and set

$$S_n := \sum_{j=1}^{n} f_j, Y_1 := S_n, Y_2 := f_{n+1}, ..., Y_m := f_{n+m-1}.$$

It is easy to see that the maps Y_j $(j = 1, ..., m)$ satisfy the hypotheses of Lemma 8.36.

For every $j = 1, ..., m$ put $T_j := \sum_{i=1}^{j} Y_i$. Set

$$C_{n,m}^{\varepsilon} := \{x \in X : | f_j(x) | < \varepsilon \ \forall \ j = n, n+1, ..., n+m \}.$$

Then we have

$$C_{n,m}^{\varepsilon} = \{x \in X : |T_j(x)| < (j+n-1)\varepsilon \ \forall \ j = 1,...,m\}.$$

Hence, by Lemma 8.36, we get:

$$P(C_{n,m}^{\varepsilon}) \geq P(X) - \sum_{j=1}^{m} \frac{E(Y_j^2)}{(j+n-1)^2 \, \varepsilon^2} = P(X) - \frac{E(S_n^2)}{n^2 \, \varepsilon^2} -$$

$$- \sum_{j=n+1}^{m-1} \frac{E(f_j^2)}{j^2 \, \varepsilon^2} \geq P(X) - \frac{E(S_n^2)}{n^2 \, \varepsilon^2} - \sum_{j=n+1}^{\infty} \frac{E(f_j^2)}{j^2 \, \varepsilon^2}.$$

So, there exists a sequence $(p_n)_n(\varepsilon)$ in R, $p_n \downarrow 0$, such that

$$P(C_{n,m}^{\varepsilon}) \geq P(X) - \frac{E(S_n^2)}{n^2 \, \varepsilon^2} - \frac{p_n}{\varepsilon^2},$$

and hence

$$P(\cap_m C_{n,m}^{\varepsilon}) = \inf_m P(C_{n,m}^{\varepsilon}) \geq P(X) - \frac{E(S_n^2)}{n^2 \, \varepsilon^2} - \frac{p_n}{\varepsilon^2}.$$

Let $E_n^{\varepsilon} := \cap_{m=1}^{\infty} C_{n,m}^{\varepsilon}$: then

$$P(\cup_n E_n^{\varepsilon}) = (O) \lim_n P(E_n^{\varepsilon}) \geq P(X) - (O) \lim_n \frac{E(S_n^2)}{n^2 \, \varepsilon^2}.$$

By hypotheses (see also Lemma 8.36), we have:

$$E(S_n^2) = \sum_{j=1}^{n} E(f_j^2), \ \forall \ n \in \mathbb{N},$$

and so

$$(O) \lim_n \frac{E(S_n^2)}{n^2 \, \varepsilon^2} = 0.$$

Thus,

$$P(X) \geq P(\cup_{n=1}^{\infty} [\cap_{j=n}^{\infty} \{x \in X : |f_j(x)| < \varepsilon\}]) \geq P(X),$$

and therefore the sequence $(f_n)_n$ converges to 0 almost everywhere.

Now we state a version of the weak law of large numbers, and we observe that the assertion still is true, even if we assume that P is only a finitely additive positive R-valued set function.

Theorem 8.38 *Let $(f_n)_n$ be a sequence of random variables, such that $E(f_n) = 0$ and f_n^2 is integrable, $\forall \ n \in \mathbb{N}$, and suppose that*

$$E((f_n + f_m)^2) = E(f_n^2) + E(f_m^2), \ \forall \ n, m \in \mathbb{N}.$$

If $(O) \lim_n \frac{1}{n^2} \sum_{i=1}^{n} E(f_i^2) = 0$, then $(O) \lim_n E(f_n^2) = 0$, where f_n is as in Theorem 8.35.

Proof: We have:

$$E(f_n^{\,2}) = \frac{1}{n^2} E\left(\left(\sum_{i=1}^{n} f_i\right)^2\right) = \frac{1}{n^2} \sum_{i=1}^{n} E(f_i^2).$$

By virtue of the hypotheses, we get:

$$(O) \lim_n E(f_n^{\,2}) = 0,$$

that is the assertion.

Remark 8.39 We consider now the case when $(X_n)_n$ is a sequence of random variables defined on a probability space (Ω, Σ, P) and $\mathcal{Z}$ is a sub-σ-algebra of Σ.

For every $A \in \Sigma$ we set $\widetilde{P}(A) = P(A \mid \mathcal{Z}) = E(1_A \mid \mathcal{Z})$. Then $\widetilde{P} : \Sigma \to L^0$. Moreover, if $X \in L^1$, we can define $\widetilde{E}(X) = E(X \mid \mathcal{Z}) \in L^1 \subset L^0$. $\widetilde{E}(X)$ is a random variable obtained by integrating X with respect to $\widetilde{P}$.

In fact, given X in L^1, we can consider a sequence of random variables $(X_n)_n$ such that $X_n(\omega) \le X_{n+1}(\omega)$ for almost every ω and $\sup_n X_n = X$ a.e. Then, using Beppo Levi's Theorem,

$$E(X \mid \mathcal{Z}) = \sup_n E(X_n \mid \mathcal{Z}) = \lim_n E(X_n \mid \mathcal{Z}).$$

So $\widetilde{P} : \Sigma \to L^1(\Omega, \mathcal{Z}, P) \subset L^0(\Omega, \mathcal{Z}, P) \subset L^0(\Omega, \Sigma, P)$ is a measure (observe in fact that if $A_n, A \in \Sigma$ such that $A_n \uparrow A$ it is $1_{A_n} \uparrow 1_A$ and so $E(1_{A_n} \mid \mathcal{Z}) \uparrow E(1_A \mid \mathcal{Z})$). Finally, if X in $L^1(\Omega, \Sigma, P)$, then $\int X d\widetilde{P} = E(X \mid \mathcal{Z})$. This is obvious for simple functions, and it is possible to obtain for the general case, using classical techniques.

If we suppose that $\widetilde{E}(X_h X_j 1_{A_i}) = E(X_h X_j 1_{A_i} \mid \mathcal{Z}) := 0$ for every $j > i \ge h$, $A_i \in \sigma(X_1, \cdots, X_i)$ (this hypothesis is stronger then H1)), in fact it means that

$$E(X_j X_h \mid \mathcal{Z} \vee \sigma(X_1, \cdots, X_i)) = X_h E(X_i \mid \mathcal{Z} \vee \sigma(X_1, \cdots, X_i)) = 0 \; ; \text{ and if}$$

$$\sum_{j=1}^{\infty} \frac{1}{j^2} E(X_j^2 \mid \mathcal{Z}) = \sum_{j=1}^{\infty} \frac{\widetilde{E}(X_j^2)}{j^2} \in L^0$$

then, using Theorem 8.35, $\overline{X}_n$ converges to 0 $\widetilde{P}$ almost everywhere, that is $P(A \mid Z) = 0$, where

$$A = \{\omega : \overline{X} \to 0\}.$$

This implies that $P(A) = 0$. Conversely if $P(A) = 0$ then $P(A \mid \mathcal{Z}) = 0$. So

$$\overline{X}_n \to 0 \;\; \widetilde{P}\text{-a.e. if and only if } \overline{X}_n \to 0 \;\; P\text{-a.e.}$$

So, we note that, in this particular case, the classical result given in Theorem 8.35, for $R = \mathbb{R}$, is obtained by modifying the hypothesis: in particular by strengthening H1) and by weakening the convergence of $\sum_{j=1}^{\infty} \frac{1}{j^2} E(X_j^2 \mid \mathcal{Z})$.

9 *(SL)*-integral

Abstract: In this chapter we deal with the strong Luzin ((*SL*)-) integral, related with the existence of primitives of functions in the weak sense. This integral is a variant of the Kurzweil-Henstock integral, which coincides with it in the real case, but is in general slightly different in the context of Riesz spaces, because some pathologies can occur. We also prove some versions of Hake and monotone convergence type theorems and of the Fundamental Theorem of Calculus, together with the basic properties.

9.1 Main Properties in the Real and Riesz Space Context

There are several generalizations of the Riemann integral, both in the classical and in abstract theory of integration. A variant of the Kurzweil-Henstock integral, which coincides with it for real-valued functions, but in general is different for Riesz space-valued mappings, is the (SL)- (strong Luzin) integral, which was investigated in the case of real-valued functions by P. Y. Lee and R. Výborný (see [169,170]). The idea related with this type of integral is the following: we introduce a condition, so called *strong Luzin condition* or *property* (SL), which lies strictly between absolute continuity and condition N (roughly speaking, we say that a function f *satisfies condition N* if it maps sets of measure zero onto sets of measure zero). We will say that a function f is (SL)-integrable if it "admits" a "weak primitive" of class (SL). In the Kurzweil-Henstock integration, condition (SL) will play, in the case of real-valued function, a role very similar to the one played by absolute continuity in the theory of the Lebesgue integral. Moreover, we will see that, for real-valued functions, the (SL)- and the Kurzweil-Henstock integral will coincide: this, in general, is true only in some particular Riesz spaces.

We begin with the real case. Sometimes, we will not report the proofs in this context because we will give them in the more sophisticated case of Riesz space-valued functions, which obviously includes the real setting as a particular case.

Definitions 9.1   Given an interval $[a,b] \subset \mathbb{R}$, we call *division of* $[a,b]$ any finite set $D = \{x_0, x_1, ..., x_n\} \subset [a,b]$, where $x_0 = a, x_n = b$ and $x_{i-1} < x_i$ for all $i = 1,...,n$. We denote by $\mathcal{T}$ the class of all divisions of $[a,b]$.

We call *partition of* $[a,b]$ a set of the type $\Pi = \{(A_i, \xi_i) : i = 1,...,n\}$, where $A_i = [x_{i-1}, x_i]$, $\{x_0, x_1, ... x_n\}$ is a division and $\xi_i \in A_i$ for all $i = 1,...,n$. A partition is said to be *special* if ξ_i is an endpoint of A_i for every $i = 1,...,n$. We call *mesh* of a partition Π the quantity $|\Pi| := \max_{i=1}^{n} (x_i - x_{i-1})$. Moreover, as no confusion can arise, given a measurable subset A of the (extended) real line, we denote by $|A|$ its Lebesgue measure (finite or $+\infty$).

Definition 9.2 A *decomposition of* $[a,b]$ is a set of the type
$$\Pi = \{(A_i, \xi_i) : i = 1,...,n\}, \tag{9.1}$$
where $\{A_i\}_{i=1}^{n}$ is a family of pairwise nonoverlapping intervals of $[a,b]$ and $\xi_i \in A_i$ for all $i = 1,...,n$.

We note that a decomposition of $[a,b]$ is not necessarily a partition of $[a,b]$.

From now on, let us denote by Δ the set of all positive real-valued functions, defined on an interval $[a,b] \subset \mathbb{R}$.

Definition 9.3   Given a partition or decomposition $\Pi = \{([x_{i-1}, x_i], \xi_i) : i = 1,...,n\}$ of $[a,b]$ and a function $\delta \in \Delta$, we say that Π is δ-*fine* if $x_i - x_{i-1} \leq \delta(\xi_i)$ for all $i = 1,...,n$.

Antonio Boccuto / Beloslav Riečan / Marta Vrábelová

Moreover, if $f:[a,b]\to R$ is a mapping, and Π is a partition or decomposition as above, we denote by $\sum_{\Pi} f$ the quantity

$$\sum_{i=1}^{n} f(\xi_i)(x_i - x_{i-1}).$$

Definition 9.4   A function $\delta:[a,b]\to \mathbb{R}_0^+$ is called *gage* if the set $Z_\delta := \{x\in [a,b]:\delta(x)=0\}$ has Lebesgue measure zero.

We now endow the set of all gages with the following ordering. Given two gages δ_1 and δ_2, we say that $\delta_1 \le \delta_2$ if and only if $\delta_1(x)\le \delta_2(x)$ for all $x\in [a,b]$.

Definition 9.5   If δ is a gage, we say that a decomposition $\Pi = \{(A_i,\xi_i):i=1,...,n\}$ is δ-*fine* if $\xi_i \notin Z_\delta$ and $|A_i|\le \delta(\xi_i)$ for all $i=1,...,n$.

Remark 9.6   We note that there exist some gages δ, without δ-fine partitions. However for all gages δ there exists a δ-fine decomposition Π (for a proof, see [212], Teorema 1, p. 259).

Definition 9.7   Let $J\subset [a,b]\subset \mathbb{R}$, and $\Delta_J = (\mathbb{R}^+)^J$. We say that a function $P:[a,b]\to \mathbb{R}$ is *of class (SL)* or has *property (SL)* on J if for every set $N\subset [a,b]$ of Lebesgue measure zero and $\forall \varepsilon >0$ there exists a map $\delta:J\to \mathbb{R}^+$ such that for any δ-fine decomposition $\Pi = \{([x_{i-1},x_i],\xi_i):i=1,...,n\}$ of $[a,b]$, with $x_i \in [a,b]$ and $\xi_i \in N\cap J$ $\forall i=1,...,n$, we have:

$$\sum_{i=1}^{n} |P(x_i)-P(x_{i-1})|\le \varepsilon. \tag{9.2}$$

We say that P is *of class (SL)* if it is of class (SL) on $[a,b]$.

The following properties hold:

Proposition 9.8

1) *The functions satisfying property (SL) form a linear space.*

2) *Property (SL) implies continuity.*

3) *If J_k, $k\in \mathbb{N}$, are such that f is of class (SL) on each J_k, then f is of class (SL) also on $\bigcup_{k=1}^{\infty} J_k$.*

Definition 9.9   Let $\Delta^{(1)}$ be the set of all gages, defined on $[a,b]$. We say that $f:[a,b]\to \mathbb{R}$ is *(SL)-integrable* on $[a,b]$ if there exists a function $P:[a,b]\to \mathbb{R}$ of class (SL) such that, for every $\varepsilon >0$, there exists $\delta\in \Delta^{(1)}$ such that

$$\left| \sum_{\Pi} f - \sum_{i=1}^{n} [P(x_i)-P(x_{i-1})] \right| \le \varepsilon \tag{9.3}$$

whenever $\Pi = \{([x_{i-1},x_i],\xi_i):i=1,...,n\}$ is a δ-fine decomposition of $[a,b]$. The function P will be called *weak primitive* of f.

In this case we put (by *definition*) $(SL)\int_a^b f = P(b)-P(a)$.

The (SL) integral is well-defined: indeed we have the following:

Proposition 9.10 *Let $N \subset [a,b]$ be a set of Lebesgue measure zero, $f_0 : [a,b] \to \mathbb{R}$ such that $f_0(x) = 0$ for all $x \notin N$ and $P_0 : [a,b] \to \mathbb{R}$ be a constant function. Then P_0 is a weak primitive of f_0.*

Proposition 9.11 *Let N and f_0 be as in Proposition 9.10, and P be a weak primitive of f_0. Then P is constant.*

The following fundamental result holds:

Theorem 9.12 *A function $f : [a,b] \to \mathbb{R}$ is Kurzweil-Henstock integrable on $[a,b]$ if and only if it is (SL)-integrable on $[a,b]$, and in this case we have*

$$(KH)\int_a^b f = (SL)\int_a^b f.$$

Proof: We begin with the "only if" part (see also [169]). Since f is (KH)-integrable on $[a,b]$, then it is (KH)-integrable even on every subinterval of $[a,b]$. Let $F(a) = 0$ and

$$F(x) = (KH)\int_a^x f, \quad \forall x \in (a,b].$$

The assertion follows because (9.3) holds when P is replaced by F and since F is of class (SL) on $[a,b]$. The last fact follows since F is of class (SL) on

$$J_k = \{x \in [a,b] : |f(x)| \le k\}$$

for each $k \in \mathbb{N}$: this is a consequence of the inequality

$$\sum_{i=1}^n |F(x_i) - F(x_{i-1})| - \sum_{i=1}^n (x_i - x_{i-1}) |f(\xi_i)|$$

$$\le \sum_{i=1}^n |[F(x_i) - F(x_{i-1})] - (x_i - x_{i-1}) f(\xi_i)|,$$

which holds for every decomposition $\Pi = \{([x_{i-1}, x_i], \xi_i) : i = 1, \dots, n\}$ of $[a,b]$: indeed, we note that, for sufficiently fine decompositions, the quantity

$$\sum_{i=1}^n (x_i - x_{i-1}) |f(\xi_i)|$$

is sufficiently small, because k is fixed, f is bounded on J_k and the involved ξ_i's belong to a fixed set of Lebesgue measure zero.

We now turn to the "if" part (see also [169] and [170], pp. 169-170). Let $f : [a,b] \to \mathbb{R}$ be (SL)-integrable, let $P : [a,b] \to \mathbb{R}$ be a weak primitive of f, and choose arbitrarily $\varepsilon > 0$. There exists a gage δ_1 such that

$$\left| \sum_{i=1}^n \{[P(x_i) - P(x_{i-1})] - (x_i - x_{i-1}) f(\xi_i)\} \right| \le \frac{\varepsilon}{3}$$

for every δ_1-fine decomposition $\Pi_1 = \{([x_{i-1}, x_i], \xi_i) : i = 1, \dots, n\}$ of $[a,b]$. Let Z_{δ_1} be the set of the zeros of δ_1, and let $f_0 := f \chi_{Z_{\delta_1}}$. We note that f_0 is zero almost everywhere, and hence f_0 is Kurzweil-Henstock integrable on $[a,b]$, and

$$(KH)\int_a^b f_0 = 0.$$

Thus there exists a function $\delta_2 : [a,b] \to \mathbb{R}^+$ such that, $\forall \delta_2$-fine partition Π_2 of $[a,b]$, one has

$$\left| \sum_{\Pi_2} f_0 \right| \le \frac{\varepsilon}{3}.$$

Since P is of class (SL) on $[a,b]$, there exists a map $\gamma : [a,b] \to \mathbb{R}^+$ such that, for each γ-fine decomposition Π_3 of $[a,b]$, $\Pi_3 = \{([y_{i-1}, y_i], \kappa_i) : i = 1, \ldots, n\}$, with $\kappa_i \in [a,b]$ and $\kappa_i \in Z_{\delta_1}$ $\forall i = 1, \ldots, n$, we have:

$$\sum_{i=1}^{n} |P(y_i) - P(y_{i-1})| \le \frac{\varepsilon}{3}. \tag{9.4}$$

Let $\delta(t) = \min\{\delta_1(t), \delta_2(t)\}$ if $t \notin Z_{\delta_1}$, and $\delta(t) = \min\{\gamma(t), \delta_2(t)\}$ if $t \in Z_{\delta_1}$.

If $\Pi = \{([w_{k-1}, w_k], \zeta_k) : k = 1, \ldots, p\}$ is any δ-fine partition of $[a,b]$, then we get:

$$\left| \sum_{k=1}^{p} (w_k - w_{k-1}) f(\zeta_k) - [P(b) - P(a)] \right|$$

$$= \left| \sum_{k=1}^{p} (w_k - w_{k-1}) f(\zeta_k) - \sum_{k=1}^{p} [P(w_k) - P(w_{k-1})] \right|$$

$$\le \left| \sum_{k=1}^{p} (w_k - w_{k-1}) f_0(\zeta_k) \right| + \sum_{k=1,\ldots,p, \zeta_k \in Z_{\delta_1}} |P(w_k) - P(w_{k-1})|$$

$$+ \left| \sum_{k=1,\ldots,p, \zeta_k \notin Z_{\delta_1}} (w_k - w_{k-1}) f(\zeta_k) - [P(w_k) - P(w_{k-1})] \right| \le \varepsilon.$$

This concludes the proof. 🍎

The following results hold (see also [169]):

Proposition 9.13 *A function f which is (SL)-integrable on $[a,b]$ is (SL)-integrable also on every subinterval $[c,d]$ of $[a,b]$.*

Moreover, if $a < c < b$, then (SL)-integrability on $[a,c]$ and on $[c,b]$ implies (SL)-integrability on $[a,b]$.

The (SL)-integral is a positive linear functional.

Theorem 9.14 *If f, $P : [a,b] \to \mathbb{R}$ are such that $P' = f$ almost everywhere on $[a,b]$ and P is of class (SL) on $[a,b]$, then*

$$(SL)\int_a^b f = P(b) - P(a).$$

Theorem 9.15 *If f, $P : [a,b] \to \mathbb{R}$ are such that $P' = f$ on $[a,b]$ except a countable set Q and P is continuous on Q, then*

$$(SL)\int_a^b f = P(b) - P(a).$$

Sketch of the proof: This follows from the fact that, if P' exists in $\mathbb{R}$ except a countable set Q, then P is of class (SL) on $[a,b] \setminus Q$, because P has property (SL) on each set of the type

$$J_k = \{x \in [a,b] : k - 1 \le |P'(x)| < k\}, \quad k \in \mathbb{N},$$

and from the fact that P is of class (SL) on every countable set on which P is continuous.

Furthermore, since absolute continuity of a real-valued function defined on $[a,b]$ implies property (SL) and existence of the derivative almost everywhere in $[a,b]$, we get the following:

Corollary 9.16 *If $P:[a,b] \to \mathbb{R}$ is absolutely continuous in $[a,b]$, then*

$$(SL)\int_a^b P' = P(b) - P(a).$$

The following results hold (see [169], Corollary 3, p. 764 and Theorem 1, p. 765):

Theorem 9.17 *If $f:[a,b] \to \mathbb{R}$ is Lebesgue integrable on $[a,b]$, then f is (SL)-integrable on $[a,b]$, and the two integrals coincide.*

Theorem 9.18 *If $f:[a,b] \to \mathbb{R}$ is (SL)-integrable and P is a weak primitive of f, then there exists a set N of Lebesgue measure zero such that*

$$P'(x) = f(x) \quad \forall\, x \in [a,b] \setminus N.$$

We will now introduce the (SL)-integral for Riesz space-valued functions. For further developments and generalizations in a more abstract setting, see also [37, 38].

Definition 9.19 Let R be any Riesz space. A *unit of R* is an element $u \in R$ such that $u \geq 0$, $u \neq 0$.

We now introduce the concept of (r)-convergence.

Definition 9.20 Let u be a unit of R. We say that the sequence $(r_n)_n$ of elements of R u-*converges* to $r \in R$ if $\forall \varepsilon > 0$, $\exists \bar{n} = \bar{n}(\varepsilon) \in \mathbb{N}$ such that $|r_n - r| \leq \varepsilon u, \forall n \geq \bar{n}$. We say that $(r_n)_n$ *converges relatively* (or (r)-*converges*) to r, if it u-converges, for some unit u. In this case, we write $(r)\lim_{n \to +\infty} r_n = r$ or $(r)\lim_n r_n = r$.

Definition 9.21 Let u be a unit of R, and $(\Lambda, \prec)$ be any directed set. We say that the net $(r_\lambda)_{\lambda \in \Lambda}$ u-*converges* to r if $\forall \varepsilon > 0$, $\exists \bar{\lambda} = \bar{\lambda}(\varepsilon) \in \Lambda$ such that

$$|r_\lambda - r| \leq \varepsilon u, \forall \lambda \in \Lambda, \lambda \prec \bar{\lambda}.$$

We say that $(r_\lambda)_{\lambda \in \Lambda}$ *converges relatively* (or (r)-*converges*) to r, if it u-converges, for some unit u. In this case, we write $(r)\lim_{\lambda \in \Lambda} r_\lambda = r$ or $(r)\lim_\lambda r_\lambda = r$.

We now recall the concept of order convergence (see also Chapter 2).

Definition 9.22 A sequence $(r_n)_n$ of elements of R is said to be *order-convergent* (or (O)-*convergent*) to $r \in R$, if there exists a sequence $(p_n)_n$ in R, such that $p_n \downarrow 0$ and $|r_n - r| \leq p_n$, $\forall n \in \mathbb{N}$ (see also [175, 273]). In this case, we write $(O)\lim_n r_n = r$.

We note that, if R is a Dedekind complete Riesz space, then (O)-convergence can be formulated in the following way (see also [273]):

Definition 9.23 We say that $(r_n)_n$ is *order convergent* (or (O)-*convergent*) to r, if

$$r = (O)\limsup_{n} r_n = (O)\liminf_{n} r_n,$$

where

$$(O)\limsup_{n} r_n = \inf_{n}[\sup_{m \geq n} r_m],$$
$$(O)\liminf_{n} r_n = \sup_{n}[\inf_{m \geq n} r_m].$$

We note that, in the context of (O)-convergence, the definition of series of the type $\sum_{n=1}^{\infty} a_n$, with $a_n \in R$ $\forall n \in \mathbb{N}$, can be formulated analogously as in the classical case.

Definition 9.24 Let $(\Lambda, \prec)$ be any directed set. A net $(r_\lambda)_{\lambda \in \Lambda}$ is said to be *order convergent* (or (O)-*convergent*) to r, if

$$\sup_{\rho \in \Lambda} \inf_{\lambda \prec \rho} r_\lambda = \inf_{\rho \in \Lambda} \sup_{\lambda \prec \rho} r_\lambda,$$

and in this case we write $(O)\lim_{\lambda \in \Lambda} r_\lambda = r$. We observe that, if a net (O)-converges both to r and to s, then necessarily $r := s$.

We note that, if R is Archimedean, then (r)-convergence implies (O)-convergence. In general, the converse is not true   (for example, in the spaces of the type $\mathcal{C}(\Omega) = \{f : \Omega \to \mathbb{R},\ f$ is continuous$\}$, where Ω is an infinite compact topological space). However, there are many cases in which these two concepts coincide, for example in the spaces of the type $L^0(X, \mathcal{B}, \mu)$, where $(X, \mathcal{B}, \mu)$ is a measure space with μ positive, σ-additive and σ-finite (see [175, 273]).

Throughout this part of this chapter, we investigate a Kurzweil-Henstock-type integral for functions, defined in $[a, b] \subset \mathbb{R}$ and taking values in a Dedekind complete Riesz space, relatively to the *order convergence,* and successively we introduce the (SL)-integral, by means of which we prove some more sophisticated versions of the Fundamental Formula of Calculus, which are similar to the classical ones existing in the literature (see also [207,212,257]).

Definition 9.25 (see also Chapter 2) A Dedekind complete Riesz space R, such that every subset $R_1 \neq \emptyset$ bounded from above contains a countable subset having the same supremum as R_1, is said to be *super Dedekind complete.*

Definition 9.26 Let R be a Riesz space. We say that R has *property σ* or *the σ property* if, for all sequences $(r_n)_n$ of positive elements of R, there exists a sequence $(\lambda_n)_n$ of strictly positive real numbers such that the sequence $(\lambda_n r_n)_n$ is bounded (see also [175], Definition 70.1., p. 478).

Remark 9.27 We observe that, if $(X, \mathcal{B}, \mu)$ is a measure space, where μ is a σ-finite countably additive $\tilde{\mathbb{R}}$-valued measure, then the spaces $L^p(X, \mathcal{B}, \mu)$, with $p \in [0, \infty]$, have property σ (see also [175]). Moreover, we note that there exist some super Dedekind complete Riesz spaces, lacking property σ (for example, we can take the space $R \subset \mathbb{R}^{\mathbb{N}}$ consisting of all sequences with only a finite number of nonzero coordinates: see [175], p. 479).

Assumptions 9.28 From now on, we always suppose that R is a Dedekind complete Riesz space. In some suitable cases, we add to R an extra element, $+\infty$, extending ordering and operations, having the same role as the element $+\infty$ with respect to the real numbers, that is in such a way that

$$\begin{cases} +\infty \geq r \quad \forall r \in R, \\ 0 \cdot (+\infty) = 0, \\ \lambda \cdot (+\infty) = +\infty \quad \forall \lambda \in \mathbb{R}^+, \\ +\infty + r = +\infty \quad \forall r \in R \cup \{+\infty\}. \end{cases}$$

We denote with the symbol $\overline{R}$ the set $R \cup \{+\infty\}$.

Definition 9.29   A net $(p_\delta)_{\delta \in \Delta}$ of elements of $\overline{R}$ is called (O)-*net* if the following conditions are fulfilled:

1) $p_{\delta_1} \leq p_{\delta_2}$ whenever $\delta_1, \delta_2 \in \Delta$, $\delta_1 \leq \delta_2$;

2) $\inf\{p_\delta : \delta \in \Delta\} = 0$.

A sequence $(p_n)_n$ in $\overline{R}$ is called (O)-*sequence* if it is decreasing and $\inf_n p_n = 0$.

Before investigating the (SL)-integral in Riesz spaces and formulating our version of the Fundamental Formula of Integral Calculus, we now introduce some concepts of continuity and differentiability and some notions related with them in the context of Riesz spaces.

Definition 9.30   A map $f : [a,b] \to R$ is said to be *continuous* at the point $x_0 \in [a,b]$ if
$$(O) \lim_{x \to x_0} f(x) = f(x_0).$$

We say that f is *continuous* in $[a,b]$ if it is continuous at every point $x_0 \in [a,b]$.

Definition 9.31   A map $f : [a,b] \to R$ is said to be *of bounded variation* on $[a,b]$ if there exists in R the quantity

$$\sup\left\{ \sum_{i=1}^{n} | f(x_i) - f(x_{i-1}) | : \right.$$

$$\left. \{x_0 := a, x_1, \ldots, x_n := b\} \text{ is a division of } [a,b]\right\}.$$

In this case, we indicate this quantity by the symbol $V(f,[a,b])$.

Remark 9.32 It is easy to see that every function of bounded variation is bounded too (see also [55]).

Let now Z be the set of all sequences $([\alpha_s, \beta_s])_s$ of nonoverlapping subintervals of $[a,b]$. A mapping $f : [a,b] \to R$ is said to be *absolutely continuous* in $[a,b]$ if there exists an (O)-sequence $(p_n)_n$ of elements of R, such that

$$\sup\left\{ \sum_{s=1}^{\infty} | f(\beta_s) - f(\alpha_s) | : ([\alpha_s, \beta_s])_s \in Z, \sum_{s=1}^{\infty} | \beta_s - \alpha_s | \leq 1/n \right\} \leq p_n \qquad (9.5)$$

for all $n \in \mathbb{N}$.

The following result holds.

Proposition 9.33 *Let $f : [a,b] \to R$ be a function, continuous in $[a,b]$. Then f is bounded on $[a,b]$.*

Proof: Let $J_0 := [a,b]$, and suppose by contradiction that f is unbounded on J_0. Then f is unbounded on at least one of the intervals

$$I_1 = \left[a, \frac{a+b}{2} \right] \text{ and } I_2 = \left[\frac{a+b}{2}, b \right];$$

denote this interval by J_1 or $[c,d]$. By proceeding as above, we get that f is unbounded either on

$$I_1 = \left[c, \frac{c+d}{2} \right] \text{ or } I_2 = \left[\frac{c+d}{2}, d \right].$$

Thus we can construct a decreasing sequence of closed subintervals of $[a,b]$, $(J_n)_{n \in \mathbb{N} \cup \{0\}}$, such that $|J_n| = (b-a)/2^n$ for all $n \geq 0$. Let now $\{x_0\} := \bigcup_{n \in \mathbb{N} \cup \{0\}} J_n$. As $(O)\lim_{x \to x_0} f(x) = f(x_0)$, then there exists an (O)-sequence $(p_k)_k$ in R, such that

$$\sup\{| f(v) - f(u) | : x_0 - 1/k \leq u \leq x_0 \leq v \leq x_0 + 1/k\} \leq p_k$$

for all $k \in \mathbb{N}$. From this it follows easily that $\exists n_0 \in \mathbb{N}$ such that f is bounded in J_{n_0}, a contradiction. $\blacksquare$

Definition 9.34 Let $J \subset [a,b]$. We say that $f : [a,b] \to R$ is (u)-*continuous in* J if there exists an (O)-net $(p_\delta)_{\delta \in \Delta}$ such that, $\forall \delta : J \to \mathbb{R}^+$,

$$| f(t) - f(x) | \leq p_\delta \quad \text{whenever } t \in [a,b], x \in J \text{ and } |t - x| \leq \delta(x).$$

We say that f is (u)-*continuous* if it is (u)-continuous in $[a,b]$.

Definition 9.35 We say that $f : [a,b] \to R$ is *uniformly continuous* (in $[a,b]$) if there exists an (O)-sequence $(p_n)_n$ such that $\forall n \in \mathbb{N}$

$$| f(t) - f(x) | \leq p_n \quad \text{whenever } t, x \in [a,b] \text{ and } |t - x| \leq 1/n.$$

Remark 9.36 We observe that:

9.36.1) (u)-continuity always implies continuity and is implied by uniform continuity.

9.36.2) If $R = \mathbb{R}$, then (u)-continuity is equivalent to (classical) continuity (and *uniform* continuity, by Heine's Theorem).

9.36.3) Absolute continuity implies uniform continuity.

In 9.38 below, we will give an example of a continuous function in $[a,b]$, which is not (u)-continuous.

Definition 9.37 A function $f : [a,b] \to R$ is said to be *differentiable at* x_0 if

$$(O)\lim_{x \to x_0} \frac{f(x) - f(x_0)}{x - x_0}$$

exists in R.

We say that f is *differentiable* in $[a,b]$ if it is differentiable at every point $x_0 \in [a,b]$. If $J \subset [a,b]$, then a mapping $f : [a,b] \to R$ is said to be (u)-*differentiable in* J if there exists a function $g : J \to R$ such that

$$(O) \lim_{\delta:J\to\mathbb{R}^+} \left(\sup\left\{ \left| \left| \frac{f(v)-f(u)}{v-u} - g(x) \right| \right| : (u,v,x) \in A_\delta \right\} \right) = 0,$$

where $A_\delta := \{(u,v,x) : u,v \in [a,b], \quad x \in J, \quad u \neq v \quad \text{and} \quad x - \delta(x) \leq u \leq x \leq v \leq x + \delta(x)\}$ for all $\delta : J \to \mathbb{R}^+$.

We say that $f : [a,b] \to R$ is (u)-*differentiable* if it is (u)-differentiable in $[a,b]$.

As a consequence of the properties of (O)-convergence, we see that such a function g is unique. In this case, the map g will be called the (u)-*derivative* of f, or simply *derivative,* when no confusion can arise. It is easy to see that every (u)-differentiable map is differentiable too, and the involved derivatives coincide. In general, the converse implication is not true (See Remark 9.38), but it is clear that, when $R = \mathbb{R}$, the concepts of differentiability and (u)-differentiability coincide.

We say that $f : [a,b] \to R$ is *Lipschitzian* if there exists $r \in R$ such that

$$\left| f(t_1) - f(t_2) \right| \leq \left| t_1 - t_2 \right| r$$

for all $t_1, t_2 \in [a,b]$. We observe that, if R is *any* Dedekind complete Riesz space, then every polynomial is (u)-differentiable (we recall that a *polynomial* is a function $p : [a,b] \to R$ of the type $p(x) = \alpha_0 + \alpha_1 x + \ldots + \alpha_n x^n$, with $n \in \mathbb{N}$ and $\alpha_i \in R, i = 0, \ldots, n$) and its (u)-derivative is $p'(x) = \alpha_1 + 2\alpha_2 x + \ldots + n\alpha_n x^{n-1}$.

Remark 9.38 We now give an example of a function $f : [a,b] \to R$, which is differentiable in $[a,b]$ but not (u)-continuous in $[a,b]$. Let $f : \mathbb{R} \to \mathbb{R}^{[0,1]}$ be defined by setting

$$f(t) = \chi_{(t-1/8,t+1/8)\cap(0,1)}$$

for all $t \in \mathbb{R}$ (where χ_A is the *characteristic function* associated with A; see also [55]). For every $t, t_0 \in \mathbb{R}$, with $0 \leq t_0 < t < t_0 + 1/8 \leq 1$, and for each $s \in [0,1]$, we have

$$\frac{f(t)-f(t_0)}{t-t_0}(s) = \begin{cases} -\dfrac{1}{t-t_0} & \text{if } s \in (t_0 - 1/8, t - 1/8), \\[2mm] \dfrac{1}{t-t_0} & \text{if } s \in (t_0 + 1/8, t + 1/8), \\[2mm] 0 & \text{otherwise.} \end{cases}$$

From this it follows easily that $f'_+(t_0) = 0$. Analogously one proves that $f'_-(t_0) = 0$. So, f is differentiable in $[0,1]$, and therefore f is continuous in $[0,1]$.

Moreover, for all s in $[0,1]$, define $f_s : [0,1] \to R$ by setting $f_s(t) := f(t)(s)$ for every $t \in [0,1]$. It is possible to see that f is not (u)-continuous. Thus f is neither (u)-differentiable nor uniformly continuous (see also [55]), and hence f is not Lipschitzian.

Definition 9.39 Let $X \neq \varnothing$ be any set, and $\Delta' \subset \Delta_0 \subset \mathbb{R}^X$. We say that Δ' is a *cofinal set* with respect to Δ_0 if for every $\delta_1, \delta_2 \in \Delta_0$ there exists $\delta_3 \in \Delta'$, such that $\delta_3 \leq \delta_1$ and $\delta_3 \leq \delta_2$.

We now introduce the definition of the Riemann integral, in the "Mengoli-Cauchy" form (see [35]).

Definition 9.40   A map $f:[a,b] \to R$ is said to be (Ri)-*integrable* on $[a,b]$ if there exists an element $Y \in R$ such that

$$(O)\lim_n \left[\sup\left\{ \left| \sum_\Pi f - Y \right| : |\Pi| \le 1/n \right\} \right] = 0. \qquad (9.6)$$

In this case, we write $Y = (Ri)\int_a^b f$.

We now give the definition of the Lebesgue integral for Riesz space-valued mappings, firstly for bounded functions and afterwards for not necessarily bounded maps.

Definition 9.41   Let R be a Dedekind complete Riesz space. We say that a map $g:[a,b] \to R$ is *simple* if there exist n measurable (disjoint) sets $E_1,...,E_n$ such that g is constant in E_i $(i=1,...,n)$. If g is a simple function, we put

$$\int_a^b g := \sum_{i=1}^n |E_i| g(\xi_i),$$

where ξ_i is an arbitrary point of E_i.

We call *upper integral* [resp. *lower integral*] of a bounded map $f:[a,b] \to R$ the element of R given by

$$\inf_{v \in V_f} \int_a^b v \quad \left[\sup_{s \in S_f} \int_a^b s \right],$$

where

$$V_f := \{v : v \text{ is a simple function}, v(t) \ge f(t) \forall t \in [a,b]\},$$
$$S_f := \{s : s \text{ is a simple function}, s(t) \le f(t) \forall t \in [a,b]\}.$$

We denote the upper and the lower integrals of f with the symbols $^*\!\int_a^b f$ and $_*\!\int_a^b f$ respectively.

We say that a bounded function $f:[a,b] \to R$ is *Lebesgue* integrable (or (L)-*integrable*) on $[a,b]$, if its lower integral coincides with its upper integral, and, in this case, we call (L)-*integral of* f their common value, and we denote it by

$$(L)\int_a^b f.$$

Definition 9.42   Let $(\psi_n : [a,b] \to R)_n$ be a sequence of functions. We say that $(\psi_n)_n$ *converges in measure* to ψ if there exist two (O)-sequences $(p_n)_n$ in R, $(q_n)_n$ in $\mathbb{R}$, such that

$$\lambda(\{x \in [a,b] : |\psi_n(x) - \psi(x)| \not\le p_n\}) \le q_n \quad \text{for all } n \in \mathbb{N}$$

(where λ is the Lebesgue measure).

Definition 9.43   We say that a sequence of bounded Lebesgue integrable functions $(\psi_n : [a,b] \to R)_n$ is *uniformly integrable* if

$$\sup_n \left[(L)\int_a^b |\psi_n| \right] \in R$$

and

$$[\lambda(B_n) \to 0] \Rightarrow \left[(O)\lim_n \left(\sup_{m \ge n} \left[(L)\int_{B_n} |\psi_m| \right] \right) = 0 \right].$$

Definition 9.44 A map $\psi : [a,b] \to R$ (not necessarily bounded) is said to be (L)-*integrable* on $[a,b]$ if there exists a uniformly integrable sequence $(\psi_n)_n$ of bounded (L)-integrable functions, convergent in measure to ψ. In this case we say that

$$(L)\int_a^b \psi := (O)\lim_n \left[(L)\int_a^b \psi_n \right].$$

Remark 9.45 One can check that Definition 9.44 makes sense, i.e. the (L)-integral of ψ does not depend on the chosen sequence $(\psi_n)_n$.

Definition 46. A map $f : [a,b] \to R$ is (KH)-*integrable* (on $[a,b]$) if there exists an element $Y \in R$ such that

$$(O)\lim_{\delta \in \Delta} \left[\sup \left\{ \left| \sum_\Pi f - Y \right| : \Pi \text{ is a } \delta-\text{fine partition of } [a,b] \right\} \right] = 0. \qquad (9.7)$$

Remark 9.47 It is easy to see that, in Definition 9.46, the element Y is unique, if it exists: in this case, we write

$$Y = (KH)\int_a^b f.$$

Remark 9.48 It is easy to check that, if α, $\beta \in \mathbb{R}$, and f and g are (KH)-integrable, then $\alpha f + \beta g$ is (KH)-integrable too, and

$$(KH)\int_a^b (\alpha f + \beta g) = \alpha (KH)\int_a^b f + \beta (KH)\int_a^b g$$

Furthermore, if $f \geq 0$ is (KH)-integrable, we have

$$(KH)\int_a^b f \geq 0.$$

We note that, without loss of generality, we can take in (9.7) the set of the δ-fine special partitions of $[a,b]$ instead of the set of all δ-fine partitions of $[a,b]$: indeed, given a partition "generated" by a division $D = \{x_0, x_1, \ldots, x_n\}$, with $\xi_i \in [x_{i-1}, x_i]$ for all $i = 1, \ldots, n$, and such that ξ_i is not an endpoint of $[x_{i-1}, x_i]$ for some i, it will be enough for our purposes to consider the division D' obtained by adding such ξ_i 's to the points of D, and to associate with D' the same ξ_i 's as in the original partition.

From now on, we denote by

$$\Psi(\delta) := \{ \Pi : \Pi \text{ is a } \delta-\text{fine partition of} [a,b] \},$$

$$a(\delta) := \sup \left\{ \sum_\Pi f : \Pi \in \Psi(\delta) \right\},$$

$$b(\delta) := \inf \left\{ \sum_{\Pi'} f : \Pi' \in \Psi(\delta) \right\},$$

$$r(\delta) := \sup \left\{ \left| \sum_\Pi f - \sum_{\Pi'} f \right| : \Pi, \Pi' \in \Psi(\delta) \right\},$$

for all $\delta \in \Delta$. Now we state the Bolzano-Cauchy condition for the (KH)-integral.

Proposition 9.49 *A necessary and sufficient condition for* (KH)*-integrability of* f *on* $[a,b]$ *is that*

$$(O)\lim_{\delta\in\Delta} r(\delta) = 0.$$

Proof: The necessary part is straightforward. ♨

We now turn to the sufficient part. We have

$$\sum_{\Pi} f \le r(\delta) + \sum_{\Pi'} f \quad \text{for all} \Pi,\Pi'\in\Psi(\delta). \tag{9.8}$$

Since by hypothesis

$$(O)\lim_{\delta\in\Delta} r(\delta) = \inf_{\delta\in\Delta} r(\delta) = 0,$$

then there exists $\delta_0\in\Delta$ such that

$$r(\delta)\in R \quad \forall\delta\in\Delta, \delta\le\delta_0.$$

Taking in (9.8) first the supremum on the left over Π, we observe that $a(\delta)\in R$ for $\delta\le\delta_0$ with δ_0 chosen above. Then we take in (9.8) the infimum over Π' to get

$$a(\delta) \le r(\delta) + b(\delta), \tag{9.9}$$

noting that $b(\delta)\in R$ for the same δ's. Moreover, still for such δ's, we have:

$$\left|\sum_{\Pi} f - \sum_{\Pi'} f\right| \le a(\delta) - b(\delta) \quad \forall\Pi,\Pi'\in\Psi(\delta), \tag{9.10}$$

and taking the supremum in (9.10) we obtain:

$$r(\delta) \le a(\delta) - b(\delta).$$

In view of (9.9) this gives

$$r(\delta) = a(\delta) - b(\delta). \tag{9.11}$$

We observe that the nets $(a(\delta))_{\delta\in\Delta}$ and $(b(\delta))_{\delta\in\Delta}$ are monotone decreasing and increasing respectively, and thus there exist in R the (O)-limits

$$(O)\lim_{\delta\in\Delta} a(\delta) \text{ and } (O)\lim_{\delta\in\Delta} b(\delta).$$

So, taking into account also of the hypothesis, we obtain:

$$(O)\lim_{\delta\in\Delta} a(\delta) \le (O)\limsup_{\delta\in\Delta} r(\delta) + (O)\lim_{\delta\in\Delta} b(\delta)$$

$$= (O)\lim_{\delta\in\Delta} b(\delta) \le (O)\lim_{\delta\in\Delta} a(\delta),$$

and thus

$$(O)\lim_{\delta\in\Delta} a(\delta) = (O)\lim_{\delta\in\Delta} b(\delta).$$

Let now

$$Y := (O)\lim_{\delta\in\Delta} a(\delta) = (O)\lim_{\delta\in\Delta} b(\delta).$$

We observe that

$$\sup\left\{\left|\sum_\Pi f - Y\right| : \Pi \in \Psi(\delta)\right\}$$

$$\leq \left(\sup\left\{\sum_\Pi f : \Pi \in \Psi(\delta)\right\} - Y\right)$$

$$+\left(Y - \inf\left\{\sum_\Pi f : \Pi \in \Psi(\delta)\right\}\right) \quad \forall \delta \in \Delta.$$

Hence,

$$(O)\lim_{\delta \in \Delta}\left(\sup\left\{\left|\sum_\Pi f - Y\right| : E \in \Psi(\delta)\right\}\right) = 0.$$

Thus the assertion follows. 🍎

We now state the following:

Theorem 9.50 *Let* $f : [a,b] \to R$, *and* $c \in (a,b)$. *Then* f *is* (KH)-*integrable on* $[a,b]$ *if and only if* f *is* (KH)-*integrable on* $[a,c]$ *and on* $[c,b]$, *and in this case we have*

$$(KH)\int_a^b f = (KH)\int_a^c f + (KH)\int_c^b f.$$

Example 9.51 Let R be a Dedekind complete Riesz space without property σ (see Remark 9.27). In R there exists a sequence $(u_n)_n$ such that for all sequences $(\lambda_n)_n$ of positive real numbers, the sequence $(\lambda_n u_n)_n$ is not bounded in R. In Chapter 7 we saw that the function $f : [0,1] \to R$, defined by setting

$$f(x) = \begin{cases} u_n & \text{if } x = 1/n, \\ 0 & \text{otherwise,} \end{cases} \tag{9.12}$$

is not (KH)-integrable. This function is (SL)-integrable with integral zero, because it coincides with the identically zero function up to the complement of a set of Lebesgue measure 0.
In Chapter 7, we saw also that, if R has property σ, then pathologies like the one in (9.12) do not hold.

However, in the case of bounded functions, we get the following result:

Proposition 9.52 *Let* R *be any Dedekind complete Riesz space,* $N \subset \mathbb{R}$ *be a set of Lebesgue measure zero, and* $f : [a,b] \to R$ *be such that* $f(x) = 0$ *for all* $x \in [a,b] \setminus N$. *Suppose that there exists* $w \in R$ *such that* $|f(x)| \leq w$ *for all* $x \in [a,b] \cap N$. *Then* f *is* (KH)-*integrable on* $[a,b]$ *and*

$$(KH)\int_a^b f = 0.$$

An immediate consequence of Proposition 9.52 is the following.

Proposition 9.53 *Let* $f, g : [a,b] \to R$ *be two maps. If* f *is* (KH)-*integrable on* $[a,b]$ *and there exists a set* N *of Lebesgue measure zero such that* $f := g$ *on* $[a,b] \setminus N$ *and* $f - g$ *is bounded on* N, *then* g *is* (KH)-*integrable on* $[a,b]$ *and*

$$(KH)\int_a^b g = (KH)\int_a^b f.$$

We now prove that every "(u)-derivative" is integrable with respect to the (KH)-integral.

Theorem 9.54 *Let R be any Dedekind complete Riesz space, and $f:[a,b] \to R$ be a (u)-differentiable map, with derivative f'. Then, f' is (KH)-integrable on $[a,b]$, and*

$$(KH)\int_a^b f' = f(b) - f(a).$$

Proof: By (u)-differentiability of f, there exists an (O)-net $(p_\delta)_{\delta \in \Delta}$, such that $\forall \delta \in \Delta$

$$\sup\left\{ \left| \frac{f(v)-f(u)}{v-u} - f'(x) \right| : (u,v,x) \in A_\delta \right\} \le p_\delta.$$

Take $\delta \in \Delta$, and choose a δ-fine partition Π of $[a,b]$, $\Pi = \{([t_{i-1},t_i],\xi_i) : i=1,\ldots,q\}$. We get:

$$0 \le \left| \sum_\Pi f' - [f(b)-f(a)] \right| \le \left| \sum_{i=1}^q [f(t_i)-f(t_{i-1}) - (t_i-t_{i-1})f'(\xi_i)] \right|$$

$$\le \sum_{i=1}^q \left\{ \left| (t_i-t_{i-1}) \frac{f(t_i)-f(t_{i-1})}{t_i-t_{i-1}} - f'(\xi_i) \right| \right\}$$

$$\le \sum_{i=1}^q (t_i-t_{i-1})p_\delta = (b-a)p_\delta.$$

Therefore the assertion follows. ✎

Remark 9.55 We note that in general, in Theorem 9.54, the hypothesis of (u)-differentiability cannot be replaced by the one of differentiability. Indeed, let f be as in Remark 9.38. We have

$$0 = (KH)\int_a^b f' \ne f(1) - f(0) = \chi_{(7/8,1)} - \chi_{(0,1/8)}.$$

We now prove a version of the Henstock Lemma for our integral (for further generalizations of this lemma to more abstract contexts, see also [37]).

Lemma 9.56 *If f is (KH)-integrable on $[a,b]$ and $F(I) = (KH)\int_I f$ for every subinterval $I \subset [a,b]$, then*

$$\inf_{\delta \in \Delta} \left\{ \sup \sum_{i=1}^q \left| |J_i| f(\xi_i) - F(J_i) \right| : \right. \qquad (9.13)$$

$$\Pi \text{ is a } \delta-\text{fine decomposition of } [a,b] \} = 0$$

(where $\Pi := \{(J_i,\xi_i) : i=1,\ldots,q\}$).

Proof: By (KH)-integrability of f, there exists an (o)-net $(p_\delta)_{\delta \in \Delta}$ such that

$$\sup\left\{ \left| \sum_{i=1}^q |J_i| f(\xi_i) - (KH)\int_a^b f \right| : \right. \qquad (9.14)$$

$$\Pi \text{ is a } \delta-\text{fine partition of } [a,b] \} \le p_\delta$$

for every $\delta \in \Delta$. Let $\delta \in \Delta$ and $\Pi = \{(J_i,\xi_i), i=1,\ldots,q\}$ be a δ-fine decomposition of $[a,b]$. Then f is integrable on J_i, $i=1,\ldots,q$. Thus for each $i=1,\ldots,q$ there exists an (o)-net $(p_{\delta_i})_{\delta_i:J_i \to \mathbb{R}^+}$ such that

$$\sup\left\{ \left| \sum_{i=1}^q |J_i| f(\xi_i) - (KH)\int_{J_i} f \right| : \right. \qquad (9.15)$$

$$\Pi \text{ is a } \delta-\text{fine partition of } J_i \} \le p_{\delta_i}.$$

Now, fix arbitrarily $\varnothing \neq L \subset \{1,\ldots,q\}$. Let Π_i be a δ-fine partition of J_i, and set

$$\Pi_0 = \{(J_i, \xi_i) \in \Pi : i \in L\} \cup (\bigcup_{i \notin L} \Pi_i).$$

Having fixed the chosen δ, we can suppose that $\delta_i(x) \leq \delta(x)$ for all $x \in J_i$. Then Π_0 is a δ-fine partition of $[a,b]$, and hence by (9.14) we get

$$\left| \sum_{\Pi_0} f - (KH) \int_a^b f \right| \leq p_\delta.$$

Thus we have

$$0 \leq \left| \sum_{i \in L} |J_i| f(\xi_i) - \sum_{i \in L} (KH) \int_{J_i} f \right|$$

$$= \left| \sum_{\Pi_0} f - (KH) \int_a^b f + \sum_{i \notin L} (KH) \int_{J_i} f - \sum_{i \notin L} \sum_{\Pi_i} f \right|$$

$$\leq \left| \sum_{\Pi_0} f - (KH) \int_a^b f \right| + \left| \sum_{i \notin L} (KH) \int_{J_i} f - \sum_{i \notin L} \sum_{\Pi_i} f \right|$$

$$\leq \left| \sum_{\Pi_0} f - (KH) \int_a^b f \right| + \sum_{i=1}^q \left| (KH) \int_{J_i} f - \sum_{\Pi_i} f \right|$$

$$\leq p_\delta + \sum_{i=1}^q p_{\delta_i}.$$

Considering this inequality for a fixed δ and for any δ_i's we can pass to the $(O)\limsup$ as the δ_i's "shrink to zero", to get

$$0 \leq \left| \sum_{i \in L} |J_i| f(\xi_i) - \sum_{i \in L} (KH) \int_{J_i} f \right| \leq p_\delta. \qquad (9.16)$$

We now observe that, since R is a Dedekind complete Riesz space, by virtue of the Maeda-Ogasawara-Vulikh representation theorem (see Chapter 2) there exists a compact extremely disconnected topological space Ω, such that R can be embedded Riesz isomorphically as a solid subset of $\mathcal{C}_\infty(\Omega) := \{f : \Omega \to \tilde{\mathbb{R}} : f$ is continuous, and the set $\{\omega \in \Omega : |f(\omega)| = +\infty\}$ is nowhere dense in $\Omega\}$ (extremely disconnected means that the closure of every open subset of Ω is still an open set of Ω). From (9.16), for all $\omega \in \Omega$ and for each δ-fine partition $\Pi = \{(J_i, \xi_i) : i = 1,\ldots,q\}$ of $[a,b]$, we have (using the same notations for elements of R and for the corresponding elements of $\mathcal{C}_\infty(\Omega)$):

$$0 \leq \left| \sum_{i \in L} |J_i| f(\xi_i) - \sum_{i \in L} (KH) \int_{J_i} f \right|(\omega) \leq p_\delta(\omega)$$

for all $L \subset \{1,2,\ldots,q\}$ (with the convention that the sum along the empty set of any quantity is zero).

Fix now $\omega \in \Omega$. If $p_\delta(\omega) = +\infty$, there is nothing to prove. Suppose that $p_\delta(\omega) \in \mathbb{R}$. Let L [resp. L'] be the sets of all indices $i \in \{1,\ldots,q\}$ such that

$$\left[\,|J_i|\,f(\xi_i)-(KH)\!\int_{J_i} f\,\right]\!(\omega)\ge 0\ \ [\,\text{resp.}<0].$$

We have

$$0\le\sum_{i=1}^{q}\left|\,\left|J_i\,|\,f(\xi_i)-(KH)\!\int_{J_i} f\,\right|\!(\omega)\right.$$

$$=\sum_{i\in L}\left[\,|J_i|\,f(\xi_i)-(KH)\!\int_{J_i} f\,\right]\!(\omega)-\sum_{i\in L'}\left[\,|J_i|\,f(\xi_i)-(KH)\!\int_{J_i} f\,\right]\!(\omega)\le 2\,p_\delta(\omega)$$

for each $\omega\in\Omega$. Coming back to the corresponding elements of R, we get (9.13). ◄

We observe that in general (KH)-integrability of a function $f:[a,b]\to R$ does not imply (KH)-integrability of the map $|f|$, even if $R=\mathbb{R}$ (see also [207]). However, this implication is true if f is bounded. This will be a consequence of the following result:

Theorem 9.57 *Let R be any Dedekind complete Riesz space, and f_1, f_2, g, $h:[a,b]\to R$ be four maps, (KH)-integrable on $[a,b]$, such that $g\le f_j\le h$ for $j=1,2$.*

Then $f_1\vee f_2$ and $f_1\wedge f_2$ are (KH)-integrable on $[a,b]$
(see also Lemma 4.2, pp. 14-15 of [168], where Theorem 9.57 is proved in the particular case $R=\mathbb{R}$).

Proof: The proof is divided in three steps.

Step 1. In this step, we always suppose that $g:=0$ and we prove (KH)-integrability of $f_1\vee f_2$ on $[a,b]$. For all u,v with $a\le u<v\le b$, define

$$\iota(u,v):=\sup\!\left((KH)\!\int_u^v f_1,(KH)\!\int_u^v f_2\right).$$

Given a division D of $[a,b]$, $D:=\{x_0:=a,x_1,\ldots,x_n:=b\}$, denote by $S^*(D)$ the quantity $\displaystyle\sum_{i=1}^{n}\iota(x_{i-1},x_i)$; moreover set $Y:=\sup_{D\in T}S^*(D)$. It is easy to see that

$$0\le S^*(D)\le(KH)\!\int_a^b h\ \ \text{for all}\ D\in\mathcal{T};$$

so, by virtue of Dedekind completeness of R, it follows that $Y\in R$.

We endow $\mathcal{T}$ with the following order

$$\text{Given}\ D_1,D_2\in\mathcal{T},\ D_1\le D_2\ \text{iff}\ D_1\ \text{is a refinement of}\ D_2. \tag{9.17}$$

As

$$[D_1\le D_2]\Rightarrow[S^*(D_1)\ge S^*(D_2)],$$

then we have

$$Y=(O)\lim_{D\in\mathcal{T}}S^*(D),$$

where the involved limit is intended with respect to the order introduced in (9.17). We now prove that $Y=(KH)\!\int_a^b (f_1\vee f_2)$.

By virtue of the Henstock Lemma 9.56, there exist two (o)-nets $(p_\delta^{(1)})_{\delta\in\Delta}$, $(p_\delta^{(2)})_{\delta\in\Delta}$, such that for every $\delta\in\Delta$ and for each δ-fine decomposition Π of $[a,b]$, $\Pi=\{([x_{i-1},x_i],\xi_i):i=1,\ldots,n\}$, we get

$$\sum_{i=1}^{n}\left|(x_i-x_{i-1})f_j(\xi_i)-(KH)\int_{x_{i-1}}^{x_i}f_j\right|\le p_\delta^{(j)}\quad(j=1,2).$$

For each $\delta\in\Delta$ and for every subinterval $[u,v]\subset[a,b]$, set

$$\mathcal{T}_{[u,v]}:=\{\Pi:\Pi\text{ is a fine decomposition of }[u,v]\}$$

and

$$\kappa_\delta^j([u,v]):=\sup_{\mathcal{T}_{[u,v]}}\sum_{i=1}^{n}\left|(x_i-x_{i-1})f_j(\xi_i)-(KH)\int_{x_{i-1}}^{x_i}f_j\right|\quad(j=1,2).$$

It is easy to check that for all $\delta\in\Delta$ and for every δ-fine decomposition Π of $[a,b]$, $\Pi=\{([x_{i-1},x_i],\xi_i):i=1,\ldots,n\}$, we have

$$(x_i-x_{i-1})(f_1\vee f_2)(\xi_i)$$
$$\le\iota(x_{i-1},x_i)+\kappa_\delta^1([x_{i-1},x_i])+\kappa_\delta^2([x_{i-1},x_i])$$
$$\le\iota(x_{i-1},x_i)+\kappa_\delta^1([a,b])+\kappa_\delta^2([a,b])$$
$$\le p_\delta^{(1)}+p_\delta^{(2)};$$

$$(x_i-x_{i-1})(f_1\vee f_2)(\xi_i)$$
$$\ge\iota(x_{i-1},x_i)-\kappa_\delta^1([x_{i-1},x_i])-\kappa_\delta^2([x_{i-1},x_i])$$
$$\ge\iota(x_{i-1},x_i)-\kappa_\delta^1([a,b])-\kappa_\delta^2([a,b])$$
$$\ge-p_\delta^{(1)}-p_\delta^{(2)};$$

therefore it follows that

$$\left|\sum_{i=1}^{n}(x_i-x_{i-1})(f_1\vee f_2)(\xi_i)-\iota(x_{i-1},x_i)\right|\le2(p_\delta^{(1)}+p_\delta^{(2)}).$$

Let now

$$\Delta^*:=\{\tau\in\Delta:\text{there exists }D\in\mathcal{T}:\qquad\qquad(9.18)$$
$$\text{every special }\tau-\text{fine partition consists in a refinement of }D\}.$$

It is readily seen that Δ^* is a cofinal subset of Δ; so, in order to prove (KH)-integrability of $f_1\vee f_2$, it will be enough to consider (O)-nets of the type $(q_\tau)_{\tau\in\Delta^*}$. Moreover, we observe in particular that for all $D\in T$ there exists $\tau\in\Delta$ such that every special τ-fine partition consists in a refinement of D: indeed, if $D:=\{y_0:=a,y_1,\ldots,y_n:=b\}$, it is enough to take

$$\tau(x)=\begin{cases}\min\{x-y_{i-1},y_i-x\}&\text{if }x\in(y_{i-1},y_i),\\b-a&\text{if }x=y_i.\end{cases}$$

Indeed, if Π is a special τ-fine partition of $[a,b]$ and $[u,v]$ is an interval of Π such that $\exists i:y_{i-1}<u<y_i<v<y_{i+1}$, then we have $v-u\le y_i-u$ or $v-u\le v-y_i$, that is $v\le y_i$ or $u\ge y_i$, which is impossible.

Put $w_\delta := 2(p_\delta^{(1)} + p_\delta^{(2)})$ for all $\delta \in \Delta$: we note that $\inf_{\delta \in \Delta} w_\delta = 0$, and hence $\inf_{\delta \in \Delta^*} w_\delta = 0$. Let now $\tau \in \Delta^*$ and $D = D(\tau)$ be associated with τ as in (9.18). For all τ-fine special partitions $\Pi = \{([x_{i-1}, x_i], \xi_i) : i = 1, \ldots, n\}$ we have:

$$0 \le \left| \sum_{i=1}^n (x_i - x_{i-1})(f_1 \vee f_2)(\xi_i) - Y \right|$$

$$\le \left| \sum_{i=1}^n \{(x_i - x_{i-1})(f_1 \vee f_2)(\xi_i) - \iota(x_{i-1}, x_i)\} \right| \qquad (9.19)$$

$$+(Y - S^*(D))$$

$$\le w_\tau + (Y - S^*(D)).$$

As τ "shrinks", D can be taken "very fine". So, the $(O)\limsup$ as τ "tends" to zero of the last line in (9.19) is zero, and this concludes the first step.

Step 2. Now we prove (KH)-integrability of $f_1 \vee f_2$ without the further assumption $g := 0$. If $g \le f_j \le h$, then

$$0 \le f_j - g \le h - g \ (j = 1, 2). \qquad (9.20)$$

We note that all functions in (9.20) are (KH)-integrable. By Step 1, $(f_1 - g) \vee (f_2 - g)$ is (KH)-integrable. As

$$f_1 \vee f_2 = [(f_1 - g) \vee (f_2 - g)] + g$$

and g is (KH)-integrable, then $f_1 \vee f_2$ is (KH)-integrable too. This concludes Step 2.

Step 3. In order to prove (KH)-integrability of $f_1 \wedge f_2$, it is enough to recall that $f_1 \wedge f_2 = -[(-f_1) \vee (-f_2)]$. Thus the theorem is completely proved. 🍎

A consequence of Theorem 9.57 is the following.

Corollary 9.58 *Let R be an arbitrary Dedekind complete Riesz space, and $f : [a,b] \to R$ be a bounded (KH)-integrable map. Then $|f|$ is (KH)-integrable.*

Proof: Let $L := \sup_{x \in [a,b]} |f(x)|$. We have that both f and $-f$ are (KH)-integrable. By Theorem 9.57 applied with $g = -L$, $h = L$, $f_1 = f$ and $f_2 = -f$, it follows that $|f| := f \vee (-f)$ is (KH)-integrable.

A consequence of the Henstock Lemma is the following theorem (in the case $R = \mathbb{R}$, see [168], Lemma 5.3., pp. 18-19).

Theorem 9.59 *Let R be an arbitrary Dedekind complete Riesz space, $f : [a,b] \to R$ be such that both f and $|f|$ are (KH)-integrable, and suppose that there exists a (u)-differentiable map $P : [a,b] \to R$ such that f is the derivative of P. Then P is of bounded variation on $[a,b]$ and*

$$(KH)\int_a^b |f| = V(P, [a,b]).$$

Proof: First of all we prove that

$$(KH)\int_a^b |f| \ge V(P, [a,b]). \qquad (9.21)$$

For each division D of $[a,b]$, $D = \{x_0 := a, x_1, \ldots, x_{n-1}, x_n := b\}$, we have

$$\sum_{i=1}^{n} |P(x_i) - P(x_{i-1})| = \sum_{i=1}^{n} \left| (KH)\int_{x_{i-1}}^{x_i} f \right|$$

$$\leq \sum_{i=1}^{n} (KH)\int_{x_{i-1}}^{x_i} |f| = (KH)\int_{a}^{b} |f|.$$

Taking the suprema, we get (9.21). We now prove that

$$(KH)\int_{a}^{b} |f| \leq V(P,[a,b]). \qquad (9.22)$$

Let Δ_0 and $(p_\delta)_{\delta \in \Delta_0}$ be a cofinal set and an (O)-net respectively, associated with integrability of f and $|f|$. For each $\delta \in \Delta_0$ and for every δ-fine partition $\Pi = \{([x_{i-1}, x_i], \xi_i) : i = 1, \ldots, n\}$, we have

$$(KH)\int_{a}^{b} |f| = (KH)\int_{a}^{b} |f| - \sum_{i=1}^{n} |(x_i - x_{i-1}) f(\xi_i)|$$

$$+ \sum_{i=1}^{n} |(x_i - x_{i-1}) f(\xi_i)|.$$

Moreover we get

$$|(x_i - x_{i-1}) f(\xi_i)| \leq |(x_i - x_{i-1}) f(\xi_i) - [P(x_i) - P(x_{i-1})]|$$

$$+ |P(x_i) - P(x_{i-1})|$$

for all $i = 1, \ldots, n$. Taking the sums, we have

$$0 \leq \sum_{i=1}^{n} |(x_i - x_{i-1}) f(\xi_i)|$$

$$\leq \sum_{i=1}^{n} |(x_i - x_{i-1}) f(\xi_i) - [P(x_i) - P(x_{i-1})]| + \sum_{i=1}^{n} |P(x_i) - P(x_{i-1})|$$

$$\leq \sum_{i=1}^{n} |(x_i - x_{i-1}) f(\xi_i) - [P(x_i) - P(x_{i-1})]| + V(P,[a,b]).$$

So we obtain

$$(KH)\int_{a}^{b} |f| \leq \left| (KH)\int_{a}^{b} |f| - \sum_{i=1}^{n} |(x_i - x_{i-1}) f(\xi_i)| \right|$$

$$+ \sum_{i=1}^{n} |(x_i - x_{i-1}) f(\xi_i) - [P(x_i) - P(x_{i-1})]| + V(P,[a,b])$$

$$\leq 3 p_\delta + V(P,[a,b])$$

by virtue of Lemma 9.56. Taking the $(O)\limsup$, we get (9.22). 🍎

Definition 9.60 Given a (KH)-integrable function $f : [a,b] \to R$, set

$$F(x) = \begin{cases} (KH)\int_{a}^{x} f & \text{if } a < x \leq b, \\ 0 & \text{if } x = a. \end{cases}$$

The map F will be called the (KH)-*integral function* associated with f.

The following result holds:

Proposition 9.61   *The map F introduced in Definition (9.60) is continuous. Moreover, if $f:[a,b] \to R$ is bounded, then F is Lipschitzian. If $f:[a,b] \to R$ is continuous at $x_0 \in [a,b]$ [(u)-continuous], then F is differentiable at x_0 [(u)-differentiable] and $F'(x_0) = f(x_0)$.*

Remark 9.62   We note that there exist Riesz spaces R and *Riemann* integrable functions $u:[a,b] \to R$, which are discontinuous at every $x \in (a,b)$.

Indeed, let $[a,b] := [0,1], R := \mathbb{R}^{[0,1]}, u(s) := \chi_{[0,s]}, \forall s \in [0,1]$. For each $x \in (0,1)$, we have:
$$\lim_{t \to x^+} u(t) = \chi_{[0,x]}, \quad \lim_{t \to x^-} u(t) = \chi_{[0,x)},$$
and hence
$$\limsup_{t \to x} u(t) - \liminf_{t \to x} u(t) = \chi_{\{x\}} \nleq \frac{1}{2}.$$
However, u is Riemann integrable, and *a fortiori* (KH)-integrable.

Put $I(s) := (Ri)\int_0^s u, s \in [0,1]$. It is easy to check that
$$I(s)(x) = \begin{cases} 0 & \text{if } x \geq s, \\ s-x & \text{if } x < s, \end{cases}$$
$\forall s, x \in [0,1]$, and that I is not differentiable at *any* point $s \in (0,1)$. So, in general, it is not true that the function F defined in 9.60 is almost everywhere differentiable. As I is Lipschitzian, this example shows also that there exist some Lipschitzian functions $F:[a,b] \to \mathbb{R}^{[0,1]}$ which are not differentiable at any point of (a,b).

We now give some versions of the Fundamental Theorem of the Calculus for Riesz space-valued functions. First of all, we introduce the concept of (SL)-integral in Riesz spaces, by means of which it is also possible to avoid some pathologies (typical of Riesz spaces without property σ), like the one in (9.12).

Definition 9.63   Let $J \subset [a,b] \subset \mathbb{R}$, and $\Delta_J = (\mathbb{R}^+)^J$. We say that a function $P:[a,b] \to R$ is *of class (SL)* or has *property (SL)* on J if for every set $N \subset [a,b]$ of Lebesgue measure zero there exists an (O)-net $(p_\delta^{(N)})_{\delta \in \Delta_J}$, such that $\forall \delta \in \Delta_J$

$$\sup\{\sum_{i=1}^n |P(x_i) - P(x_{i-1})| : \{([x_{i-1}, x_i], \xi_i) : i = 1, \dots, n\} \tag{9.23}$$

is a δ-fine decomposition of $[a,b], x_i \in [a,b]$ and $\xi_i \in N \cap J \; \forall i = 1, \dots, n, \} \leq p_\delta^{(N)}$.

We say that P is *of class (SL)* if it is of class (SL) on $[a,b]$.
We now give some examples of functions of class (SL).

Proposition 9.64   *Let $P:[a,b] \to R$ be absolutely continuous in $[a,b]$. Then, P is of class (SL) on $[a,b]$.*

Proof: Choose arbitrarily a set N of Lebesgue measure zero. For every $n \in \mathbb{N}$ there exists a countable family $\mathcal{F}$ of intervals of the type $[\alpha_j^{(n)}, \beta_j^{(n)}], j \in \mathbb{N}$, which covers N, such that every point of N is interior to at least one of the intervals of $\mathcal{F}$ and with

$$\sum_{j=1}^\infty (\beta_j^{(n)} - \alpha_j^{(n)}) \leq 1/n.$$

From this it follows that

$$\sum_{j=1}^{\infty} |P(\beta_j^{(n)}) - P(\alpha_j^{(n)})| \leq p_n,$$

where $(p_n)_n$ is a suitable (O)-sequence. Thus the assertion follows.

Proposition 9.65   *Let R be an arbitrary Dedekind complete Riesz space. If $P:[a,b] \to R$ is continuous at a point $x_0 \in [a,b]$, then P is of class (SL) on $\{x_0\}$.*

Proof: As P is continuous in x_0, we get the existence of an (O)-net $(p_\delta^{(x_0)})_{\delta \in \Delta}$ in R, such that for all $\delta \in \Delta$ we have

$$|P(v) - P(u)| \leq p_\delta^{(x_0)}$$

whenever $x_0 - \delta(x_0) \leq u \leq x_0 \leq v \leq x_0 + \delta(x_0)$. Let now $N \subset [a,b]$ be a set of Lebesgue measure zero. Then, for all $\gamma \in \Delta_{\{x_0\}}$ and for every δ-fine decomposition $\Pi = \{([x_{i-1}, x_i], \xi_i) : i = 1, ..., n\}$ of $[a,b]$, we get

$$\sum_{\xi_i \in N \cap \{x_0\}} |P(x_i) - P(x_{i-1})| = 0$$

if $x_0 \notin N$;

$$\sum_{\xi_i \in N \cap \{x_0\}} |P(x_i) - P(x_{i-1})| \leq |P(x_{\bar{i}}) - P(x_{\bar{i}-1})|,$$

where $\bar{i}$ is such that $\xi_{\bar{i}} = x_0$, if $x_0 \in N$. Thus the assertion follows. ✦

Proposition 9.66   *Let R be a Dedekind complete Riesz space, $P:[a,b] \to R$ be a map, $r \in R$ be a unit, and assume that*

$$\varnothing \neq J := \{x \in [a,b] : P'(x) \exists \text{ in } R \text{ and } |P'(x)| \leq r\},$$

where P' is the (u)-derivative of P. Then P is of class (SL) on J.

Proof: Let $(p_\delta)_{\delta \in (\mathbb{R}^+)^J}$ be an (O)-net, related with (u)-differentiability of P on J. Fix arbitrarily a set N of Lebesgue measure zero, let $\delta \in (\mathbb{R}^+)^J$ and choose a δ-fine decomposition $\Pi = \{([x_{i-1}, x_i], \xi_i) : i = 1, ..., n\}$. For all $i = 1, ..., n$, we get

$$|P(x_i) - P(x_{i-1})| - (x_i - x_{i-1}) |P'(\xi_i)|$$
$$\leq |P(x_i) - P(x_{i-1}) - (x_i - x_{i-1}) P'(\xi_i)| \leq (x_i - x_{i-1}) p_\delta,$$

and hence

$$|P(x_i) - P(x_{i-1})| \leq (x_i - x_{i-1}) |P'(\xi_i)| + (x_i - x_{i-1}) p_\delta$$
$$\leq (x_i - x_{i-1}) r + (x_i - x_{i-1}) p_\delta.$$

From this we get:

$$\sum_{\xi_i \in N} |P(x_i) - P(x_{i-1})| \leq \left(\sum_{\xi_i \in N} (x_i - x_{i-1}) \right) r + (b-a) p_\delta.$$

The assertion follows from absolute continuity of the "identity function". ✦

Definition 9.67 Let R be any Dedekind complete Riesz space, and $\Delta^{(1)}$ be the set of all gages, defined on $[a,b]$. We say that $f:[a,b] \to R$ is *(SL)-integrable* on $[a,b]$ if there exist a function $P:[a,b] \to R$ of class (SL) and an (O)-net $(p_\delta)_{\delta \in \Delta^{(1)}}$ such that $\forall \delta \in \Delta^{(1)}$ we have:

$$\sup \left\{ \left| \sum_\Pi f - \sum_{i=1}^n [P(x_i) - P(x_{i-1})] \right| : \right.$$

$$\left. \Pi = \{([x_{i-1}, x_i], \xi_i) : i = 1, \ldots, n\} \text{ is a } \delta - \text{fine decomposition of } [a,b] \right\} \le p_\delta.$$

The function P will be called *weak primitive* of f.

In this case we put (by *definition*) $(SL)\int_a^b f = P(b) - P(a)$.

We now prove that the (SL)-integral is well-defined. We begin with the following:

Proposition 9.68 *Let R be any Dedekind complete Riesz space, $N \subset [a,b]$ be a set of Lebesgue measure zero, $f_0 : [a,b] \to R$ be such that $f_0(x) = 0$ for all $x \notin N$ and $P_0 : [a,b] \to R$ be a constant function. Then P_0 is a weak primitive of f_0.*

Proof: Set $\Delta_N := \{\delta \in (\mathbb{R}_0^+)^{[a,b]} :$ there exist a gage δ_1 such that

$$\delta(x) = \begin{cases} \delta_1(x) & \text{if } x \in [a,b] \setminus N, \\ 0 & \text{if } x \in N.\} \end{cases}$$

It is clear that Δ_N is a cofinal set with respect to the set of all gages. Pick $\delta \in \Delta_N$. By definition of δ-fine decomposition and (SL)-integral, the decompositions of the type $\{([x_{i-1}, x_i], \xi_i) : i = 1, \ldots, n\}$, with $\xi_i \in N$ for at least one i, are not involved in the calculus of the "Riemann-type sums" relatively to f_0. From this we get the assertion. ✎

Proposition 9.69 *Let N and f_0 be as in Proposition 9.68, and P be a weak primitive of f_0. Then P is constant.*

Proof: Let P be a weak primitive of f_0 and δ be a gage, and denote by Z_δ the set of all zeros of δ. Denote by $(q_\delta)_\delta$ an (O)-net, related to Definition 9.67. Set $\Delta^* := \{\tau \in \Delta :$ there exist $\delta' \in \Delta$ and a gage δ such that

$$\tau(x) = \begin{cases} \delta(x) & \text{if } x \notin Z_\delta \cup N, \\ \delta'(x) & \text{if } x \in Z_\delta \cup N\} : \end{cases}$$

then Δ^* is a cofinal set with respect to Δ.

Choose now arbitrarily $[\alpha, \beta] \subset [a,b]$ and $\tau \in \Delta^*$, and pick any τ-fine partition $\Pi = \{([x_{i-1}, x_i], \xi_i) : i = 1, \ldots, n\}$ of $[\alpha, \beta]$. The decomposition $F_\delta = \{([x_{i-1}, x_i], \xi_i) \in E : \xi_i \notin Z_\delta\}$ is δ-fine. For each gage δ, let $(p_{\delta, \delta'})_{\delta' \in \Delta}$ be an (O)-net in connection with the property (SL) of P. We get:

$$0 \le |P(\beta) - P(\alpha)| = \left| \sum_{i=1}^n [P(x_i) - P(x_{i-1})] \right|$$

$$\le \left| \sum_{\xi_i \in Z_\delta} [P(x_i) - P(x_{i-1})] \right| + \left| \sum_{\xi_i \notin Z_\delta} [P(x_i) - P(x_{i-1})] \right| \qquad (9.24)$$

$$\leq p_{\delta,\delta'} + q_\delta.$$

As

$$(O)\limsup_{\delta}\, [(O)\limsup_{\delta'}\, (p_{\delta,\delta'} + q_\delta)] = \inf_{\delta}\, [\inf_{\delta'}\, (p_{\delta,\delta'} + q_\delta)] = 0,$$

then it follows that $P(\beta) = P(\alpha)$. By arbitrariness of α and β, we get that P is constant. 🍎

Some immediate consequences of Propositions 9.68 and 9.69 are the following:

Proposition 9.70 *Let R be any Dedekind complete Riesz space, $N \subset [a,b]$ be a set of Lebesgue measure zero, and $f_0 : [a,b] \to R$ such that $f_0(x) = 0$ for all $x \notin N$. Then*

$$(SL)\int_a^b f_0 = 0.$$

Proposition 9.71 *Let R be as in Proposition 9.70, and $f, g : [a,b] \to R$ two functions, which differ only on a set of Lebesgue measure zero. Then f is (SL)-integrable on $[a,b]$ if and only if g does, and in this case*

$$(SL)\int_a^b f = (SL)\int_a^b g.$$

Proposition 9.72 *Let $f : [a,b] \to R$ be (SL)-integrable on $[a,b]$, and P_1, P_2 be two weak primitives of f. Then*

$$P_1(\beta) - P_1(\alpha) = P_2(\beta) - P_2(\alpha)$$

for all α, β with $[\alpha, \beta] \subset [a,b]$.

This shows that the (SL)-integral is well-defined:

Theorem 9.73 *Let $f, P : [a,b] \to R$ be such that P is of class (SL) on $[a,b]$. Suppose that there exists a set $N \subset [a,b]$ of Lebesgue measure zero, such that $f|_{[a,b]\setminus N}$ is the (u)-derivative of $P|_{[a,b]\setminus N}$. Then f is (SL)-integrable on $[a,b]$, and $(SL)\int_a^b f = P(b) - P(a)$.*

Proof: Let $\Delta_N := \mathbb{R}^{+([a,b]\setminus N)}$. As f is the (u)-derivative of P in $[a,b] \setminus N$, then we get the existence of an (O)-net $(p_\sigma)_{\sigma \in \Delta_N}$, such that, if $\Pi = \{([x_{i-1}, x_i], \xi_i) : i = 1, \ldots, n\}$ is a σ-fine decomposition (with $\xi_i \notin N$ for all i), then

$$|(x_i - x_{i-1})f(\xi_i) - [P(x_i) - P(x_{i-1})]| \leq p_\sigma (x_i - x_{i-1})$$

for all $i = 1, \ldots, n$. Define now $f_1 : [a,b] \to R$ by setting

$$f_1(x) = \begin{cases} |f(x)| & \text{if } x \in N, \\ 0 & \text{if } x \notin N. \end{cases}$$

As f_1 is zero almost everywhere, then f_1 is (SL)-integrable on $[a,b]$; let $(p'_\delta)_{\delta \in \Delta^{(1)}}$ be an (O)-net, associated with (SL)-integrability of f_1.

Set $\Delta_* := \{\delta_* \in (\mathbb{R}_0^+)^{[a,b]}$: there exist a gage δ, two functions $\theta \in \Delta$ and $\sigma \in \Delta_N$ such that

$$\delta_*(x) = \begin{cases} \min\{\delta(x), \sigma(x)\} & \text{if } x \in [a,b] \setminus N, \\ \min\{\delta(x), \theta(x)\} & \text{if } x \in N\}. \end{cases}$$

Then Δ_* is a cofinal set with respect to the set $\Delta^{(1)}$ of all gages. Pick $\delta_* \in \Delta_*$ and $n \in \mathbb{N}$. If $\Pi = \{([x_{i-1}, x_i], \xi_i) : i = 1, \dots, n\}$ is a δ_*-fine decomposition of $[a,b]$, then we have

$$0 \le \left| \sum_{i=1}^{n} \{(x_i - x_{i-1}) f(\xi_i) - [P(x_i) - P(x_{i-1})]\} \right|$$

$$\le \sum_{i=1}^{n} \left| (x_i - x_{i-1}) f(\xi_i) - [P(x_i) - P(x_{i-1})] \right|$$

$$= \sum_{\xi_i \notin N} \left| (x_i - x_{i-1}) f(\xi_i) - [P(x_i) - P(x_{i-1})] \right|$$

$$+ \sum_{\xi_i \in N} \left| (x_i - x_{i-1}) f(\xi_i) - [P(x_i) - P(x_{i-1})] \right| \qquad (9.25)$$

$$\le (b-a) p_\sigma + \sum_{\xi_i \in N} (x_i - x_{i-1}) |f(\xi_i)| + \sum_{\xi_i \in N} |P(x_i) - P(x_{i-1})|$$

$$\le (b-a) p_\sigma + \sum_{\xi_i \in N} (x_i - x_{i-1}) f_1(\xi_i) + v_\theta$$

$$\le (b-a) p_\sigma + p'_\delta + v_\theta,$$

where $(v_\theta)_{\theta \in \Delta}$ is a suitable (O)-net, existing because P satisfies property (SL). The quantity in the last line of (27) tends to zero as δ_* decreases to zero. Thus the theorem is proved.

It is easy to check that the (SL)-integral is a monotone linear functional. We now give a first result of comparison between the (KH)- and the (SL)-integral.

Theorem 9.74   *Let R be an arbitrary Dedekind complete Riesz space, and $f : [a,b] \to R$ be bounded. Then the following are equivalent:*

9.74.1) f is (KH)-integrable on $[a,b]$;
9.74.2) f is (SL)-integrable on $[a,b]$.

In this case we have

$$(KH) \int_a^b f = (SL) \int_a^b f.$$

Proof: $[9.74.1)] \Rightarrow [9.74.2)]$

Suppose that f is (KH)-integrable, and let F be the integral function associated with f, defined in 9.60. As f is bounded, then F is Lipschitzian and hence absolutely continuous in $[a,b]$, and so F is of class (SL) on $[a,b]$. For each $\delta \in \Delta^{(1)}$, let

$$E_\delta := \{\Pi = \{([x_{i-1}, x_i], \xi_i) : i = 1, \dots, n\} :$$
$$\Pi \text{ is a } \delta - \text{fine decomposition of } [a,b]\}.$$

By virtue of the Henstock Lemma, we have

$$0 = (O) \lim_{\delta \in \Delta^{(1)}} \left[\sup \left\{ \left| \sum_{\Pi} f - \sum_{i=1}^{n} [F(x_i) - F(x_{i-1})] \right| : E \in E_\delta \right\} \right].$$

This concludes the proof of the considered implication.

$[9.74.2)] \Rightarrow [9.74.1)]$

Let f be bounded and (SL)-integrable, P be a weak primitive of f, and $(r_\delta)_{\delta \in \Delta^{(1)}}$ be an (O)-net, related with (SL)-integrability of f. For each $\delta \in \Delta^{(1)}$, let Z_δ be the set of its zeros. Let $f_1 := f \chi_{Z_\delta}$: as f_1 is bounded and zero almost everywhere, then f_1 is (KH)-integrable. Thus there exists an (O)-net $(q_{\delta''})_{\delta'' \in \Delta}$ such that

$$\sup \left\{ \left| \sum_{\Pi} f_1 \right| : \Pi = \{([x_{i-1}, x_i], \xi_i) : i = 1, \ldots, n\} \text{ is a} \right.$$

δ''-fine partition of $[a,b]\} \le q_{\delta''}$.

For all $\delta \in \Delta^{(1)}$, in correspondence with Z_δ there exists an (O)-net $(p_{\delta,\delta'})_{\delta' \in \Delta}$, related to property (SL) of P. Let now $\Delta_* := \{\tau \in \Delta : \text{there exist } \delta \in \Delta^{(1)}, \delta', \delta'' \in \Delta \text{ such that}$

$$\tau(x) = \begin{cases} \min\{\delta(x), \delta''(x)\} & \text{if } x \notin Z_\delta, \\ \delta'(x) & \text{if } x \in Z_\delta.\} \end{cases}$$

It is easy to check that Δ_* is a cofinal set with respect to Δ. Let Π be any τ-fine partition of $[a,b]$, $\Pi = \{([x_{i-1}, x_i], \xi_i) : i = 1, \ldots, n\}$. The decomposition $F_\delta = \{([x_{i-1}, x_i], \xi_i) \in \Pi : \xi_i \notin Z_\delta\}$ is δ-fine, and

$$0 \le \left| \sum_{i=1}^{n} (x_i - x_{i-1}) f(\xi_i) - [P(b) - P(a)] \right|$$
$$= \left| \sum_{i=1}^{n} (x_i - x_{i-1}) \{f(\xi_i) - [P(x_i) - P(x_{i-1})]\} \right|$$
$$\le \left| \sum_{i=1}^{n} (x_i - x_{i-1}) f_1(\xi_i) \right| + \left| \sum_{\xi_i \in Z_\delta} [P(x_i) - P(x_{i-1})] \right| \qquad (9.26)$$
$$+ \left| \sum_{\xi_i \notin Z_\delta} \{(x_i - x_{i-1}) f(\xi_i) - [P(x_i) - P(x_{i-1})]\} \right|$$
$$\le q_{\delta''} + p_{\delta,\delta'} + r_\delta,$$

where the quantity in the last line of (9.26) tends to zero as τ decreases. Therefore the assertion follows. ◉

9.2 Convergence theorems

We begin with the following convergence theorem for the real case, whose proof we will give in a (more general) Riesz space version in Theorem 9.76 (see also [169], Theorem 2, pp. 766-767).

Theorem 9.75 *Assume that* $f:[a,b] \to R$ *satisfies the following properties:*

9.75.1) f *is bounded and* (SL)*-integrable on* $[a,c]$ *for every* $c \in (a,b)$;

9.75.2) $\displaystyle \lim_{c \to b^-}\left[(SL)\int_a^c f\right] = l \in R$.

Then f *is* (SL)*-integrable on* $[a,b]$ *and*

$$(SL)\int_a^b f = l.$$

We now turn to a convergence theorem in the context of Riesz spaces.

Let $R = L^0(X, \mathcal{B}, \mu)$, where $(X, \mathcal{B}, \mu)$ is a measure space, with μ positive, σ-finite, σ-additive and $\tilde{\mathbb{R}}$-valued. From now on, we will write simply $R = L^0(X, \mathcal{B}, \mu)$.

In the following convergence theorem, we prove that, under these assumptions, in a certain sense, the (SL)-integral contains the improper integral (on $[a,b]$).

Theorem 9.76 *Let* $f:[a,b] \to R$ *satisfy the following properties:*

9.76.1) f *is bounded and* (SL)*-integrable on* $[a,c]$ *for every* $c \in (a,b)$;

9.76.2) $(O)\displaystyle\lim_{c \to b^-}\left[(SL)\int_a^c f\right] = l \in R$.

Then f *is* (SL)*-integrable on* $[a,b]$ *and*

$$(SL)\int_a^b f = l.$$

Proof: Let us define $P:[a,b] \to R$ as follows:

$$P(x) = (SL)\int_a^x f \text{ if } x \in [a,b); \quad P(b) = l.$$

Let $(y_n)_{n \in \mathbb{N} \cup \{0\}}$ be a strictly increasing sequence such that $y_0 = a$ and $\lim_{n \to +\infty} y_n = b$. Since f is bounded in $[a,x]$ for each $x \in [a,b)$, then P is absolutely continuous in $[y_n, y_{n+1}]$ for every $n \in \mathbb{N} \cup \{0\}$, and hence P is there of class (SL). From this it follows that f is of class (SL) on $[a,b)$. Moreover, by 9.76.2), P is continuous at b, and hence P is of class (SL) on $[a,b]$. Since f is (SL)-integrable on *every* interval of the type $[y_n, y_{n+1}]$ and, in our space R, (O)-convergence coincides with (r)-convergence, then, for each $n \in \mathbb{N}$, there exists a unit w_n such that, for all $\varepsilon > 0$, there exists a gage $\delta_n^* : [a,b] \to \mathbb{R}_0^+$ such that, for every δ_n^*-fine decomposition $\Pi_n = \{([x_{i-1}, x_i], \xi_i) : i = 1, 2, \ldots, k_n\}$ of $[y_n, y_{n+1}]$, we have:

$$\left|\sum_{i=1}^{k_n} (x_i - x_{i-1})f(\xi_i) - [P(x_i) - P(x_{i-1})]\right| \le \varepsilon\, w_n. \tag{9.27}$$

By virtue of property σ, in correspondence with the sequence $(w_n)_n$ there exist a sequence of positive real numbers $(\lambda_n)_n$ and a unit w, such that

$$\lambda_n w_n \le w \quad \forall n \in \mathbb{N}.$$

By applying the previous argument as in (9.27) when we replace ε with $\dfrac{\varepsilon \lambda_n}{2^n}$, we get that, for each $n \in \mathbb{N}$, there exists a unit w_n such that, for all $\varepsilon > 0$, there exists a gage $\delta_n : [a,b] \to \mathbb{R}_0^+$ such that, for every δ_n-fine decomposition $\Pi_n = \{([x_{i-1}, x_i], \xi_i) : i = 1, 2, \ldots, k_n\}$ of $[y_n, y_{n+1}]$, we have

$$\left|\sum_{i=1}^{k_n}(x_i-x_{i-1})f(\xi_i)-[P(x_i)-P(x_{i-1})]\right|\leq\frac{\varepsilon\,\lambda_n}{2^n}w_n\leq\frac{\varepsilon}{2^n}w.\qquad(9.28)$$

Let now

$$\delta(x)=\begin{cases}0 & \text{if } x\in\{y_n:n\in\mathbb{N}\cup\{0\}\},\\ \min\{x-y_n,y_{n+1}-x,\delta_n(x)\} & \text{if } x\in(y_n,y_{n+1}).\end{cases}$$

It is easy to check that $\delta:[a,b]\to\mathbb{R}_0^+$ is a gage. We note that every δ-fine decomposition must be such that its intervals are contained in some interval of the type $[y_n,y_{n+1}]$. Indeed, if there would exist an element $([u,v],\xi)$ of such a decomposition (where $[u,v]$ is a subinterval of $[a,b]$ and $\xi\in[u,v]$) such that $\exists n\in\mathbb{N}$ with $u<y_n<v$, then $v-u\leq\delta(\xi)$. Since $\delta(y_n)=0$, then ξ cannot coincide with y_n. If $u\leq\xi<y_n$, then

$$v-u\leq\delta(\xi)\leq y_n-\xi\leq y_n-u,$$

and hence $v\leq y_n$, a contradiction. Moreover, if $y_n<\xi\leq v$, then

$$v-u\leq\delta(\xi)\leq\xi-y_n\leq v-y_n,$$

and hence $u\geq y_n$, a contradiction.

Let now Π be any δ-fine decomposition of $[a,b]$: by virtue of what previously seen, Π "induces" on every interval $[y_n,y_{n+1}]$ a decomposition Π_n of $[y_n,y_{n+1}]$: thus

$$\Pi=\bigcup_{n\in T^*}\Pi_n,\qquad(9.29)$$

where T^* is a suitable subset of $\mathbb{N}$. We observe that T^* is a *finite* set: indeed the involved decompositions are formed by a finite number of intervals and $\lim_{n\to+\infty}y_n=b$. Moreover, by δ-fineness, the endpoint of the last interval of the involved decomposition must be strictly smaller than y_{n_0} for some positive integer n_0: in particular, our decomposition Π cannot have *any* interval of the type $[t,b]$, and hence no δ-fine decomposition can be a partition. Finally we have:

$$0\leq\left|\sum_{i\in T^*}(x_i-x_{i-1})f(\xi_i)-[P(x_i)-P(x_{i-1})]\right|$$

$$\leq\sum_{\Pi_n}\left|\sum_{i=1}^{k_n}(x_i-x_{i-1})f(\xi_i)-[P(x_i)-P(x_{i-1})]\right|\qquad(9.30)$$

$$\leq\sum_{n=1}^{\infty}\frac{\varepsilon\,\lambda_n}{2^n}w_n\leq\varepsilon\,w.$$

The assertion follows from (9.30). This concludes the proof. ◼

We now formulate some convergence theorems in the case of Riesz spaces R of the type $R=L^0(X,\mathcal{B},\mu)$: these results are generalizations of corresponding theorems proved for real-valued functions by P. Y. Lee and R. Výborný (see [169] and [170], pp. 162-169). We begin with the following preliminary definitions.

Definition 9.77   Let $J \subset [a,b] \subset \mathbb{R}$, and $\Delta_J = (\mathbb{R}^+)^J$. We say that a sequence of functions $(P_n : [a,b] \to R)_n$ is *of class (USL)* or has *property (USL)* (uniform *(SL)*) on J if for every set $N \subset [a,b]$ of Lebesgue measure zero there exists a unit w such that, $\forall \varepsilon > 0$, there exist $n_0 = n_0(\varepsilon) \in \mathbb{N}$ and a map $\delta : J \to \mathbb{R}^+$ such that, for every $n \geq n_0$ and for any δ-fine decomposition $\Pi = \{([x_{i-1}, x_i], \xi_i) : i = 1, \ldots, p\}$ of $[a,b]$ with $x_i \in [a,b]$ and $\xi_i \in N \cap J$ $\forall i = 1, \ldots, p$, we have

$$\sum_{i=1}^{p} |P_n(x_i) - P_n(x_{i-1})| \leq \varepsilon \, w. \qquad (9.31)$$

We say that $(P_n)_n$ is *of class (USL)* [has *property (USL)*] if it is of class *(USL)* on $[a,b]$.

Definition 9.78   Let $\Delta^{(1)}$ be the set of all gages, defined on $[a,b]$. We say that the sequence $(f_n : [a,b] \to R)_n$ is *(SL)-equiintegrable* (or simply *equiintegrable*) on $[a,b]$ if there exist a sequence of functions $(P_n : [a,b] \to R)_n$ of class *(USL)* and a unit w such that, for every $\varepsilon > 0$, there exist $\delta \in \Delta^{(1)}$ and $n_0 \in \mathbb{N}$, such that

$$\left| \sum_{\Pi} f_n - \sum_{i=1}^{p} [P_n(x_i) - P_n(x_{i-1})] \right| \leq \varepsilon \, w \qquad (9.32)$$

for every $n \geq n_0$ and whenever $\Pi = \{([x_{i-1}, x_i], \xi_i) : i = 1, \ldots, p\}$ is a δ-fine decomposition of $[a,b]$.

We now prove the following:

Theorem 9.79   *Let R be any Dedekind complete Riesz space, $(f_n : [a,b] \to R)_n$ be (SL)-equiintegrable, F_n be a weak primitive of f_n on $[a,b]$ $\forall n \in \mathbb{N}$, such that the sequence $(F_n)_n$ is of class (USL) on $[a,b]$, and assume that there exists a function $f : [a,b] \to R$, such that*

$$(r) \lim_n f_n(x) = f(x)$$

for almost all $x \in [a,b]$ and with respect to the same unit y. Then there exists a map $F : [a,b] \to R$, such that

$$(r) \lim_n [F_n(x) - F_n(a)] = F(x) \quad \forall x \in [a,b]$$

with respect to a same unit, f is (SL)-integrable on $[a,b]$ and F is a weak primitive of f on $[a,b]$.

Proof: (see also [169], Theorem 3, pp. 767-768 and [170], Theorem 4.3.3, pp. 163-165) Let N be that set (of Lebesgue measure zero) where the sequence $(f_n)_n$ does not converge to f, $\varepsilon > 0$, $z \in R$ be a unit and δ be a gage chosen according to *(SL)*-equiintegrability of the sequence $(f_n)_n$. Let Z_δ be the set of all zeros of δ, $\gamma : [a,b] \to \mathbb{R}^+$ be associated with ε and $N \cup Z_\delta$, according to property *(USL)* of the sequence $(F_n)_n$. Put

$$\delta^*(x) = \begin{cases} \delta(x) & \text{if } x \notin N \cup Z_\delta, \\ \gamma(x) & \text{if } x \in N \cup Z_\delta. \end{cases}$$

Fix $x \in (a,b]$, and let $\Pi = \{([u_i, v_i], x_i) : i = 1, \ldots, p\}$ be a δ^*-fine partition of $[a,x]$. Let Π_0 be that part of Π for which $x_i \in N \cup Z_\delta$ $\forall i = 1, \ldots, p$, and $\Pi_1 = \Pi \setminus \Pi_0$. For every $m, n \in \mathbb{N}$ we have:

$$0 \leq |[F_m(x) - F_m(a)] - [F_n(x) - F_n(a)]|$$

$$\leq \left| \sum_{\Pi} \{[F_m(v_i) - F_m(u_i)] - [F_n(v_i) - F_n(u_i)]\} \right|$$

$$\leq \sum_{\Pi_0} [F_m(v_i) - F_m(u_i)] + \sum_{\Pi_0} [F_n(v_i) - F_n(u_i)]$$

$$\leq \sum_{\Pi_0} [F_m(v_i) - F_m(u_i)] + \sum_{\Pi_0} [F_n(v_i) - F_n(u_i)]$$

$$+ \left| \sum_{\Pi_1} \{[F_m(v_i) - F_m(u_i)] - (v_i - u_i) f_m(x_i)\} \right|$$

$$+ \left| \sum_{\Pi_1} \{[F_n(v_i) - F_n(u_i)] - (v_i - u_i) f_n(x_i)\} \right|$$

$$+ \sum_{\Pi_1} (v_i - u_i) |f_m(x_i) - f_n(x_i)|. \qquad (9.33)$$

For m, n larger than a suitable positive integer n^* depending only on ε, the first two terms in the last member of inequality (9.33) are less than εw, where w is a suitable unit, existing by property (USL), and the third and the fourth term are less than εz, where z is as above. We now estimate the last term.

By hypothesis, the unit y is such that, for every $x \in [a,b] \setminus N$ and for each $\varepsilon > 0$, there is $n_0 \in \mathbb{N}$ such that, $\forall n, m \geq n_0$,

$$|f_m(x) - f_n(x)| \leq \varepsilon y.$$

From this it follows that the last term in the inequality (9.33) is less than $\varepsilon(b-a)y$ for n large enough (depending on the involved partition Π). From this it follows that the sequence $(F_n(x) - F_n(a))_n$ is Cauchy, and hence there exists in R the limit

$$F(x) := (r) \lim_n [F_n(x) - F_n(a)],$$

for every $x \in (a,b]$ (relatively to the same unit as the one with respect to which the involved sequence is Cauchy), because every Dedekind complete Riesz space is complete with respect to (r)-convergence (see also [175]). Moreover, set $F(a) = 0$. We have that F is of class (SL), since the sequence $(F_n)_n$ is of class (USL). Now change δ, if necessary, by defining it equal to zero on the set N. So, $\forall n \geq n^*$ (where n^* is as above) and for every δ-fine decomposition $\Pi = \{([s_i, t_i], \xi_i) : i = 1, \ldots, q\}$ of $[a,b]$, we obtain:

$$\left| \sum_{i=1}^{q} \{[F_n(t_i) - F_n(s_i)] - (t_i - s_i) f_n(\xi_i)\} \right| \leq \varepsilon z, \qquad (9.34)$$

where z is as above. Taking in (9.34) the (r)-limit as n tends to $+\infty$, we get:

$$\left| \sum_{i=1}^{q} \{[F(t_i) - F(s_i)] - (t_i - s_i) f(\xi_i)\} \right| \leq \varepsilon z$$

whenever $\Pi = \{([s_i, t_i], \xi_i) : i = 1, \ldots, q\}$ is a δ-fine decomposition of $[a,b]$. Thus we get (SL)-integrability of f on $[a,b]$ with weak primitive F. This concludes the proof. 🍎

We now turn to the following version of the monotone convergence theorem.

Theorem 9.80 *Let* $R = L^0(X, \mathcal{B}, \mu)$, *let* $f_n : [a, b] \to R$, $n \in \mathbb{N}$, *be* (SL)-*integrable on* $[a, b]$, *and suppose that the sequence* $\left((SL) \int_a^b f_n \right)_n$ *is bounded in* R. *Moreover, assume that*

$$f_n(x) \le f_{n+1}(x) [\text{or } f_n(x) \ge f_{n+1}(x)] \quad \forall n \in \mathbb{N}$$

for almost all $x \in [a, b]$, *and that there exists a function* $f : [a, b] \to R$ *such that*

$$(r) \lim_n f_n(x) = f(x)$$

for almost all $x \in [a, b]$ *and with respect to the same unit* z.
Then f *is* (SL)-*integrable on* $[a, b]$ *and*

$$(SL) \int_a^x f = (r) \lim_n \left[(SL) \int_a^x f_n \right] \quad \forall x \in [a, b]$$

with respect to a same unit.

Proof: By virtue of Theorem 9.79, it will be enough to prove equiintegrability of the sequence $(f_n)_n$. By considering $-f_n$ or $f_n - f_1$ respectively instead of f_n, we can suppose, without loss of generality, that $(f_n)_n$ is increasing and $f_n \ge 0$. Put

$$F_n(x) := (SL) \int_a^x f_n, \quad n \in \mathbb{N}, \ x \in [a, b].$$

We have:

$$F_n(x) \le F_n(b) \quad \forall n \in \mathbb{N}, \forall x \in [a, b],$$

the sequence $(F_n(b))_n$ is increasing and bounded, and thus there exists in R the following limit:

$$F(x) := (O) \lim_n F_n(x), \quad x \in [a, b].$$

So, there is a unit w such that, $\forall \varepsilon > 0$, $\exists n_0 \in \mathbb{N}$ such that

$$F(b) - F_n(b) \le \varepsilon w \quad \forall n \ge n_0.$$

Let N be the set (of Lebesgue measure zero) where the sequence $(f_n)_n$ does not converge to f. By hypothesis, the unit z is such that, $\forall \varepsilon > 0$ and $\forall x \in [a, b] \setminus N$, there exists $n(x) \ge n_0$ such that

$$f(x) - f_n(x) \le \varepsilon z \quad \forall n \ge n(x).$$

Since f_n is (SL)-integrable $\forall n \in \mathbb{N}$, then for every $n \in \mathbb{N}$ there exists a unit w_n such that, $\forall \varepsilon > 0$, there exists a gage δ_n^* such that, for every δ_n^*-fine decomposition Π_n of $[a, b]$, $\Pi_n = \{([u_i, v_i], x_i) : i = 1, \dots, q\}$, we have:

$$\left| \sum_{i=1}^q [F_n(v_i) - F_n(u_i) - (v_i - u_i) f_n(x_i)] \right| \le \varepsilon w_n.$$

Since L^0 has property σ, there exist a sequence $(\lambda_n)_n$ of positive real numbers and a unit y such that

$$\lambda_n w_n \le y \quad \forall n \in \mathbb{N}.$$

Thus the unit y is such that, $\forall \varepsilon > 0$, $\forall n \in \mathbb{N}$, there exists a gage δ_n such that, for every δ_n-fine decomposition $\Pi_n = \{([u_i, v_i], x_i) : i = 1, \dots, q\}$, we have:

$$\left| \sum_{i=1}^q [F_n(v_i) - F_n(u_i) - (v_i - u_i) f_n(x_i)] \right| \le \frac{\varepsilon \lambda_n}{2^n} w_n \le \frac{\varepsilon}{2^n} y.$$

Without loss of generality, we can assume

$$\delta_n(x) \ge \delta_{n+1}(x) \quad \forall n \in \mathbb{N}, \quad \forall x \in [a, b].$$

Moreover, let

$$\delta(x) = \begin{cases} \delta_{n(x)}(x) & \text{if } x \notin N, \\ 0 & \text{if } x \in N. \end{cases} \qquad (9.35)$$

We now prove that the sequence $(F_n)_n$ has property (USL) on $[a,b]$. For every decomposition $\Pi = \{([u_i, v_i], x_i) : i = 1, \ldots, q\}$ we get:

$$\sum_{i=1}^{q} |F_n(v_i) - F_n(u_i)| \leq \sum_{i=1}^{q} |F_{n_0}(v_i) - F_{n_0}(u_i)|$$

$$+ \sum_{i=1}^{q} [F_n(v_i) - F_n(u_i) - (F_{n_0}(v_i) - F_{n_0}(u_i))] \qquad (9.36)$$

$$\leq \sum_{i=1}^{q} |F_{n_0}(v_i) - F_{n_0}(u_i)| + [F(b) - F_{n_0}(b)]$$

$$\leq \sum_{i=1}^{q} |F_{n_0}(v_i) - F_{n_0}(u_i)| + \varepsilon w \quad \forall n \geq n_0.$$

There exists a gage γ_{n_0} from the fact that F_{n_0} is of class (SL) on $[a,b]$. By virtue of (9.36), γ_{n_0} is the required gage, in order that the sequence $(F_n)_n$ satisfies property (USL) on $[a,b]$.

Finally we prove (SL)-equiintegrability of the sequence $(f_n)_n$ on $[a,b]$. Let n_0 be as above, and fix arbitrarily $m, n \geq n_0$. Let δ be as in (9.35), $\Pi' = \{([u_i, v_i], x_i) : i = 1, \ldots, q\}$ be a δ-fine decomposition of $[a,b]$, Π be that part of Π' whose x_i's are such that $m > n(x_i)$ and $\Pi_0 = \Pi' \setminus \Pi$. Since the sequence $(\delta_n)_n$ is decreasing, we have that Π_0 is δ_m-fine, and

$$\left| \sum_{\Pi'} \{(v_i - u_i) f_m(x_i) - [F_m(v_i) - F_m(u_i)]\} \right|$$

$$\leq \left| \sum_{\Pi} \{(v_i - u_i) f_m(x_i) - [F_m(v_i) - F_m(u_i)]\} \right|$$

$$+ \left| \sum_{\Pi_0} \{(v_i - u_i) f_m(x_i) - [F_m(v_i) - F_m(u_i)]\} \right|$$

$$\leq \left| \sum_{\Pi} \{(v_i - u_i) f_m(x_i) - [F_m(v_i) - F_m(u_i)]\} \right| + \frac{\varepsilon}{2^m} y.$$

Moreover we get

$$\left| \sum_{\Pi} \{(v_i - u_i) f_m(x_i) - [F_m(v_i) - F_m(u_i)]\} \right|$$

$$\leq \sum_{\Pi} (v_i - u_i)| f_m(x_i) - f_{n(x_i)}(x_i)|$$

$$+ \left| \sum_{\Pi} \{[F_{n(x_i)}(v_i) - F_{n(x_i)}(u_i)] - (v_i - u_i) f_{n(x_i)}(x_i)\} \right|$$

$$+ \left| \sum_{\Pi} \{[F_{n(x_i)}(v_i) - F_{n(x_i)}(u_i)] - [F_m(v_i) - F_m(u_i)]\} \right|.$$

We have

$$0 \leq \sum_{\Pi} (v_i - u_i)| f_m(x_i) - f_{n(x_i)}(x_i)| \leq \varepsilon z;$$

$$0 \leq \sum_{\Pi} \{[F_m(v_i) - F_m(u_i)] - [F_{n(x_i)}(v_i) - F_{n(x_i)}(u_i)]\}$$

$$\leq \sum_{\Pi} \{[F(v_i) - F(u_i)] - [F_{n_0}(v_i) - F_{n_0}(u_i)]\}$$

$$\leq F(b) - F_{n_0}(b) \leq \varepsilon w.$$

Furthermore we get

$$\left| \sum_{\Pi} \{[F_{n(x_i)}(v_i) - F_{n(x_i)}(u_i)] - (v_i - u_i) f_m(x_i)\} \right|$$

$$\leq \sum_{\kappa=1}^{\infty} \left| \sum_{n(x_i)=\kappa} [F_\kappa(v_i) - F_\kappa(u_i)] - (v_i - u_i) f_\kappa(x_i)] \right| \leq \sum_{\kappa=1}^{\infty} \frac{\varepsilon}{2^\kappa} y = \varepsilon y.$$

Finally we obtain

$$\left| \sum_{\Pi'} \{(v_i - u_i) f_m(x_i) - [F_m(v_i) - F_m(u_i)]\} \right| \leq \varepsilon (z + w + 2y).$$

This concludes the proof.

Remark 9.81 We note that the theory of *(SL)*-integration can be "naturally" generalized to the case of functions, defined on unbounded subintervals of the (extended) real line (for example, property *(SL)* on $[-\infty, +\infty]$ of a function f there defined can be formulated by requiring property *(SL)* on R and continuity at the points $+\infty$ and $-\infty$).

10 Pettis-type Approach

Abstract: In this chapter we begin with investigating the Pettis, Bochner, Gelfand, Dunford, McShane and Kurzweil-Henstock integrals in the context of Banach spaces, and give some comparison results.

Furthermore, we introduce the Pettis-Kurzweil-Henstock integral for Riesz space-valued functions, giving a Hake-type convergence theorem and a version of the Levi monotone convergence theorem.

10.1 Banach space Valued Case

In this paragraph we will deal in short with the main elementary properties of the Pettis integral for Banach space-valued functions. We essentially follow [197], [196], [131] and [134].

Let $(X, \mathcal{B}, \mu)$ be a measure space. We say that μ is *complete* if every subset F of any set $E \in \mathcal{B}$ with $\mu(E) = 0$ belongs to $\mathcal{B}$. Let $\mathcal{S}$ be a Banach space, $\mathcal{S}^*$ be its topological dual and $(X, \mathcal{B}, \mu)$ be a finite complete measure space. We denote by $\mathcal{N}(\mu)$ the σ-ideal of all sets of measure μ zero.

Definition 10.1 A function $f : X \to \mathcal{S}$ is said to be μ-*measurable* if there exists a sequence of simple functions $(f_n : X \to \mathcal{S})_n$ such that

$$\lim_n \| f_n(x) - f(x) \| = 0$$

for almost all $x \in X$ (with respect to μ).

The following characterization of μ-measurability holds (see [75], Corollary 3, p. 42):

Proposition 10.2 *A function* $f : X \to \mathcal{S}$ *is* μ-*measurable if and only if* f *is the* μ-*almost everywhere uniform limit of a sequence of countably valued* μ-*measurable functions.*

Definition 10.3 We say that $f : X \to \mathcal{S}$ is *scalarly* μ-*measurable* if $g \circ f$ is μ-measurable for each $g \in \mathcal{S}^*$.

A map $f : X \to \mathcal{S}^*$ is said to be *weak* scalarly* μ-*measurable* if $f(\cdot)(s)$ is μ-measurable for each $s \in \mathcal{S}$.

The following result holds (Pettis' measurability theorem):

Theorem 10.4 *A function* $f : X \to \mathcal{S}$ *is* μ-*measurable if and only if it is scalarly* μ-*measurable and there exists a set* $E \in \mathcal{N}(\mu)$ *such that* $f(X \setminus E)$ *is a separable subset of* $\mathcal{S}$.

We say that $f : X \to \mathcal{S}$ is *scalarly* μ-*bounded* if there exists a positive real number M such that

$$\left|(g \circ f)(x)\right| \le M \|g\|$$

for almost all $x \in X$ and for each $g \in \mathcal{S}^*$. A map $f : X \to \mathcal{S}^*$ is said to be *weak* scalarly μ-bounded* if there exists a positive real number L such that

$$\left|f(x)(s)\right| \le L \|s\|$$

for almost all $x \in X$ and for each $s \in \mathcal{S}$. We say that $f : X \to \mathcal{S}$ is *μ-bounded* if there exists a positive real number M such that

$$\|f(x)\| \le M$$

for almost all $x \in X$.

Proposition 10.5 *If $f : X \to \mathcal{S}$ is μ-measurable and scalarly μ-bounded, then f is μ-bounded.*

We now turn to the Pettis integral.

Definition 10.6 A function $f : X \to \mathcal{S}$ is said to be *scalarly μ-integrable* if $g \circ f \in L^1 = L^1(X, \mathcal{B}, \mu)$ for each $g \in \mathcal{S}^*$.

A map $f : X \to \mathcal{S}^*$ is said to be *weak* scalarly μ-integrable* if $|f(\cdot)(s)| \in L^1$ for each $s \in \mathcal{S}$.

Definition 10.7 If $f : X \to \mathcal{S}$ is scalarly μ-integrable, then we define the operator $T_f : \mathcal{S}^* \to L^1$ by setting

$$T_f(g) = g \circ f, \quad g \in \mathcal{S}^*.$$

Definition 10.8 A scalarly μ-integrable function $f : X \to \mathcal{S}$ is said to be *Pettis μ-integrable* if for every $E \in \mathcal{B}$ there exists $v_f(E) \in \mathcal{S}$ such that

$$g(v_f(E)) = \int_E g \circ f \, d\mu \quad \forall g \in \mathcal{S}^*.$$

The set function $v_f : \mathcal{B} \to \mathcal{S}$ is called the *indefinite Pettis integral* of f with respect to μ.

A weak* scalarly μ-integrable function $f : X \to \mathcal{S}^*$ is said to be *Gelfand μ-integrable* if for every $E \in \mathcal{B}$ there exists $\xi_f(E) \in \mathcal{S}^*$ such that

$$\xi_f(E)(s) = \int_E f(\cdot)(s) \, d\mu \quad \forall s \in \mathcal{S}.$$

The Gelfand integral of f will be denoted by $^*v(f)$. If $f : X \to \mathcal{S}$ is considered as a $\mathcal{S}^{**}$-valued function, then its Gelfand integral in $\mathcal{S}^{**} = (\mathcal{S}^*)^*$ is called the *Dunford integral* and is denoted by v_f^{**}.

The following results hold:

Proposition 10.9 *Every scalarly μ-integrable function $f : X \to \mathcal{S}$ is Dunford μ-integrable.*

Proposition 10.10 *Every weak* scalarly μ-integrable function $f : X \to \mathcal{S}^*$ is Gelfand μ-integrable.*

If $\mathcal{S}$ is reflexive, then the Dunford and Pettis integrals coincide. When $\mathcal{S}$ is not reflexive, this, in general, is not true (see [197]). However, the following result holds:

Proposition 10.11 *If $\mathcal{S}$ is a separable Banach space without an isomorphic copy of c_0, then every Dunford integrable function $f : X \to \mathcal{S}$ is Pettis integrable.*

We now state some characterizations of Pettis integrability.

Proposition 10.12 *A scalarly μ-integrable function $f : X \to \mathcal{S}$ is Pettis μ-integrable if and only if the set*

$$\{g \in \mathcal{S}^* : g \circ f = 0 \ \mu-\text{almost everywhere}\}$$

is closed with respect to the weak* topology.

Proposition 10.13 *If $f : X \to \mathcal{S}$ is Pettis integrable, then the operator T_f defined in 10.7 is weakly compact and the set*

$$Z_f = \{g \circ f : g \in \mathcal{S}^*, \ \|g\| \le 1\}$$

is weakly closed in L^1.

We now state some convergence theorems for the Pettis integral in the context of Banach spaces (for the proofs, see also [197], pp. 550-552 and [196], pp. 221-223). We begin with the following:

Proposition 10.14 *Let $\mathcal{S}$ be any Banach space. A bounded set $\varnothing \ne W \subset \mathcal{S}$ is relatively weakly compact if and only if, for every two sequences $(s_n)_n$ in W and $(g_n)_n$ in the unit ball of $\mathcal{S}^*$, one has*

$$\lim_n [\lim_m g_m(s_n)] = \lim_m [\lim_n g_m(s_n)],$$

provided that all the involved limits exist.

Now we formulate the following Vitali-type theorem:

Theorem 10.15 *Let $f : X \to \mathcal{S}$ be a map, and $(f_n : X \to \mathcal{S})_n$ be a sequence of Pettis μ-integrable functions, such that $\lim_n (g \circ f_n) = g \circ f$ in μ-measure for every $g \in \mathcal{S}^*$, and the set*

$$\{g \circ f_n : g \in \mathcal{S}^*, \ \|g\| \le 1, n \in \mathbb{N}\}$$

is uniformly μ-integrable, that is, $\forall \varepsilon > 0$, $\exists \delta(\varepsilon) > 0$ such that

$$\int_E |g \circ f_n| d\mu \le \varepsilon,$$

$\forall n \in \mathbb{N}$, $\forall g \in \mathcal{S}^*$ with $\|g\| \le 1$ and $\forall E \in \mathcal{B}$ with $\mu(E) < \delta$.

Then f is Pettis μ-integrable, and

$$\lim_n \int_E f_n d\mu = \int_E f d\mu.$$

weakly in S, *for every* $E \in B$.

A consequence of Theorem 15 is the following version of the Lebesgue dominated convergence theorem (see [197], pp. 551-552 and [196], p. 224):

Theorem 10.16 *Let* $f : X \to S$ *be a map, and* $(f_n : X \to S)_n$ *be a sequence of Pettis* μ-*integrable functions, such that* $\lim_n (g \circ f_n) = g \circ f$ *in* μ-*measure for every* $g \in S^*$. *Moreover, suppose that there exists* $h \in L^1$ *such that, for every* g *belonging to the unit ball of* S^* *and* $\forall n \in \mathbb{N}$, *one has* $|g \circ f_n| \leq h$ μ-*almost everywhere (the exceptional set depending on* g). *Then* f *is Pettis* μ-*integrable, and*

$$\lim_n \int_E f_n \, d\mu = \int_E f \, d\mu$$

weakly in S, *for every* $E \in B$.

Moreover, in the following result, it is shown that c_0 is the only "exceptional" Banach space (see [197], pp. 552-553 and [196], p. 226-227):

Theorem 10.17 *Let* S *be a Banach space without any isomorphic copy of* c_0. *If* $f : X \to S$ *is scalarly* μ-*integrable and there are Pettis* μ-*integrable functions* $f_n : X \to S$, $n \in \mathbb{N}$, *such that*

$$\lim_n \int_E g \circ f_n \, d\mu = \int_E g \circ f \, d\mu$$

for every $E \in B$ *and* $g \in S^*$, *then* f *is Pettis* μ-*integrable and*

$$\lim_n \int_E f_n \, d\mu = \int_E f \, d\mu$$

weakly in S, *for every* $E \in B$.

We now introduce the Bochner, McShane and Kurzweil-Henstock integrals for functions, defined in $[a, b] \subset \mathbb{R}$ and taking values in a Banach space S. We begin with the Bochner integral (see [75], p. 44-45):

Definition 10.18 A μ-*measurable function* $f : X \to S$ *is called Bochner integrable if there exists a sequence of simple functions* $(f_n : X \to S)_n$ *such that*

$$\lim_n \int_X \|f_n - f\| \, d\mu = 0.$$

In this case, $\int_E f \, d\mu$ is defined $\forall E \in B$ by setting:

$$\int_E f \, d\mu = \lim_n \int_E f_n \, d\mu,$$

where the involved limit is intended with respect to the norm.

The following characterization of the Bochner integral holds (see also [75]):

Proposition 10.19 A μ-*measurable function* $f : X \to S$ *is Bochner integrable if and only if*

$$\int_X \|f\| \, d\mu < +\infty.$$

We note that Bochner integrability implies Pettis integrability. However, if $\mathcal{S}$ is an infinite dimensional Banach space, then there exists a $\mathcal{S}$-valued μ-measurable function which is Pettis but not Bochner integrable (see [196], pp. 203-204).

We now turn to the McShane and Kurzweil-Henstock integrals.

Definitions 10.20   Given an interval $[a,b] \subset \mathbb{R}$, we call *division of* $[a,b]$ any finite set $D = \{x_0, x_1, \ldots, x_n\} \subset [a,b]$, *where* $x_0 = a, x_n = b$ and $x_{i-1} < x_i$ for all $i = 1, \ldots, n$ (see also Chapter 9).

We call *McShane partition of* $[a,b]$ a set of the type $\Pi = \{(A_i, \xi_i) : i = 1, \ldots n\}$, where $A_i = [x_{i-1}, x_i]$ and $\{x_0, x_1, \ldots x_n\}$ is a division. A McShane partition is said to be *partition* if $\xi_i \in A_i$ for all $i = 1, \ldots, n$. A partition is said to be *special* if ξ_i is an endpoint of A_i for every $i = 1, \ldots, n$. We call *mesh* of a McShane partition Π the quantity $|\Pi| := \max_i (x_i - x_{i-1})$.

Definition 10.21   Given a McShane partition $\Pi = \{([x_{i-1}, x_i], \xi_i), i = 1, \ldots, n\}$ and a function $\delta : [a,b] \rightarrow \mathbb{R}^+$, we say that Π is δ-*fine* if $x_i - x_{i-1} \leq \delta(\xi_i)$ for all $i = 1, \ldots, n$.

Moreover, if $\mathcal{S}$ is a Banach space, $f : [a,b] \rightarrow \mathcal{S}$ is a mapping, and Π is a McShane partition as above, we denote by $\sum_{\Pi} f$ (Riemann sum) the quantity

$$\sum_{i=1}^{n} (x_i - x_{i-1}) f(\xi_i).$$

We now are ready to define the McShane and Kurzweil-Henstock integrals. For the sake of simplicity, we assume $[a,b] = [0,1]$.

Definition 10.22 A function $f : [0,1] \rightarrow \mathcal{S}$ is said to be *McShane integrable,* with *McShane integral* w, if for every $\varepsilon > 0$ there exists a map $\delta : [0,1] \rightarrow \mathbb{R}^+$ such that

$$\left\| \sum_{\Pi} f - w \right\| \leq \varepsilon$$

for every δ-fine McShane partition of $[0,1]$.

Moreover, we say that a mapping $f : [0,1] \rightarrow \mathcal{S}$ is *Kurzweil-Henstock integrable,* with *Kurzweil-Henstock integral* w, if for every $\varepsilon > 0$ there exists $\delta : [0,1] \rightarrow \mathbb{R}^+$ such that

$$\left\| \sum_{\Pi} f - w \right\| \leq \varepsilon$$

for each δ-fine partition of $[0,1]$.

The following results hold (for a proof, see [131,134,142]):

Theorem 10.23 *A function $f:[0,1] \to S$ is McShane integrable if and only if it is Kurzweil-Henstock integrable and Pettis integrable.*

Moreover, if S is a separable Banach space, then $f:[0,1] \to S$ is Pettis integrable if and only if it is McShane integrable. In general, this is not true in an arbitrary Banach space.

Theorem 10.24 *A measurable function $f:[0,1] \to S$ (with respect to the Lebesgue measure) is Pettis integrable if and only if it is McShane integrable.*

We note that there are even non-separable Banach spaces S, such that every S-valued function defined on $[0,1]$ is Pettis integrable if and only if it is McShane integrable (see also [76]).

D. H. Fremlin ([132]) generalized Theorems 10.23 and 10.24 for functions defined not necessarily on a closed bounded subinterval of the real line, but even in a suitable abstract space.

We begin with the following (see [129], pp. 148, 172):

Definition 10.25 Let $(X, \mathcal{B}, \mu)$ be a measure space. We write

$$\mathcal{B}^f = \{E \in \mathcal{B} : \mu(E) < +\infty\}.$$

We say that μ is *semifinite* if, whenever $E \in \mathcal{B}$ and $\mu(E) = +\infty$, there is $F \in \mathcal{B}$, $F \subset E$ such that $0 < \mu(F) < +\infty$. We say tat μ is *locally determined* if it is semifinite and, for any set $E \subset X$, the following implication holds:

$$[E \cap F \in \mathcal{B} \quad \forall F \in \mathcal{B}^f] \Rightarrow [E \in \mathcal{B}].$$

Finally, we are ready to give the definitions of quasi-Radon and Radon measure space (see also [129], p. 202, 210).

Definition 10.26 A *quasi-Radon measure space* is a quadruple $(X, \mathcal{T}, \mathcal{B}, \mu)$, where $(X, \mathcal{B}, \mu)$ is a measure space and $\mathcal{T}$ is a topology on X such that:

(i) μ is complete and locally determined;

(ii) $\mathcal{T} \subset \mathcal{B}$ (that is, every open set is measurable);

(iii) if $E \in \mathcal{B}$ and $\mu(E) > 0$, then $\exists G \in \mathcal{T}$ such that $\mu(G) < +\infty$ and $\mu(E \cap G) > 0$;

(iv) $\mu(E) = \sup\{\mu(F) : F \in \mathcal{B}, F \subset E, F \text{ closed}\} \quad \forall E \in \mathcal{B}$;

(v) if G is an upward directed set in $\mathcal{T}$, then

$$\mu(\cup G) = \sup\{\mu(G) : G \in \mathcal{G}\}.$$

We note that the Lebesgue measure on $\mathbb{R}^n$ is quasi-Radon (see also [132]).

Definition 10.27 A *Radon measure space* is a quadruple $(X, \mathcal{T}, \mathcal{B}, \mu)$, where $(X, \mathcal{B}, \mu)$ is a measure space and $\mathcal{T}$ is a Hausdorff topology on X such that:

(i) μ is complete and locally determined;

(ii) $\mathcal{T} \subset \mathcal{B}$ (that is, every open set is measurable);

(iii) if $t \in X$, then there exists $G \in \mathcal{T}$ such that $t \in G$ and $\mu(G) < +\infty$;

(iv) $\mu(E) = \sup\{\mu(F) : F \in B, F \subset E, F \text{ compact}\} \quad \forall E \in \mathcal{B}$.

We now introduce the McShane integral for Banach space-valued functions, defined in quasi-Radon measure spaces.

Definition 10.28 Let $(X, \mathcal{T}, \mathcal{B}, \mu)$ be a σ-finite quasi-Radon measure space which is *outer regular,* that is such that

$$\mu(E) = \inf\{\mu(G) : E \subset G \in \mathcal{T}\} \quad \forall E \in \mathcal{B}.$$

A *generalized McShane partition* on X is a sequence $(E_i, t_i)_{i \in \mathbb{N}}$ such that $(E_i)_{i \in \mathbb{N}}$ is a disjoint family of measurable sets of finite measure, $\mu(X \setminus \bigcup_{i \in \mathbb{N}} E_i) = 0$ and $t_i \in X \; \forall i$. A *gauge* on X is a function $\gamma : X \to \mathcal{T}$ such that $\xi \in \gamma(\xi) \; \forall \xi \in X$. A generalized McShane partition $(E_i, t_i)_{i \in \mathbb{N}}$ is said to be γ-*fine* if $E_i \subset \gamma(t_i) \; \forall i \in \mathbb{N}$.

Definition 10.29 Let $\mathcal{S}$ be a Banach space. We say that a function $f : X \to \mathcal{S}$ is *McShane integrable,* with *McShane integral* w, if for every $\varepsilon > 0$ there exists a gauge γ such that

$$\limsup_{n \to +\infty} \left\| \sum_{i=1}^{n} \mu(E_i) f(t_i) - w \right\| \leq \varepsilon$$

for every γ-fine generalized McShane partition $(E_i, t_i)_{i \in \mathbb{N}}$ of X.

The following result holds (see [132], Theorem 1Q, p. 50 and Corollary 4C, p. 64):

Theorem 10.30 *Let* $(X, \mathcal{T}, \mathcal{B}, \mu)$ *be a* σ-*finite outer regular quasi-Radon measure space and* $\mathcal{S}$ *be a Banach space. If* $f : X \to \mathcal{S}$ *is McShane integrable, then* f *is Pettis integrable, and the two integrals coincide.*

Moreover, if $\mathcal{S}$ *is separable, then* $f : X \to \mathcal{S}$ *is McShane integrable if and only if it is Pettis integrable.*

We now state another condition under which Pettis integrability implies McShane integrability (see also [132], Corollary 4D, p. 64).

Theorem 10.31 *Let* $(X, \mathcal{T}, \mathcal{B}, \mu)$ *be a* σ-*finite outer regular Radon measure space and* $\mathcal{S}$ *be a Banach space. If* $f : X \to \mathcal{S}$ *is Pettis integrable and* $f^{-1}(A) \in \mathcal{B}$ *for each norm-open set* $A \subset \mathcal{S}$, *then* f *is McShane integrable.*

10.2 Riesz Space Valued Case

We now introduce and investigate a Pettis-type integral (p-integral) for functions with values in a Dedekind complete π-space R (see also Chapter 2). Our results about this kind of Pettis integral were proved in [25].

Throughout this paragraph, we always suppose that R is a Dedekind complete π-space. Recall that $R^{\times}$ is the space of all order continuous functionals on R (see Chapter 2), that $g_1 \geq g_2$ in $R^{\times}$ means

$g_1(r) \geq g_2(r)$ $\forall r \in R$, $r \geq 0$, and that an element $r \in R$ satisfies $r \geq 0$ if and only if $g(r) \geq 0$ holds for each $0 \leq g \in R^\times$ (see also [4], Theorem 5.1, p. 55). In economic models, a way to describe the commodity-price system is the pair $(R, R^\times)$, in which the hypothesis that R is a π-space is essential (see [5], pp. 100 and 115).

Let X be a Hausdorff compact topological space. If $A \subset X$, then the interior of the set A is denoted by $int\, A$.

We shall work with a family $\mathcal{F}$ of compact subsets of X such that $X \in \mathcal{F}$ and closed under the intersection and finite union and a monotone additive mapping $\lambda : \mathcal{F} \to [0, +\infty)$. The additivity means that

$$\lambda(A \cup B) + \lambda(A \cap B) = \lambda(A) + \lambda(B)$$

whenever $A, B \in \mathcal{F}$.

By a *partition* (detaily, $(\mathcal{F}, \lambda)$-*partition*) of a nonempty set $A \in \mathcal{F}$ we mean a finite collection $\Pi = \{(A_1, \xi_1), \ldots, (A_q, \xi_q)\}$ such that

(i) $A_1, \ldots, A_q \in \mathcal{F}$, $\displaystyle\bigcup_{i=1}^{q} A_i = A$,

(ii) $\lambda(A_i \cap A_j) = 0$ whenever $i \neq j$,

(iii) $\xi_j \in A_j$ $(j = 1, \ldots, q)$.

Sometimes, as no confusion can arise, we will indicate by *partition of A* a finite collection $\{A_j : j = 1, \ldots, q\}$, satisfying conditions (i) and (ii). The concepts of partition, separation of points, gauge, γ-fineness etc. are intended as in Chapter 7.

Let us turn to the definition of p-integral on X. If $\Pi = \{(A_1, \xi_1), \ldots, (A_q, \xi_q)\}$ is a partition of a set $A \in \mathcal{F}$, and $f : X \to R$, then we define the Riemann sum as follows:

$$\sum_{\Pi} f = \sum_{j=1}^{q} \lambda(A_j) f(\xi_j).$$

Definition 10.32 A function $f : X \to R$ is *p-integrable* on a set $A \in \mathcal{F}$, if there exists $I \in R$ such that $\forall \varepsilon > 0$ and $\forall g \in R^\times$ there exists a gauge γ on A such that

$$\left| g(\sum_{\Pi} f) - g(I) \right| \leq \varepsilon \qquad (10.1)$$

whenever Π is a γ-fine partition of A. We denote

$$I = \int_A f.$$

Remark 10.33 We note that, if $I_1, I_2 \in R$ satisfy (10.1), then $I_1 = I_2$: indeed, for all $g \in R^\times$, $g \geq 0$, and for fine enough partitions Π, we get

$$\left| g(I_1) - g(I_2) \right| \le \left| g(I_1) - g(\sum_{\Pi} f) \right| + \left| g(\sum_{\Pi} f) - g(I_2) \right| \le 2\varepsilon. \qquad (10.2)$$

Since R is a π-space, then we get $I_1 - I_2 = 0$, that is the assertion.

Remark 10.34 It is easy to check that, in the case $R = \mathbb{R}$, the p-integral coincides with the classical Kurzweil-Henstock integral, as introduced in [23]. In this case, we often will use the term "integrable" instead of "p-integrable".

We now state the main properties of the p-integral.

Proposition 10.35 *If f_1, f_2 are p-integrable on $A \in \mathcal{F}$ and $c_1, c_2 \in \mathbb{R}$, then $c_1 f_1 + c_2 f_2$ is p-integrable on A and*

$$\int_A (c_1 f_1 + c_2 f_2) = c_1 \int_A f_1 + c_2 \int_A f_2.$$

The proof is similar to the one of [170], Theorems 2.5.1 and 2.5.3.

Proposition 10.36 *If $f : X \to R$ is p-integrable on $A \in \mathcal{F}$, then for every $g \in R^\times$ the real-valued map $g \circ f$ is integrable on A, and*

$$\int_A g \circ f = g \left(\int_A f \right)$$

Conversely, if $f : X \to R$ is such that $g \circ f$ is integrable on $A \in \mathcal{F}$ for each $g \in R^\times$ and there exists $I \in R$ such that

$$\int_A g \circ f = g(I) \quad \forall g \in R^\times,$$

then f is p-integrable on A, and $\int_A f = I$.

Proof: Fix arbitrarily $g \in R^\times$ and a partition Π of A, $\Pi = \{(A_i, \xi_i) : i = 1, \ldots, q\}$. We have

$$g \left(\sum_{\Pi} f \right) = g \left(\sum_{i=1}^{q} \lambda(A_i) f(\xi_i) \right)$$
$$= \sum_{i=1}^{q} \lambda(A_i) g(f(\xi_i)) = \sum_{\Pi} (g \circ f). \qquad (10.3)$$

The assertion follows from (10.3) and definitions of integrability and p-integrability.

Proposition 10.37 *If f_1 and f_2 are p-integrable on $A \in \mathcal{F}$ and $f_1 \le f_2$, then*

$$\int_A f_1 \le \int_A f_2.$$

Proof: Fix arbitrarily $g \in R^\times$, $g \ge 0$. Then $g \circ f_1 \le g \circ f_2$. By the first part of Proposition 10.36 and Proposition 1.4 of [217] we get that $g \circ f_1$ and $g \circ f_2$ are integrable on A, and

$$\int_A g \circ f_1 \le \int_A g \circ f_2. \tag{10.4}$$

Again by Proposition 10.36, we have

$$\int_A g \circ f_l = g\left(\int_A f_l\right), \quad l = 1, 2. \tag{10.5}$$

From (10.4) and (10.5) it follows:

$$g\left(\int_A f_1\right) \le g\left(\int_A f_2\right). \tag{10.6}$$

The assertion follows from (10.6) and arbitrariness of $g \in R^\times$. ❦

A simple consequence of Proposition 10.37 is the following:

Corollary 10.38 *If both f and $|f|$ are p-integrable on $A \in \mathcal{F}$, then*

$$\left|\int_A f\right| \le \int_A |f|.$$

We now state the following results:

Proposition 10.39 *Let $u \in R$, $u \ge 0$. For every $E \in \mathcal{F}$, the function $f = \chi_E u : X \to R$ satisfies the following condition:*

$$\exists I \in R : \forall \varepsilon > 0, \exists \text{ gauge } \gamma :$$

$$\left|\sum_\Pi f - I\right| \le \varepsilon u \tag{10.7}$$

for all γ-fine partitions Π of X.

Proof: It is enough to take into account of Proposition 1.5., pp. 155-156, of [217], and to use the same technique as in Theorem 3.18 of [26]. ❦

Proposition 10.40 *Let $f : X \to R$ satisfy condition (10.7) for suitable I and $u \in R$, $u \ge 0$. Then f is p-integrable on X, and $\int_X f = I$.*

Proof: Let I and u be as in the hypothesis of the proposition. Fix arbitrarily $\varepsilon > 0$ and $g \in R^\times$, $g \ge 0$. Then there exists $\eta > 0$:

$$\eta \, g(u) \le \varepsilon. \tag{10.8}$$

Moreover, by condition (10.7), in correspondence with η there exists a gauge γ such that

$$\left|\sum_\Pi f - I\right| \le \eta u \tag{10.9}$$

for all γ-fine partitions Π of X. From (10.8) and (10.9) it follows that

$$\left|g(\sum_\Pi f) - g(I)\right| \le g(\eta u) = \eta \, g(u) \le \varepsilon \tag{10.10}$$

for all γ-fine partitions Π of X. The assertion follows from (10.10).

Proposition 10.41 *For every $E \in F$ and $u \in R$ the function $\chi_E u$ is p-integrable on X and*

$$\int_X \chi_E u = \lambda(E)u.$$

Proof: Since R is a Riesz space, we have $u = u^+ - u^-$, where $u^+, u^- \in R$, $u^+ \geq 0$, $u^- \geq 0$. So, we can suppose, without loss of generality, that $u \geq 0$. The assertion follows from Propositions 10.39 and 10.40.

The following theorem generalizes to the context of Riesz spaces and our Pettis-type integral Theorem 3.1 of [23], which was formulated for real-valued functions.

Theorem 10.42 *Let $X = X_0 \cup \{x_0\}$ be the one-point compactification of a locally compact space X_0. Let $f : X \to R$ be a function such that $f(x_0) = 0$. Let $(A_n)_n$ be a sequence of sets, such that*

$$A_n \in F, \quad A_n \subset \text{int } A_{n+1}, \quad A_{n+1} \setminus \text{int } A_n \in F, \quad \lambda(A_n \setminus \text{int } A_n) = 0 \quad (n \in \mathbb{N}), \quad \bigcup_{n=1}^{\infty} A_n = X_0.$$ *Let f be p-integrable on every $A \in F$, with $A \subset X_0$, and let there exist in R an element I such that, $\forall \varepsilon > 0$, $\forall g \in R^\times$, there exists an integer n_0 such that*

$$\left| \int_A g \circ f - g(I) \right| \leq \varepsilon \quad \forall A \in F, X_0 \supset A \supset A_{n_0}.$$

Then f is p-integrable on X and $\int_X f = I$.

Proof: By hypothesis and the first part of Proposition 10.36, we get that $g \circ f$ is integrable on A for all $g \in R^\times$ and $A \in F$, $A \subset X_0$. Moreover, by Theorem 3.1 of [23], $g \circ f$ is integrable on X and

$$\int_X g \circ f = g(I). \tag{10.11}$$

The assertion follows by (10.11) and the second part of Proposition 10.36. ◼

We now state a monotone convergence Levi-type theorem.

Theorem 10.43 *Let $f_n : X \to R$, $n \in \mathbb{N}$ be p-integrable, let the sequence $\left(\int_X f_n \right)_n$ be bounded, and suppose that for every $g \in R$, $g \geq 0$, and $\forall x \in X$, $g(f_n(x)) \uparrow g(f(x))$.*

Then f is p-integrable and

$$\sup_n \int_X f_n = \int_X f.$$

Proof: Fix arbitrarily $g \in R^\times$, $g \geq 0$. By hypothesis, we get that the sequence $\left(g\left(\int_X f_n \right) \right)_n$ is bounded. Thus, by [217], Theorem 2.2, pp. 159-162 and the first part of Proposition 10.36, the real-valued function $g \circ f$ is integrable and

$$\int_X g \circ f = \lim_n \int_X g \circ f_n = \sup_n \int_X g \circ f_n$$

$$= \sup_n \left[g \left(\int_X f_n \right) \right] = g \left(\sup_n \int_X f_n \right). \qquad (10.12)$$

The assertion follows from (10.12) and the second part of Proposition 10.36.

11 Applications in Multivalued Logic

Abstract: This chapter contains an introduction to the theory of MV-algebras and states, together with the notion of observable.

It is proved that *every* probability MV-algebra is weakly σ-distributive and some applications to intuitionistic fuzzy sets (IF-sets) are given.

Furthermore we show that the probability theory of IF-sets can be considered as a particular case of the probability theory on a suitable MV-algebra.

11.1 MV-algebras

While in the classical two-valued logic any assertion can be evaluated by two numbers $0,1$, in multivalued logic there is used the whole interval $[0,1]$. The disjunction, conjunction are considered as binary operations on $[0,1]$ and negation as a unary operation. These operations have been introduced by Łukasiewicz.

Definition 11.1 If $a,b \in [0,1]$, then we define

$$a \oplus b = \min(a+b,1),$$

$$a \odot b = \max(a+b-1,0),$$

$$\neg a = 1 - a.$$

If $a,b \in \{0,1\}$, then actually $a \oplus b$ corresponds to the disjunction, $a \odot b$ to the conjunction and $\neg a$ to the negation. Moreover, if A,B are subsets of a set Ω, and $\omega \in \Omega$, then

$$\chi_A(\omega) \oplus \chi_B(\omega) = \chi_{A \cup B}(\omega),$$

$$\chi_A(\omega) \odot \chi_B(\omega) = \chi_{A \cap B}(\omega),$$

$$\neg \chi_A(\omega) = \chi_{A'}(\omega).$$

The unit interval $[0,1]$ with the Łukasiewicz connectives $\oplus, \odot, \neg$ and two fixed elements $0,1$ is a prototype of the notion of MV-algebra.

Definition 11.2 An MV-algebra $(M,0,1,\neg,\oplus,\odot)$ is a system satisfying the following conditions:

(i) $\oplus$ is a commutative and associative binary operation;

(ii) $\neg 0 = 1, \neg 1 = 0$;

(iii) $x \oplus 1 = 1$ for any $x \in M$;

(iv) $x \odot y = \neg(\neg x \oplus \neg y)$ for any $x,y \in M$;

(v) $y \oplus \neg(y \oplus \neg x) = x \oplus \neg(x \oplus \neg y)$ for any $x,y \in M$.

Antonio Boccuto / Beloslav Riečan / Marta Vrábelová

The properties (i) - (iv) are natural, the property (v) is sophisticated, of course, in the example of characteristic functions it means that

$$B \cup (B \cup A')' = A \cup (A \cup B')',$$

where the symbols A', B' mean the *complement* of the involved sets A, B, and hence

$$B \cup (A \cap B') = A \cup (B \cap A').$$

In general, the property (v) makes possible to define a partial ordering on M. If we want to have $a \vee b = b$, i.e. $a \oplus \neg (a \oplus \neg b) = b \oplus \neg (b \oplus \neg a)$ equals b, then the equality $a \odot \neg b = 0$ will characterize the relation $a \leq b$.

Definition 11.3 We define $a \leq b$, if $a \odot \neg b = 0$.

Proposition 11.4 *The relation $\leq$ is a partial ordering, M is a distributive lattice with the least element 0 and the greatest element 1.*

Proof: See [64], Chapters 1, 3, and Lemma 6.6.4.

Definition 11.5 An MV-algebra M is said to be σ-*complete* or σ-*MV-algebra* if its underlying lattice is σ-complete, i.e. every non-empty countable subset of M has a supremum in M. Analogously we can give the definition of σ-complete l-group.

We say that a lattice (or MV-algebra, or l-group) is *weakly* σ-*complete* if it can be expressed as the union of σ-complete lattices (MV-algebras, l-groups respectively).

Example 11.6 Consider a family $\mathcal{T}$ of functions $f : \Omega \to [0,1]$ satisfying the following properties:

(i) $1_\Omega \in \mathcal{T}$;

(ii) if $f \in \mathcal{T}$, then $1 - f \in \mathcal{T}$;

(iii) if $f, g \in \mathcal{T}$, then $f \oplus g \in \mathcal{T}$;

(iv) if $f_n \in \mathcal{T}, f_n \uparrow f$, then $f \in \mathcal{T}$.

Then $\mathcal{T}$ is an example of a σ-complete MV-algebra.

A very important example of an MV-algebra is the MV-algebra induced by an l-group. Recall (see Chapter 2) that an l-*group* (*lattice ordered group*) is a structure $(G, +, \leq)$, where

(i) $(G, +)$ is an Abelian group;

(ii) $(G, \leq)$ is a lattice;

(iii) if $a \leq b$, then $a + c \leq b + c$, $\forall a, b, c \in G$.

Example 11.7 Let G be an l-group, 0 be its neutral element (i.e. $x + 0 = x$ for any $x \in G$), u be an element of G such that $u > 0$. Put $M = [0, u] = \{x \in G : 0 \leq x \leq u\}$ and define the following operations on M:

$$a \oplus b = (a + b) \wedge u,$$

$$a \odot b = (a + b - u) \vee 0,$$

$$\neg a = u - a.$$

Then $(M, 0, u, \neg, \oplus, \odot)$ is an MV-algebra. We shall denote it by $M = \Gamma(G, u)$.

Theorem 11.8 *(Mundici) Every MV-algebra M, up to isomorphisms, can be identified with the unit interval $[0, u]$ of a unique l-group G with a strong unit u (i.e. to any $a \in G$ there exists a natural number n such that $nu \geq a$).*

Proof: See [64], Chapter 7 and [193]. 🍎

The Mundici theorem stated above has some practical consequences because one can use group operations instead of quite complicated axioms. Moreover, many known results of the theory of l-groups can be used in the MV-algebra theory. We shall illustrate the advantage of the approach on the notion of state.

Definition 11.9 Let M be a σ-complete MV-algebra. A *state* is a map $m : M \to [0, 1]$ satisfying the following conditions:

(i) $m(1) = 1$;

(ii) whenever $b, c \in M$ and $b \odot c = 0$, it follows that $m(b \oplus c) = m(b) + m(c)$;

(iii) if $a_n \uparrow a$, then $m(a_n) \uparrow m(a)$.

Proposition 11.10 A map $m : M \to [0, 1]$ is a state if and only if it satisfies (i), (iii), and

(ii') if $a, b, c \in M, a = b + c$, then $m(a) = m(b) + m(c)$.

Proof: $\Rightarrow$

If $a = b + c$, then

$$b \odot c = (b + c - u) \vee 0 = (a - u) \vee 0 = 0,$$

since $a \leq u$, hence $a - u \leq 0$. Moreover

$$b \oplus c = (b + c) \wedge u = a = b + c,$$

hence

$$m(a) = m(b \oplus c) = m(b) + m(c).$$

$\Leftarrow$

If $b \odot c = 0$, then $(b + c - u) \vee 0 = 0$, hence $b + c \leq u$. Put $a = b + c$. Then

$$m(a) = m(b) + m(c).$$

Of course,

$$a = b + c = (b + c) \wedge u = b \oplus c,$$

hence

$$m(b \oplus c) = m(b) + m(c). \quad \text{🍎}$$

11.2 Group-valued Measures

In the mathematical models of quantum systems the Hilbert space model plays an important role as well as quantum logics (orthomodular lattices), its natural algebraization. Recently a similar role plays a tribe of fuzzy sets and MV-algebra as its natural algebraization.

Let M be a given σ-complete MV-algebra. In our quantum model there are two fundamental notions: state and observable. They corresponds to the notions of probability and random variable in the Kolmogorov model. The notion of a state has been exposed in the previous section. Now we shall introduce the notion of observable.

Definition 11.11 Let M be an MV-algebra. An *observable* is a mapping $x : \mathcal{B}(\mathbb{R}) \to M(\mathcal{B}(\mathbb{R})$ being the σ-algebra of the Borel subsets of the real line $\mathbb{R}$) satisfying the following conditions:

  (i) $x(\mathbb{R}) = 1$;
  (ii) $A \cap B = \varnothing \Rightarrow x(A) \odot x(B) = 0, \quad x(A \cup B) = x(A) \oplus x(B)$;
  (iii) $A_n \uparrow A \Rightarrow x(A_n) \uparrow x(A)$.

Theorem 11.12 *Let M be a σ-complete MV-algebra, $x : \mathcal{B}(\mathbb{R}) \to M$ be an observable, G be the unique l-group including M given by Theorem 11.8. Then $x : \mathcal{B}(\mathbb{R}) \to G$ is a group-valued measure.*

Proof: Let us use the group operations. Then $x(A) \odot x(B) = 0$ means $(x(A) + x(B) - u) \vee 0 = 0$, hence $x(A) + x(B) \leq u$. Therefore

$$x(A) \oplus x(B) = (x(A) + x(B)) \wedge u = 0 = x(A) + x(B),$$

and property (ii) can be rewritten

$$A \cap B = \varnothing \Rightarrow x(A \cup B) = x(A) + x(B).$$

If $A_n \uparrow A$, then $x(A_n) \uparrow x(A)$, hence x is continuous too. 🍎

Of course, to use the measure extension theorem, we need to have a weakly σ-distributive group (see Chapter 2). The positive answer is given by the following three assertions.

We shall work with a probability MV-algebra, i.e. with a couple (M, m), where M is a σ-complete MV-algebra and $m : M \to [0,1]$ is a state such that

$$m(a) = 0 \Leftrightarrow a = 0.$$

Such a state is usually called *faithful state*.

Theorem 11.13 *Any probability MV-algebra (M, m) is weakly σ-distributive (Here, weak σ-distributivity is intended analogously as in Chapter 2).*

Proof: Let $(b_{i,j})_{i,j}$ in M be a double sequence having the property that, for each $i \in \mathbb{N}$, $b_{i,j} \geq b_{i,j+1}$ and $\bigwedge_{j=1}^{\infty} b_{i,j} = 0$, that is a (D)-sequence or regulator, according to Chapter 2, and set $\Phi = \mathbb{N}^{\mathbb{N}}$. We now prove that

$$0 = \bigwedge_{\varphi \in \Phi} \bigvee_{i=1}^{\infty} b_{i,\varphi(i)}.$$

Assume (absurdum hypothesis) that $0 < b \in M$ is a lower bound for each element $\bigvee_{i=1}^{\infty} b_{i,\varphi(i)}$, where φ ranges over Φ. From the assumed faithfulness of m it follows that $m(b) > 0$. Since, for each fixed $i, b \wedge b_{i,j}$ tends to 0 as j tends to infinity, then so does $m(b \wedge b_{i,j})$. There is $j^* = \psi(i)$ such that

$$m(\bigvee_{i=1}^{\infty} (b \wedge b_{i,\psi(i)})) \le m(b)/2^{i+1}.$$

By the absurdum hypothesis, we get

$$b \le \bigvee_{i=1}^{\infty} b_{i,\psi(i)}.$$

Then, from the distributivity law, we get the contradiction

$$m(b) = m\left(b \wedge \left(\bigvee_{i=1}^{\infty} b_{i,\psi(i)}\right)\right) \le \sum_{i=1}^{\infty} m(b \wedge b_{i,\psi(i)}) < m(b).$$

Theorem 11.14 $M_0 = \Gamma(G,u)$ *is* σ *-complete if and only if* G *is weakly* σ *-complete.*

Proof: $\Leftarrow$

If $t_i \in M_0$ $(i \in \mathbb{N})$, then $(t_i)_i$ is bounded in G, hence the element $\bigvee_{i=1}^{\infty} t_i$ is in G. Of course $t_i \le u$ $(i \in \mathbb{N})$ implies $\bigvee_{i=1}^{\infty} t_i \le u$, hence $\bigvee_{i=1}^{\infty} t_i \in [0,u] = M_0$. Similarly $\bigwedge_{i=1}^{\infty} t_i \in M_0$.

$\Rightarrow$

Since $M_0 = [0,u]$, the interval $[0,u]$ is σ-complete. Since u is a strong unit of G, it suffices to prove that $[0,nu]$ is σ-complete for every n. We shall prove it by induction. It holds for $n=1$. Assume that $n > 1$ and $[0,(n-1)u]$ is σ-complete. We have to prove that $[0,nu]$ is σ-complete.

Let $\{t_i : i \in \mathbb{N}\} \subset [0,nu]$. Put

$$v_i = t_i \wedge (n-1)u, w_i = t_i \vee (n-1)u.$$

Evidently $v_i \in [0,(n-1)u], w_i \in [(n-1)u,nu]$. Since the last interval is isomorphic to $[0,u]$, it is σ-complete, hence there exist the elements

$$v = \bigvee_{i=1}^{\infty} v_i, w = \bigvee_{i=1}^{\infty} w_i.$$

Since $t_i + (n-1)u = v_i + w_i$, we obtain

$$t_i = v_i + w_i - (n-1)u, \qquad i \in \mathbb{N}.$$

If we put
$$t = v + w - (n-1)u,$$
then $t \geq t_i (i \in \mathbb{N})$. Let $s \geq t_i (i \in \mathbb{N})$. Then $s \wedge (n-1)u \geq v_i, s \vee (n-1)u \geq w_i (i \in \mathbb{N})$, hence
$$s \wedge (n-1)u \geq v, s \vee (n-1)u \geq w.$$
Therefore
$$s + (n-1)u + s \vee (n-1)u \geq u + w,$$
$$s \geq u + w - (n-1)u = t,$$
hence $t = \bigvee_{i=1}^{\infty} t_i$. Similarly the existence of $\bigwedge_{i=1}^{\infty} t_i$ can be proved. 🍎

Theorem 11.15 *The MV- algebra $\mathcal{M}_0 = \Gamma(G,u)$ is weakly σ-distributive if and only if the l-group G is weakly σ-distributive.*

Proof: $\Rightarrow$

If $(a_{i,j})_{i,j}$ is a regulator, then, by weak σ-distributivity of G, we have

$$\bigwedge_{\varphi \in \Phi} \bigvee_{i=1}^{\infty} a_{i,\varphi(i)} = 0.$$

$\Leftarrow$

Let $(a_{i,j})_{i,j}$ be a regulator in G. Put

$$b_{i,j} = u \wedge a_{i,j}.$$

Then $(b_{i,j})_{i,j}$ is a regulator.

By weak σ-distributivity of $\mathcal{M}_0 = \Gamma(G,u)$, we obtain

$$\bigwedge_{\varphi \in \Phi} \bigvee_{i=1}^{\infty} b_{i,\varphi(i)} = 0.$$

Since G is a distributive lattice, we have:

$$0 = \bigwedge_{\varphi \in \Phi} \bigvee_{i=1}^{\infty} (u \wedge a_{i,\varphi(i)}) = u \wedge \left(\bigwedge_{\varphi \in \Phi} \bigvee_{i=1}^{\infty} a_{i,\varphi(i)} \right).$$

Put

$$v = \bigwedge_{\varphi \in \Phi} \bigvee_{i=1}^{\infty} a_{i,\varphi(i)},$$

hence

$$u \wedge v = 0.$$

Since every strong unit in G is a weak unit (see [13], Chapter XIII, § 11, Lemma 4), the relation $u \wedge v = 0$ implies $v = 0$, hence

$$\bigwedge_{\varphi \in \Phi} \bigvee_{i=1}^{\infty} a_{i,\varphi(i)} = 0.$$

This concludes the proof. 🍎

11.3 Intuitionistic Fuzzy Sets

We shall consider a measurable space (Ω, S), where $\Omega \neq \emptyset$ and S is a σ-algebra of subsets of Ω. By an *intuitionistic fuzzy event* (shortly *IF-event*) we denote any pair $A = (\mu_A, \nu_A)$ of S-measurable functions $\mu_A, \nu_A : \Omega \to [0,1]$ such that

$$\mu_A + \nu_A \leq 1.$$

The function μ_A is called *membership function*, and ν_A is the non-membership function.

The theory of IF-events has been developed by K. Atanassov and it has many important applications (see [9]). Here we show that the probability theory on the family $\mathcal{F}$ of all IF-events can be considered as a particular case of the probability theory on a suitable MV-algebra.

Example 11.16 Consider $G = \mathbb{R}^2$,

$$(a,b) \hat{+} (c,d) = (a+c, b+d-1), \ (a,b) \leq (c,d) \Longleftrightarrow a \leq c, b \leq d .$$

Then $(\mathbb{R}^2, \hat{+}, \leq)$ is a lattice ordered group. Evidently the operation $\hat{+}$ is commutative and associative, $(0,1)$ is the neutral element, since

$$(0,1) \hat{+} (a,b) = (a+0, b+1-1) = (a,b),$$

and $(-a, 2-b)$ is the inverse element, since

$$(a,b) \hat{+} (-a, 2-b) = (0,1).$$

Furthermore $\leq$ is a partial ordering,

$$(a,b) \vee (c,d) = (\max(a,c), \min(b,d));$$
$$(a,b) \wedge (c,d) = (\min(a,c), \max(b,d)).$$

Finally $(a,b) \leq (c,d) \Rightarrow a \leq c, b \geq d$, and hence

$$a+e \leq c+e;$$

$$b+f-1 \geq d+f-1;$$

$$(a,b) + (e,f) = (a+e, b+f-1) \leq (c+e, d+f-1) = (c,d) + (e,f).$$

Theorem 11.17 *Let (Ω, S) be a measurable space, $\mathcal{M}$ be the family of all pairs $A = (\mu_A, \nu_A)$, where $\mu_A, \nu_A : \Omega \to [0,1]$ are S-measurable functions,*

$$A \leq B \Leftrightarrow \mu_A \leq \mu_B, \nu_A \geq \nu_B,$$
$$A \oplus B = ((\mu_A + \mu_B) \wedge 1, (\nu_A + \nu_B - 1) \vee 0),$$

$$\neg A = (1 - \mu_A, 1 - \nu_A).$$

Then the system $(\mathcal{M}, \oplus, \neg, (0,1), (1,0))$ *is an MV-algebra, and* $\mathcal{F} \subset \mathcal{M}$.

Proof: To see that $\mathcal{M}$ is an MV-algebra, let us consider the group G of all mappings from Ω to $(\mathbb{R}^2, \hat{+}, \leq)$ (see Example 11.16), where

$$(\mu_A, \nu_A) \hat{+} (\mu_B, \nu_B)(\omega) = (\mu_A(\omega), \nu_A(\omega)) \hat{+} (\mu_B(\omega), \nu_B(\omega)),$$
$$(\mu_A, \nu_A) \leq (\mu_B, \nu_B) \Leftrightarrow \mu_A \leq \mu_B, \nu_A \geq \nu_B.$$

Now

$$\mathcal{M} = \{(\mu_A, \nu_A) : 0 \leq \mu_A \leq 1, 0 \leq \nu_A \leq 1\}$$
$$= \{A : (0,1) \leq A = (\mu_A, \nu_A) \leq (1,0)\}. \quad \text{\tiny\Apple}$$

Definition 11.18 By *L-state on* $\mathcal{F}$ we denote any mapping $m : \mathcal{F} \to [0,1]$, satisfying the following conditions:

(i) $m((1,0)) = 1$, $m((0,1)) = 0$;

(ii) $a \odot b = (0,1) \Rightarrow m(a \oplus b) = m(a) + m(b)$;

(iii) $a_n \uparrow a \Rightarrow m(a_n) \uparrow m(a)$.

Theorem 11.19 *For any L-state on* $\mathcal{F}$, $m : \mathcal{F} \to [0,1]$, *there exists exactly one state* $\overline{m} : \mathcal{M} \to [0,1]$ *such that* $\overline{m}|_F = m$.

Proof: Let $A \in \mathcal{M}$. Define

$$\overline{m}((\mu_A, \nu_A)) = m((\mu_A, 0)) - m((0, 1 - \nu(A))).$$

It is not difficult to prove that $\overline{m}$ is a state. If $A = (\mu_A, \nu_A) \in \mathcal{F}$, then

$$(\mu_A, \nu_A) \odot (0, 1 - \nu_A) = (0,1);$$

$$(\mu_A, \nu_A) \oplus (0, 1 - \nu_A) = (\mu_A, 0),$$

hence

$$m((\mu_A, 0)) = m((\mu_A, \nu_A)) + m((0, 1 - \nu_A)).$$

Therefore

$$m((\mu_A, \nu_A)) = m((\mu_A, 0)) - m((0, 1 - \nu_A)) = \overline{m}((\mu_A, \nu_A)). \quad \text{\tiny\Apple}$$

Definition 11.20 By *L-observable on* $\mathcal{F}$ we denote any mapping $x : \mathcal{B}(\mathbb{R}) \to \mathcal{F}$ satisfying the following conditions:

(i) $x(\mathbb{R}) = (1,0)$, $x(\varnothing) = (0,1)$;

(ii) if $A \cap B = \varnothing$, then $x(A) \odot x(B) = (0,1)$ and $x(A \cup B) = x(A) \oplus x(B)$;

(iii) if $A_n \uparrow A$, then $x(A_n) \uparrow x(A)$.

Remark 11.21 It is evident that any L-observable $x : \mathcal{B}(\Omega) \to \mathcal{F} \subset \mathcal{M}$ is an observable according Definition 11.11 This is, together with Theorem 11.19, the key for applications of MV-algebra probability theory to the IF-probability theory.

12 Applications in Probability Theory

Abstract: In this chapter we introduce the concept of independence of states, and give a version of the weak law of large numbers. Moreover, we deal with conditional expectation in the context of Riesz spaces, and give three constructions.

Finally we present further results about probability theory in the context of IF-sets and in particular we deal with joint observables.

12.1 Independence

Recall (see Chapter 11) that a *state* (= probability measure) on an MV-algebra M is a mapping $m : M \to [0,1]$ such that

$m(1) = 1$;

$m(a) = m(b) + m(c)$, whenever $a = b + c$;

$m(a_n) \uparrow m(a)$, whenever $a_n \uparrow a$.

Let $\mathcal{B}(\mathbb{R})$ be as in Chapter 11. A mapping $x : \mathcal{B}(\mathbb{R}) \to M$ is an *observable,* if

$x(\mathbb{R}) = 1$;

$x(A) \odot x(B) = 0$, and $x(A \cup B) = x(A) + x(B)$, whenever $A \cap B = \varnothing$;

$x(A_n) \uparrow x(A)$, whenever $A_n \uparrow A$.

It is easy to see that the composite map $m_x = m \circ x : \mathcal{B}(\mathbb{R}) \to [0,1]$ is a probability measure for any state m and any observable x. If $\xi : \Omega \to \mathbb{R}$ is a random variable on a probability space (Ω, S, P), then its probability distribution $P_\xi : \mathcal{B}(\mathbb{R}) \to [0,1]$ is defined by the formula

$$P_\xi(B) = P(\xi^{-1}(B)), \quad B \in \mathcal{B}(\mathbb{R}).$$

Hence, if we define an observable x on

$$\mathcal{T} = \{ f : \Omega \to [0,1] : f \text{ is } S-\text{measurable function} \}$$

by the formula

$$x(B) = \chi_{\xi^{-1}(B)}$$

and a state $m : \mathcal{T} \to [0,1]$ by the formula

$$m(f) = \int_\Omega f \, dP,$$

then $m_x = m \circ x$ can be presented in the form

Antonio Boccuto / Beloslav Riečan / Marta Vrábelová

$$m_x(B) = \int_\Omega \chi_{\xi^{-1}(B)} dP = P(\xi^{-1}(B)) = P_\xi(B).$$

We see that m_x corresponds with the notion of the probability distribution. Therefore the following definition is natural.

Definition 12.1 An observable $x : \mathcal{B}(\mathbb{R}) \to M$ is called *integrable* if there exists the integral

$$E(x) = \int_{-\infty}^{\infty} t\, dm_x(t).$$

The quantity $E(x)$ is called the *expectation* of x.

We shall be interested in a sequence of independent observables.

Definition 12.2 Let M be an MV-algebra, $m : M \to [0,1]$ a state. We say that the observables $x_1, ..., x_n : \mathcal{B}(\mathbb{R}) \to M$ are *independent* (with respect to m) if there exists an n-dimensional observable $h : \mathcal{B}(\mathbb{R}^n) \to M$ such that

$$m(h(C_1 \times ... \times C_n)) = m(x_1(C_1)) \cdot ... \cdot m(x_n(C_n))$$

for any $C_1, C_2, ..., C_n \in \mathcal{B}(\mathbb{R})$. The mapping h is called the *joint observable* of $x_1, x_2, ..., x_n$.

Example 12.3 Consider a probability space (Ω, S, P) and a sequence of independent random variables $\xi_1, ..., \xi_n : \Omega \to \mathbb{R}$. The independence means that

$$P(\bigcap_{i=1}^n \xi_i^{-1}(C_i)) = \Pi_{i=1}^n P(\xi_i^{-1}(C_i))$$

for any $C_1, ..., C_n \in \mathcal{B}(\mathbb{R})$. Put $T = (\xi_1, ..., \xi_n) : \Omega \to \mathbb{R}^n$. Then

$$\bigcap_{i=1}^n \xi_i^{-1}(C_i) = \{\omega \in \Omega : \xi_1(\omega) \in C_1, ..., \xi_n(\omega) \in C_n\}$$

$$= \{\omega \in \Omega : T(\omega) \in C_1 \times ... \times C_n\} =$$

$$= T^{-1}(C_1 \times ... \times C_n).$$

Put furthermore

$M = \{f : \Omega \to [0,1] : f \text{ is measurable}\}$,

$m : M \to [0,1]$, $m(f) = \int_\Omega f\, dP$, $x_1, ..., x_n : \mathcal{B}(\mathbb{R}) \to M$, $x_i(C) = \chi_{\xi_i^{-1}(C)}$,

$h : \mathcal{B}(\mathbb{R}^n) \to M$, $h(A) = \chi_{T^{-1}(A)}$.

Then the independence of $\xi_1, ..., \xi_n$ can be expressed by the formula

$$m(h(C_1 \times ... \times C_n)) = m(x_1(C_1)) \cdot ... \cdot m(x_n(C_n)).$$

Indeed,

$$m(h(C_1 \times ... \times C_n)) = \int_\Omega \chi_{T^{-1}(C_1 \times ... \times C_n)} dP$$

$$= P(T^{-1}(C_1 \times ... \times C_n))$$

$$= P(\bigcap_{i=1}^{n} \xi_i^{-1}(C_i)) = \Pi_{i=1}^{n} P(\xi_i^{-1}(C_i))$$

$$= \Pi_{i=1}^{n} \int_\Omega \chi_{\xi_i^{-1}(C_i)} dP$$

$$= \Pi_{i=1}^{n} m(x_i(C_i)).$$

The mapping $h : \mathcal{B}(\mathbb{R}^n) \to M$ makes possible to define some functions of $x_1,..., x_n$.

Definition 12.4 Let $g : \mathbb{R}^n \to \mathbb{R}$ be a Borel measurable function, $x_1,..., x_n : \mathcal{B}(\mathbb{R}) \to M$ be independent observables. The mapping $y = g(x_1,..., x_n) : \mathcal{B}(\mathbb{R}) \to M$ is defined by the formula

$$y(B) = h(g^{-1}(B)), \quad B \in \mathcal{B}(\mathbb{R}),$$

where h is the joint observable of $x_1,..., x_n$.

Example 12.5 Let $(\Omega, \mathcal{S}, P)$ be a probability space, $\xi_1,..., \xi_n : \Omega \to \mathbb{R}$ be observables, $T = (\xi_1,..., \xi_n)$, $\eta = g(\xi_1,..., \xi_n)$. Then

$$\eta = g(\xi_1,..., \xi_n) = g \circ T : \Omega \to \mathbb{R},$$

hence

$$(g \circ T)^{-1}(B) = T^{-1}(g^{-1}(B)), \quad B \in \mathcal{B}(\mathbb{R}),$$

and again (putting $h(A) = \chi_{T^{-1}(A)}, A \in \mathcal{B}(\mathbb{R}^n), x_i(C) = \chi_{\xi_i^{-1}(C)}, C \in \mathcal{B}(\mathbb{R}), y(B) = h(g^{-1}(B)), B \in \mathcal{B}(\mathbb{R})$)

$$y(B) = \chi_{T^{-1}(g^{-1}(B))} = \chi_{\eta^{-1}(B)}.$$

We see that the observable $y = g(x_1,..., x_n)$ actually corresponds with the random variable $\eta = g(\xi_1,..., \xi_n)$. If we put

$$g(u_1,..., u_n) = \frac{1}{n}(u_1 + ... + u_n) - a,$$

then we obtain the observable

$$y = \frac{1}{n}(x_1 + ... + x_n) - a.$$

Theorem 12.6 *(Weak law of large numbers). Let* $(x_n)_n$ *be an independent sequence of integrable observables having the same probability distribution* $m_{x_1} = m_{x_2} = ...$ *Let* $a = E(x_1) = E(x_2) = ...$ *Put*

$$y_n = \frac{x_1 + ... + x_n}{n} - a \ (n \in \mathbb{N}).$$

Then, for every $\varepsilon > 0$, *we get:*

$$\lim_{n \to +\infty} m(y_n((-\varepsilon, \varepsilon))) = 1.$$

Proof: We shall construct a special probability space $(\mathbb{R}^{\mathbb{N}}, \sigma, P)$. Here σ is the σ-algebra generated by the family of all cylinders

$$\{u \in \mathbb{R}^{\mathbb{N}} : u_1 \in A_1, ..., u_n \in A_n\},$$

where $n \in \mathbb{N}$ and $A_i \in \mathcal{B}(\mathbb{R})$ $(i = 1, 2, ..., n)$. The probability P is the infinite product of measures $m_{x_1}, m_{x_2}, ...$, hence

$$P(\{u \in \mathbb{R}^{\mathbb{N}} : u_1 \in A_1, ..., u_n \in A_n\}) = m_{x_1}(A_1) \cdot ... \cdot m_{x_n}(A_n)$$

for any cylinder. Define $\xi_n : \mathbb{R}^{\mathbb{N}} \to \mathbb{R}$ by the formula

$$\xi_n((u_i)_i) = u_n.$$

Then ξ_n is a random variable,

$$P_{\xi_n}(A) = P(\xi_n^{-1}(A)) = P(\{u : u_n \in A\}) = m_{x_n}(A),$$

hence

$$E(\xi_n) = \int_{\mathbb{R}} t \, dP_{\xi_n}(t) = \int_{\mathbb{R}} t \, dm_{x_n}(t) = E(x_n)$$

for any n. Moreover $\xi_1, \xi_2, ..., \xi_n, ...$ are independent. Indeed,

$$P(\xi_1^{-1}(A_1) \cap ... \cap \xi_n^{-1}(A_n)) =$$

$$= P(\{t \in \mathbb{R}^{\mathbb{N}} : t_1 \in A_1, ..., t_n \in A_n\}) =$$

$$= m_{x_1}(A_1) \cdot ... \cdot m_{x_n}(A_n) =$$

$$= P_{\xi_1}(A_1) \cdot ... \cdot P_{\xi_n}(A_n) =$$

$$= P(\xi_1^{-1}(A_1)) \cdot ... \cdot P(\xi_n^{-1}(A_n))$$

for any $n \in \mathbb{N}$, and any $A_1, ..., A_n \in \mathcal{B}(\mathbb{R})$. By the classical law of large numbers we have:

$$\lim_{n \to +\infty} P\left(\left\{ u : \left| \frac{\xi_1(u) + \dots + \xi_n(u)}{n} - a \right| < \varepsilon \right\} \right) = 1.$$

Of course,

$$m(y_n((-\varepsilon, \varepsilon))) = m(h_n(g_n^{-1}((-\varepsilon, \varepsilon)))),$$

where

$$g_n(u_1, \dots, u_n) = \frac{1}{n}(u_1, \dots, u_n) - a$$

and

$$m(h_n(C)) = P(\{u : (\xi_1(u), \dots, \xi_n(u)) \in C\})$$

for any $C \in \mathcal{B}(\mathbb{R}^n)$. Therefore

$$m(y_n((-\varepsilon, \varepsilon))) = P(\{u : (\xi_1(u), \dots, \xi_n(u)) \in g_n^{-1}((-\varepsilon, \varepsilon))\}) =$$

$$P(\{u : g_n((\xi_1(u), \dots, \xi_n(u))) \in (-\varepsilon, \varepsilon)\}) =$$

$$= P\left(\left\{ u : \left| \frac{1}{n}(\xi_1(u) + \dots + \xi_n(u)) - a \right| < \varepsilon \right\} \right).$$

This concludes the proof.

12.2 Conditional Probability

Everybody knows the definition of conditional probability

$$P(A \mid B) = \frac{P(A \cap B)}{P(B)}$$

defined under the assumption that $P(B) > 0$, or

$$P(A \cap B) = P(A \mid B)P(B)$$

generally. The next step is the definition of the conditional probability in the case that there are given disjoint sets $B_1, \dots, B_k$ of positive probability such that

$$P\left(\Omega \setminus \bigcup_{i=1}^{k} B_i \right) = 0.$$

Let $\mathcal{S}_0$ be the σ-algebra generated by $B_1, \dots, B_k$. Then the conditional probability $P(A \mid \mathcal{S}_0)$ is defined by the formula

$$P(A \mid \mathcal{S}_0) = \sum_{i=1}^{k} P(A \mid B_i)\chi_{B_i} + c\chi_B,$$

where $c \in \mathbb{R}$ is arbitrary and $B = \Omega \setminus \bigcup_{i=1}^{k} B_i$. Evidently

1) $P(A \mid \mathcal{S}_0) : \Omega \to \mathbb{R}$ is an $\mathcal{S}_0$-measurable function

and

2) $\int_C P(A\,|\,S_0)\,dP = P(A\cap C)$

for any $C \in S_0$.

Consider now a random variable $\xi : \Omega \to \mathbb{R}$ and a sub-σ-algebra $S_0 \subset S$. Then the conditional expectation $E(\xi\,|\,S_0)$ is defined by the following two properties:

(i) $E(\xi\,|\,S_0): \Omega \to \mathbb{R}$ is S_0-measurable,

(ii) $\int_C E(\xi\,|\,S_0)\,dP = \int_C \xi\,dP$ for any $C \in S_0$.

Again $E(\xi\,|\,S_0)$ is not defined uniquely, but any two functions satisfying (i) and (ii) coincide almost everywhere. Also we get

$$E(\chi_A\,|\,S_0) = P(A\,|\,S_0)$$

for any $A \in S$. The existence of $E(\xi\,|\,S_0)$, and in particular of $P(A\,|\,S_0)$, can be guaranteed by the Radon - Nikodým theorem. Namely, if we put

$$\nu(C) = \int_C \xi\,dP, \quad \mu(C) = P(C), \quad C \in S_0,$$

then $\nu \ll \mu$, and there exists an S_0-measurable function $f : \Omega \to \mathbb{R}$ such that

$$\nu(C) = \int_C f\,d\mu, \quad C \in S_0.$$

The problem is the construction of the conditional expectation in the case that $\xi : \Omega \to R$, where R is a Riesz space or a similar structure. We shall present here three constructions. The first one was proposed by P. Maličký (see [180]).

Let R be a Dedekind complete Riesz space, and let $f : \Omega \to R$ be simple, i.e. $\xi = \sum_{i=1}^{n} \chi_{B_i} a_i$, where the B_i's are disjoint sets from S, and $a_i \in R\,(i = 1,2,...,n)$. Then he puts

$$E(\xi\,|\,S_0) = \sum_{i=1}^{n} E(\chi_{B_i}\,|\,S_0)a_i.$$

Then $E(.\,|\,S_0)$ has all expected properties. In the next step the function $E(.\,|\,S_0)$ is extended from the set F of all simple functions to the set $\overline{F}$ of all uniform limits of elements of F.

The second construction has been described in [226] and it can be realized in MV-algebras with product.

Definition 12.7 An *MV-algebra with product* is a pair $(M,.)$ where M is an MV-algebra and . is a commutative and associative binary operation on M satisfying the following conditions:

(i) $1.a = a$ for all $a \in M$;

(ii) $a.(b+c)=(a.c)+(b.c)$ whenever $a,b,c\in M, b+c\leq 1$.

The second limitation of the theory is the equality $\mathcal{S}_0 = y(\mathcal{B}(\mathbb{R}))$ where $y: \mathcal{B}(\mathbb{R}) \to M$ is an observable. Recall that in the case $\mathcal{S}_0 = \eta^{-1}(\mathcal{B}(\mathbb{R}))$ the expected value $E(\xi\,|\,\eta)$ can be defined as a Borel function such that

$$\int_B E(\xi\,|\,\eta)dP_\eta = \int_{\eta^{-1}(B)} \xi\,dP, \quad \forall\, B\in \mathcal{B}(\mathbb{R}).$$

Namely, if $\mathcal{S}_0 = \eta^{-1}(\mathcal{B}(\mathbb{R}))$, then any $\mathcal{S}_0$-measurable function f can be written in the form $f = g\circ\eta$, where g is a Borel function, and

$$\int_{\eta^{-1}(B)} \xi\,dP = \int_{\eta^{-1}(B)} E(\xi\,|\,\mathcal{S}_0)dP = \int_{\eta^{-1}(B)} g\circ\eta\,dP = \int_B g\,dP_\eta.$$

Therefore we can define directly $g = E(\xi\,|\,\eta) : E(\xi\,|\,\eta)$ is in one - to - one correspondence with $E(\xi\,|\,\mathcal{S}_0)$.

Of course, in the general case one can define also $\int_c x\,dm$. We have defined only

$$\int x\,dm = \int_{\mathbb{R}} t\,dm_x(t).$$

The value $\int_c x\,dm$ corresponds to $\int_c \xi\,dP = \int_\Omega \xi\chi_C\,dP$. Inspiring by $(\xi\chi_C)^{-1}(B)$ we first define x_c.

Definition 12.8 Let x be an integrable observable. Then we define

$x_c(A) = c.x(A)$, if $0\notin A$;

$x_c(A) = c.x(A) + \neg A$, if $0\in A$.

Definition 12.9 If x is an integrable observable, $c\in M$, then we define

$$\int_c x\,dm = \int_{\mathbb{R}} t\,dm_{x_c}(t).$$

Definition 12.10 If x, y are integrable observables, then $E(x\,|\,y): \mathbb{R} \to \mathbb{R}$ is a Borel function such that

$$\int_{y(B)} x\,dm = \int_B E(x\,|\,y)dm_y$$

for any $B\in \mathcal{B}(\mathbb{R})$.

The third construction has been recently suggested by Dvurečenskij and Pulmannová. It is based on the analogous of the Loomis - Sikorski theorem given by Dvurečenskij and Mundici (see [102, 103, 104, 194]).

Theorem 12.11 *Let M be a σ-complete MV-algebra. Let Ω be the set of maximal ideals of M with the spectral topology. Then there exists a tribe $\mathcal{T}$ over Ω and a σ-homomorphism η of $\mathcal{T}$ onto M satisfying the following conditions:*

For each $a\in M$ there is exactly one continuous function $a^*\in \mathcal{T}$ such that $\eta(a^*)=a$. A function in $\mathcal{T}$ has the same image as a^* if and only if it differs from a^* on a meager subset of Ω.

If we define $m^* : T \to [0,1]$ by the formula $m^*(a^*) = m(a)$, we obtain a state. By a theorem of Butnariu and Klement (see [51]), on the σ-algebra $S = \{A \subset \Omega : \chi_A \in T\}$ there exists a probability measure $P : S \to [0,1]$ such that

$$m^*(f) = \int_\Omega f \, dP$$

for any $f \in T$, hence $m(a) = m^*(a^*) = \int_\Omega a^* dP$.

Suppose $A \in M$, $M_0 \subset M$ is a σ-complete sub-MV-algebra of M. Consider $B(M_0)$ the family of all idempotents, $B(M_0) = \{a \in M : a \oplus a = a\}$. $B(M_0)$ is a Boolean σ-algebra, hence

$$S_0 = \{B \subset \Omega : \eta(\chi_B) \in B(M_0)\}$$

is a σ-algebra of subsets of Ω. Define $\mu, \nu : S_0 \to [0,1]$ by the formulas

$$\nu(B) = \int_B a^* dP,$$
$$\mu(B) = P(B).$$

Evidently $\nu \ll \mu$, hence there exists a function (denote it by $m(a \mid B(M_0))$) from Ω to $\mathbb{R}$ such that

$$\int_B m(a \mid B(M_0)) dP = \nu(B) = \int_B a^* dP \quad \forall B.$$

Of course, if $b = \eta(\chi_B) \in B(M_0)$, then

$$m(a \wedge b) = m^*(a^* \wedge \chi_B) = \int_\Omega a^* \wedge \chi_B \, dP.$$

By Lemma 4.2 of [104], $a^* \wedge \chi_B = a^*.\chi_B$, hence

$$m(a \wedge b) = \int_\Omega a^* \wedge \chi_B dP = \int_\Omega a^* \chi_B dP = \int_B a^* dP.$$

This leads to the following definition ([104]):

Definition 12.12 Given $a \in M$ and $M_0 \subset M$, a *conditional expectation* $m(a \mid B(M_0)) : \Omega \to \mathbb{R}$ is an S_0−measurable function such that

$$\int_B m(a \mid B(M_0)) dP = m(a \wedge b)$$

whenever $b = \eta(\chi_B) \in B(M_0)$.

12.3 Probability Theory on IF-events

We have seen in Section 11.3 that the probability theory on the set $\mathcal{F}$ of all IF-events can be considered as a particular case of the MV-algebra probability theory. Moreover, we shall show that there exist always joint observables.

Theorem 12.13 *The operation $\cdot$ defined on $\mathcal{F}$ by*

$$(x_1, y_1) \cdot (x_2, y_2) = (x_1 \cdot x_2, 1 - (1 - y_1) \cdot (1 - y_2)) = (x_1 \cdot x_2, y_1 + y_2 - y_1 \cdot y_2)$$

for each $(x_1, y_1), (x_2, y_2) \in \mathcal{F}$ satisfies the following conditions:

(i) $(1,0) \cdot (a_1, a_2) = (a_1, a_2)$ for each $(a_1, a_2) \in \mathcal{F}$;

(ii) the operation $\cdot$ is commutative and associative;

(iii) if $(a_1, a_2) \odot (b_1, b_2) = (0,1)$ and $(a_1, a_2), (b_1, b_2) \in \mathcal{F}$, then

$$(c_1, c_2) \cdot ((a_1, a_2) \oplus (b_1, b_2)) = ((c_1, c_2) \cdot (a_1, a_2)) \oplus ((c_1, c_2) \cdot (b_1, b_2))$$

and

$$((c_1, c_2) \cdot (a_1, a_2)) \oplus ((c_1, c_2) \cdot (b_1, b_2)) = (0,1)$$

for each $(c_1, c_2) \in \mathcal{F}$.

Proof: (i) Let (a_1, a_2) be an element of the family $\mathcal{F}$. Then

$$(1,0) \cdot (a_1, a_2) = (1 \cdot a_1, 0 + a_2 - 0 \cdot a_2) = (a_1, a_2).$$

(ii) Straightforward.

(iii) Let $(a_1, a_2) \odot (b_1, b_2) = (0,1)$. Then

$$
\begin{aligned}
(c_1, c_2) \cdot ((a_1, a_2) \oplus (b_1, b_2)) &= (c_1, c_2) \cdot (a_1 \oplus b_1, a_2 \odot b_2) \\
&= (c_1, c_2) \cdot ((a_1 + b_1) \wedge 1, (a_2 + b_2 - 1) \vee 0) \\
&= (c_1, c_2) \cdot (a_1 + b_1, a_2 + b_2 - 1) \\
&= (a_1 c_1 + b_1 c_1, a_2 + b_2 + 2c_2 - 1 - a_2 c_2 - b_2 c_2)
\end{aligned}
$$

and

$$
\begin{aligned}
((c_1, c_2) \cdot (a_1, a_2)) \oplus ((c_1, c_2) \cdot (b_1, b_2)) &= \\
&= (c_1 a_1, c_2 + a_2 - c_2 a_2) \oplus (c_1 b_1, c_2 + b_2 - c_2 b_2) \\
&= (c_1 a_1 \oplus c_1 b_1, (c_2 + a_2 - c_2 a_2) \odot (c_2 + b_2 - c_2 b_2)) \\
&= ((c_1 a_1 + c_1 b_1) \wedge 1, (a_2 + b_2 + 2c_2 - 1 - a_2 c_2 - b_2 c_2) \vee 0) \\
&= (a_1 c_1 + b_1 c_1, a_2 + b_2 + 2c_2 - 1 - a_2 c_2 - b_2 c_2).
\end{aligned}
$$

Hence

$$(c_1, c_2) \cdot ((a_1, a_2) \oplus (b_1, b_2)) = ((c_1, c_2) \cdot (a_1, a_2)) \oplus ((c_1, c_2) \cdot (b_1, b_2)).$$

Moreover

$$
\begin{aligned}
((c_1, c_2) \cdot (a_1, a_2)) \odot ((c_1, c_2) \cdot (b_1, b_2)) &= \\
&= (c_1 a_1, c_2 + a_2 - c_2 a_2) \odot (c_1 b_1, c_2 + b_2 - c_2 b_2) \\
&= (c_1 a_1 \odot c_1 b_1, (c_2 + a_2 - c_2 a_2) \oplus (c_2 + b_2 - c_2 b_2)) = (0,1). \quad \blacksquare
\end{aligned}
$$

The next important notion is the one of joint observable.

Definition 12.14 Let $x, y : \mathcal{B}(\mathbb{R}) \to \mathcal{F}$ be two L-observables. The joint L-observable of the L-observables x, y is the mapping $h : \mathcal{B}(\mathbb{R}^2) \to \mathcal{F}$ satisfying the following conditions:

(i) $h(\mathbb{R}^2) = (1, 0)$, $h(\emptyset) = (0, 1)$;

(ii) if $A, B \in \mathcal{B}(\mathbb{R}^2)$ and $A \cap B = \emptyset$, then $h(A \cup B) = h(A) \oplus h(B)$ and $h(A) \odot h(B) = (0, 1)$;

(iii) if $A, A_1, \ldots \in \mathcal{B}(\mathbb{R}^2)$ and $A_n \uparrow A$, then $h(A_n) \uparrow h(A)$;

(iv) $h(C \times D) = x(C) \cdot y(D)$ for each $C, D \in \mathcal{B}(\mathbb{R})$.

Theorem 12.15 *For each two L-observables $x, y : \mathcal{B}(\mathbb{R}) \to \mathcal{F}$ there exists their joint L-observable.*

Proof: Put $x(A) = (x^\flat(A), 1 - x^\sharp(A))$, $y(B) = (y^\flat(B), 1 - y^\sharp(B))$ We want to construct $h(C) = (h^\flat(C), 1 - h^\sharp(C))$. Fix $\omega \in \Omega$ and put

$$\mu(A) = x^\flat(A)(\omega), \quad \nu(B) = y^\flat(B)(\omega).$$

It is not difficult to prove that $\mu, \nu : \mathcal{B}(\mathbb{R}) \to [0, 1]$ are probability measures. Let $\mu \times \nu : \mathcal{B}(\mathbb{R}^2) \to [0, 1]$ be the product of measures and define

$$h^\flat(A)(\omega) = \mu \times \nu(A).$$

Then $h^\flat : \mathcal{B}(\mathbb{R}^2) \to \mathcal{T}$, where $\mathcal{T}$ is the family of all $\mathcal{S}$-measurable functions from Ω to $[0, 1]$. If $C, D \in \mathcal{B}(\mathbb{R})$, then

$$h^\flat(C \times D)(\omega) = \mu \times \nu(C \times D) = \mu(C) \cdot \nu(D) = x^\flat(C)(\omega) \cdot y^\flat(D)(\omega).$$

Hence

$$h^\flat(C \times D) = x^\flat(C) \cdot y^\flat(D).$$

Similarly we can construct $h^\sharp : \mathcal{B}(\mathbb{R}^2) \to \mathcal{T}$, such that

$$h^\sharp(C \times D) = x^\sharp(C) \cdot y^\sharp(D).$$

Put

$$h(A) = (h^\flat(A), 1 - h^\sharp(A)), \quad A \in \mathcal{B}(\mathbb{R}^2).$$

By Theorem 12.13, for any $C, D \in \mathcal{B}(\mathbb{R})$ we get:

$$
\begin{aligned}
x(C) \cdot y(D) &= (x^\flat(C), 1 - x^\sharp(C)) \cdot (y^\flat(D), 1 - y^\sharp(D)) \\
&= \left(x^\flat(C) \cdot y^\flat(D), 1 - (1 - (1 - x^\sharp(C))) \cdot (1 - (1 - y^\sharp(D))) \right) \\
&= (x^\flat(C) \cdot y^\flat(D), 1 - x^\sharp(C) \cdot y^\sharp(D)) \\
&= (h^\flat(C \times D), 1 - h^\sharp(C \times D)) = h(C \times D).
\end{aligned}
$$

We have proved that the MV-algebra probability can be applied to L-states on the family $\mathcal{F}$ of all IF-events.

Moreover, we proved that M endowed with the product defined in Theorem 12.13 is an MV-algebra with product (Definition 12.7), hence also the theory of MV-algebra conditional probability and expectation can be applied to the family $\mathcal{F}$ (see [226], Part 3.2).

13 Integration in Metric Semigroups

Abstract: In this chapter we present a Kurzweil-Henstock-type integral for metric semigroup-valued functions, defined on (possibly unbounded) subintervals of the extended real line. An example of a metric semigroup which is not a group is the set of all fuzzy numbers.

Besides the elementary properties we prove a version of the Henstock lemma and some convergence theorems in this setting.

13.1 Elementary Properties

Although the book is devoted to measure and integration theory on ordered spaces, this chapter contains some result concerning structures without ordering. More precisely, it deals with Kurzweil-Henstock integration for functions, defined on a (not necessarily bounded) interval $[A, B]$ of the extended real line and with values in a metric semigroup. The basic example is the set $L(\mathbb{R})$ of all fuzzy numbers. The idea of investigating integration in the context of metric semigroups arises from two papers of M. Matloka ([185, 186]; see also [218]).

The Kurzweil-Henstock integral for functions defined in unbounded intervals was introduced and investigated in [36] for Banach space-valued maps and in [26] for Riesz space-valued mappings (for real-valued functions, see [170]). In [282, 283] the "classical" Kurzweil-Henstock integral for $L(\mathbb{R})$-valued maps was studied (see also [139, 140, 141, 153, 163, 198, 281, 284, 285, 286]).

In this chapter some basic properties of the KH-integral for metric semigroup-valued functions defined in (possibly unbounded) subintervals of the extended real line are investigated, and some convergence theorems are proved (see [28]).

Definition 13.1 A *metric semigroup* is a structure $(X, \rho, +, \cdot)$, where $\rho : X \times X \to \mathbb{R}$, $+ : X \times X \to X$, $\cdot : \mathbb{R} \times X \to X$ satisfy the following conditions:

(i) (X, ρ) is a complete metric space;

(ii) $(X, +)$ is a commutative semigroup endowed with a neutral element 0;

(iii) $\rho(w + y, z + t) \leq \rho(w, z) + \rho(y, t)$ for any $w, y, z, t \in X$;

(iv) $\rho(\alpha w, \alpha y) \leq |\alpha| \rho(w, y)$ for all $\alpha \in \mathbb{R}$ and $w, y \in X$;

(v) $\alpha(w + y) = \alpha w + \alpha y$ for each $\alpha \in \mathbb{R}$, $w, y \in X$;

(vi) $(\alpha + \beta)w = \alpha w + \beta w$ for every $\alpha, \beta \in \mathbb{R}_0^+$, $w \in X$, $0 \cdot w = 0$ and $1 \cdot w = w$ for each $w \in X$.

A metric semigroup $(X, \rho, +, \cdot)$ is called *invariant*, if $\rho(w + z, y + z) = \rho(w, y)$ for any $w, y, z \in X$.

Observe that a consequence of invariance and the triangular property is the following condition:

(vii) $\qquad \rho(w + y, z) \leq \rho(w, t) + \rho(y + t, z)$ whenever $x, y, z, t \in X$.

An example of metric semigroup is the set of all fuzzy numbers (see also [28, 282]).

Definition 13.2 A *fuzzy number* or *fuzzy set* is a function $\mu : \mathbb{R} \to [0,1]$ satisfying the following conditions:

(j) there exists $x_0 \in \mathbb{R}$ such that $\mu(x_0) = 1$;

(jj) the α-*cut* set $\mu_\alpha = \{x \in \mathbb{R} : \mu(x) \geq \alpha\}$ is convex for $\alpha \in]0,1]$;

(jjj) μ is upper semi-continuous, i.e. any α-cut μ_α is a closed subset of $\mathbb{R}$;

(jv) the support $\overline{\{x \in \mathbb{R} : \mu(x) > 0\}}$ of the function μ is a compact set.

Any real number u_0 can be identified with a fuzzy number μ_0 in the following way:

$$\mu_0(x) = \chi_{\{u_0\}}(x),$$

i.e. $\mu_0(u_0) = 1$, and $\mu_0(x) = 0$, if $x \neq u_0$.

The set of all fuzzy numbers is denoted by $L(\mathbb{R})$.

We now endow $L(\mathbb{R})$ with a metric and a linear structure (see also [28, 282]). We define the *Hausdorff distance* H on the set of all compact possibly degenerate intervals in $\mathbb{R}$:

$$H([a,b],[c,d]) = \max(|c-a|,|d-b|).$$

Let $\mu, v \in L(\mathbb{R})$. It is easy to check that, for every $\alpha \in (0,1]$, there exist a, b, c, $d \in \mathbb{R}$ (depending on α) such that $\mu_\alpha = [a,b]$, $v_\alpha = [c,d]$. So, for $\mu, v \in L(\mathbb{R})$, set

$$\rho(\mu,v) = \sup\{H(\mu_\alpha, v_\alpha) : \alpha \in (0,1]\}.$$

Using this definition $(L(\mathbb{R}), \rho)$ becomes a complete metric space.

To define a linear structure on $L(\mathbb{R})$, recall that every fuzzy number is completely determined by its α-cuts. Hence, for any $\mu, v \in L(\mathbb{R})$, $\alpha \in \mathbb{R}^+$ and $\lambda \in \mathbb{R}$, set

$$(\mu + v)_\alpha = \mu_\alpha + v_\alpha,$$
$$(\lambda\mu)_\alpha = \lambda\mu_\alpha$$

(here, $V + Z = \{v + z : v \in V, z \in Z\}; \lambda V = \{\lambda v : v \in V\}$).

Finally, we note that $(L(\mathbb{R}), +)$ is not a group, but only a semigroup (see also [28]), in fact let $\mu \in L(\mathbb{R})$ be defined by the formula:

$$\mu(x) = \begin{cases} x, & \text{if } x \in [0,1]; \\ 2-x, & \text{if } x \in [1,2]; \\ 0, & \text{otherwise.} \end{cases}$$

Then $-\mu = (-1) \cdot \mu$ is given by

$$-\mu(x) = \begin{cases} -x, & \text{if } x \in [-1,0]; \\ 2+x, & \text{if } x \in [-2,-1]; \\ 0, & \text{otherwise.} \end{cases}$$

Note that $\mu(x)+(-\mu(x))$ is not the zero element $0 := \chi_{\{0\}}(x)$, but

$$\mu(x)+(-\mu(x)) = \begin{cases} 1-\dfrac{x}{2}, & \text{if } x \in [0,2]; \\[2mm] 1+\dfrac{x}{2}, & \text{if } x \in [-2,0]; \\[2mm] 0, & \text{otherwise.} \end{cases}$$

On the other hand the subset $R_0 \subset L(\mathbb{R})$ consisting of all functions $\chi_{\{a\}}$, $a \in \mathbb{R}$, is group isomorphic to the commutative group $(\mathbb{R},+)$. There are many applications of the space $L(\mathbb{R})$ in the fuzzy set theory. Let us mention an application from the probability theory.

A *random variable* is a measurable map from the probability space (X,B,P) to $\mathbb{R}$, a *fuzzy random variable* is a measurable mapping from the probability space $(X,\mathcal{B},P)$ to $L(\mathbb{R})$. Since expectation in any probability model of the Kolmogorov type coincides with an abstract integral of the Lebesgue type, the theory of the Kurzweil-Henstock integral with values in $L(\mathbb{R})$ could be useful in the theory of fuzzy random variables. Moreover, the axiomatic approach presented in Definition 13.1 could be simpler for exposition and possibly useful for applications.

Of course, every random variable with values in a Banach space is also a special kind of function with values in a metric semigroup.

We now introduce the Kurzweil-Henstock integral for functions with values in a metric semigroup X. From now on, we denote by $[A,B]$ a closed interval or halfline contained in $\tilde{\mathbb{R}}$, or the whole of $\tilde{\mathbb{R}}$. Moreover, given a measurable set $E \subset \tilde{\mathbb{R}}$, we denote by $|E|$ its Lebesgue measure (this quantity can be finite or $+\infty$). Our integral deals with X-valued functions defined on $[A,B]$, but it can be investigated analogously if we take functions defined on $\mathbb{R}$ or on halflines of the type $[a,+\infty)$ or $(-\infty,a]$, with $a \in \mathbb{R}$. The concepts of *partition, decomposition, gauge, γ-, δ-fineness* and *Riemann sum* are as the ones formulated in Chapter 7.

We now formulate our definition of Kurzweil-Henstock integral for functions defined on $[A,B]$ and with values in a metric semigroup X.

Definition 13.3 We say that a function $f:[A,B] \to X$ is *Kurzweil-Henstock integrable* (in short *KH-integrable* or simply *integrable*) on $[A,B]$ if there exists an element $I \in X$ such that $\forall \varepsilon > 0$ there exist a function $\delta:[A,B] \to \mathbb{R}^+$ and a positive real number P such that

$$\rho(\sum_{\Pi} f, I) \le \varepsilon \qquad\qquad (13.1)$$

whenever $\Pi = \{(I_k, t_k) : k = 1,\dots,p\}$ is a δ-fine partition of any bounded interval $[a,b]$ with $[a,b] \supset [A,B] \cap [-P,P]$ and $[a,b] \subset [A,B]$. In this case we say that I is the *KH-integral of f*, and we denote the element I by the symbol $\int_A^B f$.

It is easy to see that the element I is uniquely determined: this follows from the inequality

$$\rho(I',I'') \le \rho(I',\sum_{\Pi} f) + \rho(\sum_{\Pi} f,I'').$$

It is easy to check that, if we take $[A,B]=[a,b]$, where $[a,b]$ is a closed bounded subinterval of $\mathbb{R}$, then our definition of KH-integrability is equivalent to the following, similar to the classical one:

Definition 13.4 A map $f:[a,b] \to X$ is said to be KH-integrable on $[a,b]$ if there exists $I \in X$ such that to every $\varepsilon > 0$ there exists a function $\delta:[a,b] \to \mathbb{R}^+$ such that

$$\rho(\sum_{\Pi} f, I) \le \varepsilon$$

for any δ-fine partition Π of $[a,b]$.

We now prove the following characterization of KH-integrability when $[A,B] \subset \tilde{\mathbb{R}}$ is not necessarily bounded.

Theorem 13.5 *A function $f:[A,B] \to X$ is KH-integrable on $[A,B]$ if and only if there exists $J \in X$ such that $\forall \varepsilon > 0$ there exists a gauge γ on $[A,B]$ such that*

$$\rho(\sum_{\Pi} f, J) \le \varepsilon \qquad (13.2)$$

whenever $\Pi = \{(I_k, t_k) : k = 1, \ldots, p\}$ is a γ-fine partition of $[A,B]$, and in this case we have $\int_A^B f = J$.

Proof: Though the proof is similar to the analogous one of Chapter 7, for the sake of clearness we report it.

We begin with the "only if" part. By hypothesis, $\forall \varepsilon > 0$ there exist a function $\delta:[A,B] \to \mathbb{R}^+$ and a positive real number P such that (13.1) holds. We now define on $[A,B]$ a gauge γ in the following way:

$$\gamma(t) = \begin{cases} (t - \delta(t), t + \delta(t)) & \text{if } t \in [A,B] \cap \mathbb{R}, \\ [-\infty, -P) & \text{if } t = -\infty \text{ and } A = -\infty, \\ (P, +\infty] & \text{if } t = +\infty \text{ and } B = +\infty. \end{cases}$$

We observe that every γ-fine partition $\Pi = \{(I_k, t_k) : k = 1, \ldots, p\}$ of $[A,B]$ is such that $I_k \subset \gamma(t_k)$ $\forall k = 1, \ldots, p$. In the case $A = -\infty$, $B = +\infty$, the partition Π contains two unbounded intervals, which we call J and K: of course, if $\inf J = -\infty$ and $\sup K = +\infty$, then the t_k's associated with J and K are $-\infty$ and $+\infty$ respectively. Then, since Π is γ-fine, we have $J \subset \gamma(-\infty)$ and $K \subset \gamma(+\infty)$. Then $J \subset [-\infty, -P)$ and $K \subset (P, +\infty]$. So, if $a = \sup J$ and $b = \inf K$, then $[a,b]$ is a bounded interval, containing $[-P,P]$. If Π' is the restriction of Π to $[a,b]$, then Π' is δ-fine, and by construction we get

$$\sum_{\Pi'} f = \sum_{\Pi} f. \qquad (13.3)$$

In this case, the assertion follows from (13.1) and (13.3).

In the case $A \in \mathbb{R}$, $B = +\infty$, the partition Π contains only an unbounded interval K, with $\sup K = +\infty$. Let P be associated with K as above, and $b = \inf K$: we have $P \leq b$. We note that, without loss of generality, P can be taken greater than $|A|$. Thus, $[A,b]$ is a bounded interval, containing $[-P,P]$, and the assertion follows by proceeding as in the previous case. The case $A = -\infty$, $B \in \mathbb{R}$ is analogous to the previous one. Finally, if $[A,B]$ is bounded, then the assertion is straightforward, because in this case the number P can be taken greater than $\max(|A|,|B|)$ and, of course, (13.1) holds even in the case $[a,b] = [A,B]$. This concludes the proof of the "only if" part.

We now turn to the "if" part. By hypothesis, we know that $\forall \varepsilon > 0$ there exists a gauge γ on $[A,B]$, satisfying (13.2). By definition of gauge, there exist $\delta_1, \delta_2 : [A,B] \to \mathbb{R}^+$ such that

$$\gamma(t) = (t - \delta_1(t), t + \delta_2(t)) \quad \forall t \in [A,B] \cap \mathbb{R}.$$

For such t's, let $\delta(t) = \min\{\delta_1(t), \delta_2(t)\}$. Moreover, if $+\infty$ and $-\infty$ belong to $[A,B]$, and $\gamma(-\infty) = [-\infty, P_1^*)$, $\gamma(+\infty) = (P_2^*, +\infty]$, put $P_1 = \min\{P_1^*, -1\}$, $P_2 = \max\{P_2^*, 1\}$, $P = \max\{-P_1, P_2\}$: we note that, in the case $A \in \mathbb{R}$ (resp. $B \in \mathbb{R}$), P can be chosen greater than $|A|$ (resp. $|B|$); moreover, set $\delta(-\infty) = \delta(+\infty) = P$. Let now $[a,b] \subset [A,B]$ be any bounded interval, containing $[A,B] \cap [-P,P]$, and $\Pi = \{(I_k, t_k) : k = 1, \dots, p\}$ be a δ-fine partition of $[a,b]$. Let Π' be that partition of $[A,B]$, whose elements are the ones of Π with the addition of $([A,a], A)$, if $A = -\infty$, and $([b,B], B)$, if $B = +\infty$: we note that Π' is γ-fine. This follows from the fact that, if (I_k, t_k) is any element of Π, then

$$I_k \subset (t_k - \delta(t_k), t_k + \delta(t_k)) \subset (t_k - \delta_1(t_k), t_k + \delta_2(t_k)) = \gamma(t_k),$$

and from the following inclusions:

$$(b, +\infty] \subset (P, +\infty] \subset (P_2, +\infty] \subset (P_2^*, +\infty] = \gamma(+\infty),$$
$$[-\infty, a) \subset [-\infty, P) \subset [-\infty, P_1) \subset [-\infty, P_1^*) = \gamma(-\infty).$$

Then, taking into account that the Riemann sum concerning the partition Π' is done without considering the unbounded intervals (see Chapter 7), we get

$$\sum_{\Pi'} f = \sum_{\Pi} f.$$

From this and (13.2) the assertion follows, by proceeding analogously as at the end of the proof of the converse implication. This concludes the proof of the theorem. $\blacklozenge$

Proposition 13.6 *A function* $f : [A,B] \to X$ *is KH-integrable on* $[A,B]$ *if and only if the Bolzano-Cauchy condition is satisfied:*

For every $\varepsilon > 0$ *there exists a gauge* γ *on* $[A,B]$ *such that*

$$\rho(\sum_{\Pi_1} f, \sum_{\Pi_2} f) \leq \varepsilon$$

whenever Π_1, Π_2 *are* γ*-fine partitions.*

Proof: The integrability of f implies evidently the Bolzano-Cauchy condition: this follows from the inequality

$$\rho(\sum_{\Pi_1} f, \sum_{\Pi_2} f) \leq \rho(\sum_{\Pi_1} f, I) + \rho(I, \sum_{\Pi_2} f).$$

Let f satisfy the Bolzano-Cauchy condition. Put $\varepsilon = \dfrac{1}{n}$. Then, for all $n \in \mathbb{N}$, there exists a gauge γ_n on $[A, B]$ such that

$$\rho(\sum_{\Pi_1} f, \sum_{\Pi_2} f) \leq \frac{1}{n}$$

whenever Π_1, Π_2 are γ_n-fine partitions. Put $\eta_n = \gamma_1 \cap \gamma_2 \cap ... \cap \gamma_n$, $\forall n \in \mathbb{N}$. Then

$$\gamma_1 = \eta_1 \supset \eta_2 \supset \eta_3 \supset ... \supset \eta_n.$$

Put

$$A_n = \{x \in X : \exists \text{ an } \eta_n - \text{fine partition} \, \Pi_1 : x = \sum_{\Pi_1} f\}, \quad \forall n \in \mathbb{N}.$$

If $x, y \in A_n$, then $\rho(x, y) \leq \dfrac{1}{n}$, hence

$$\text{diam} \, \overline{A_n} = \text{diam} \, A_n \leq \frac{1}{n}.$$

Since $\eta_{n+1} \subset \eta_n$, we obtain $\overline{A_{n+1}} \subset \overline{A_n}$. Since X is complete, there exist exactly one element

$$I \in \cap_{n=1}^{\infty} \overline{A_n}.$$

Let $\varepsilon > 0$. Choose $n \in \mathbb{N}$ such that $\dfrac{1}{n} < \varepsilon$. Consider now η_n. If Π is any η_n-fine partition, then

$$\sum_{\Pi} f \in A_n.$$

Since $I \in \overline{A_n}$, we obtain

$$\rho(\sum_{\Pi} f, I) \leq \frac{1}{n} < \varepsilon.$$

Therefore f is integrable and $I = \int_A^B f$. 

Proposition 13.7 *If f, g are integrable, then $f + g$ is integrable too, and*

$$\int_A^B (f + g) = \int_A^B f + \int_A^B g.$$

Proof: Put $I = \int_A^B f + \int_A^B g$. Let $\varepsilon > 0$. By integrability of f and g there are two gauges γ_1, γ_2 on $[A, B]$ such that

$$\rho(\sum_{\Pi_1} f, \int_A^B f) \leq \varepsilon, \qquad \rho(\sum_{\Pi_2} g, \int_A^B g) \leq \varepsilon$$

whenever Π_i is any γ_i-fine partition, $i = 1, 2$. Put $\gamma = \gamma_1 \cap \gamma_2$. Then, for any γ-fine partition Π, we have:

$$\rho(\sum_{\Pi} f, \int_A^B f) \leq \varepsilon, \qquad \rho(\sum_{\Pi} g, \int_A^B g) \leq \varepsilon.$$

Evidently

$$\sum_{\Pi} (f + g) = \sum_{\Pi} f + \sum_{\Pi} g,$$

therefore

$$\rho(\sum_{\Pi} (f + g), I) = \rho(\sum_{\Pi} f + \sum_{\Pi} g, I)$$

$$\leq \rho(\sum_{\Pi} f, \int_A^B f) + \rho(\sum_{\Pi} g, \int_A^B g) \leq 2\varepsilon.$$

Hence $f + g$ is integrable, and

$$\int_A^B (f + g) = \int_A^B f + \int_A^B g. \qquad \text{\fontspec{Apple Color Emoji}🍎}$$

Proposition 13.8 *Let f be integrable, $\alpha \in \mathbb{R}$. Then αf is integrable, and*

$$\int_A^B \alpha f = \alpha \int_A^B f.$$

Proof: If $\alpha = 0$, then the assertion is straightforward.

Let $\alpha \neq 0$. Then, for sufficiently fine partitions Π, we have

$$\rho(\alpha \int_A^B f, \sum_{\Pi} (\alpha f)) \leq |\alpha| \rho(\int_A^B f, \sum_{\Pi} f) \leq |\alpha| \varepsilon. \qquad \text{\fontspec{Apple Color Emoji}🍎}$$

Remark 13.9 If f is integrable, then by Propositions 13.7 and 13.8 we obtain also

$$\int_A^B (f + (-f)) = \int_A^B f + \int_A^B (-f) = \int_A^B f + (-\int_A^B f).$$

Of course, in general, $f + (-f)$ need not be $\chi_{\{0\}}$ as well as

$$\int_A^B f + (-\int_A^B f)$$

need not be 0.

Proposition 13.10 *Let X be invariant. If f is integrable on $[A, B]$ and $A < c < d < B$, then f is integrable on $[c, d]$ too.*

Proof: We will use the Bolzano-Cauchy condition (Proposition 13.6).

Let $\varepsilon > 0$. There exists a gauge γ on $[A, B]$ such that

$$\rho(\sum_{\Pi'_1} f, \sum_{\Pi'_2} f) \le \varepsilon$$

whenever Π'_1 and Π'_2 are any two γ-fine partitions. Put $\gamma_1 = \gamma_{[c,d]}$. Let Π_1, Π_2 be any two γ_1-fine partitions of $[c,d]$. Let Π_{00} and Π_{01} be any $\gamma_{[A,c]}$- and $\gamma_{[d,B]}$-fine partition of $[A,c]$ and $[d,B]$ respectively. Put $\Pi = \Pi_{00} \cup \Pi_{01}$. Then

$$\Pi'_1 = \Pi_1 \cup \Pi, \quad \Pi'_2 = \Pi_2 \cup \Pi$$

are γ-fine, and

$$\rho(\sum_{\Pi_1} f, \sum_{\Pi_2} f)$$
$$= \rho(\sum_{\Pi_1} f + \sum_{\Pi} f, \sum_{\Pi_2} f + \sum_{\Pi} f)$$
$$= \rho(\sum_{\Pi'_1} f, \sum_{\Pi'_2} f) \le \varepsilon.$$

Since Π_1, Π_2 were *arbitrary* γ_1-fine partitions of $[c,d]$, then f is integrable on $[c,d]$. 🍎

Remark 13.11 We observe that Proposition 13.10 holds even if $c = A$ or $d = B$ (the technique is analogous).

Proposition 13.12 *Let X be invariant and $A < c < B$. Then f is integrable on $[A,B]$ if and only if f is integrable on $[A,c]$ and $[c,B]$.*

Proof: If f is integrable on $[A,B]$, then it is integrable on $[A,c]$ and $[c,B]$ (see Proposition 13.10 and Remark 13.11).

We now turn to the converse implication (the technique is similar to the one used in Chapter 7; however, for the sake of clearness, we report it). In correspondence with KH-integrability of f on $[A,c]$ and $[c,B]$, $\forall \varepsilon > 0$ there exist two mappings $\underline{\delta}:[A,c] \to \mathbb{R}^+$, $\overline{\delta}:[c,B] \to \mathbb{R}^+$, and two positive real numbers $\underline{P}$ and $\overline{P}$ (without loss of generality, $\underline{P} > |c|$, $\overline{P} > |c|$) such that, if $\underline{\Pi}$ is any $\underline{\delta}$-fine partition of any bounded interval $[a_1,b_1] \subset [A,c]$, $[a_1,b_1] \supset [A,c] \cap [-\underline{P},\underline{P}]$ and $\overline{\Pi}$ is any $\overline{\delta}$-fine partition of any bounded interval $[a_2,b_2] \subset [c,B]$, $[a_2,b_2] \supset [c,B] \cap [-\overline{P},\overline{P}]$, then

$$\rho(\sum_{\underline{\Pi}} f, \int_{a_1}^{b_1} f) \le \frac{\varepsilon}{2}$$

and

$$\rho(\sum_{\overline{\Pi}} f, \int_{a_2}^{b_2} f) \le \frac{\varepsilon}{2}.$$

If $A = -\infty$, let $\delta(-\infty) = \underline{\delta}(-\infty)$; if $B = +\infty$, let $\delta(+\infty) = \overline{\delta}(+\infty)$. Moreover, set

$$\delta(x) = \begin{cases} \min\left\{\underline{\delta}(x), \dfrac{1}{2}(c-x)\right\} & \text{if } x \in [A,c) \cap \mathbb{R}, \\[2mm] \min\left\{\overline{\delta}(x), \dfrac{1}{2}(x-c)\right\} & \text{if } x \in (c,B] \cap \mathbb{R}, \\[2mm] \min\{\underline{\delta}(c), \overline{\delta}(c)\} & \text{if } x = c, \end{cases}$$

and $P = \max\{\underline{P}, \overline{P}\}$. Take now any arbitrary bounded interval $[a,b] \subset [A,B]$, $[a,b] \supset [A,B] \cap [-P,P]$, and any δ-fine partition $\Pi = \{([u_k,v_k],t_k) : k=1,\ldots,p\}$ of $[a,b]$. Then necessarily $c \in (a,b)$. We now claim that there exists $k \in \{1,\ldots,p\}$ such that $c = t_k$, or $c = u_k$, or $c = v_k$. Otherwise there would be an interval $[u_j, v_j]$ such that $u_j < c < v_j$ and either $c < t_j < v_j$ or $u_j < t_j < c$. Since Π is δ-fine, we should get $[u_j, v_j] \subset (t_j - \delta(t_j), t_j + \delta(t_j))$ and thus $v_j - u_j < 2\delta(t_j)$. So $v_j - u_j < t_j - c$ if $t_j > c$ or $v_j - u_j < c - t_j$ if $t_j < c$. This would imply that t_j is outside the interval (u_j, v_j), contradiction. Thus we have:

$$\sum_{\Pi} f = \sum_{l=1}^{j-1}(v_l - u_l)f(t_l) + (v_j - u_j)f(t_j) + \sum_{l=j+1}^{p}(v_l - u_l)f(t_l)$$

$$= \sum_{l=1}^{j-1}(v_l - u_l)f(t_l) + (t_j - u_j)f(t_j) + (v_j - t_j)f(t_j) \qquad (13.4)$$

$$+ \sum_{l=j+1}^{p}(v_l - u_l)f(t_l).$$

The quantity $S_a^c = \sum_{l=1}^{j-1}(v_l - u_l)f(t_l) + (t_j - u_j)f(t_j)$ is a Riemann sum for a suitable $\underline{\delta}$-fine partition of $[a,c]$, which is a bounded interval contained in $[A,c]$ and containing $[A,c] \cap [-\underline{P},\underline{P}]$, by construction.

Analogously, the quantity $S_c^b = (v_j - t_j)f(t_j) + \sum_{l=j+1}^{p}(v_l - u_l)f(t_l)$ is a Riemann sum for a suitable $\overline{\delta}$-fine partition of $[c,b]$, which is a bounded interval contained in $[c,B]$ and containing $[c,B] \cap [-\overline{P},\overline{P}]$. Thus we have:

$$\rho\left(S_a^c, \int_A^c f\right) \le \frac{\varepsilon}{2}, \qquad \rho\left(S_c^b, \int_c^B f\right) \le \frac{\varepsilon}{2},$$

and hence

$$\rho\left(\sum_{\Pi} f, \int_A^c f + \int_c^B f\right)$$

$$= \rho\left(S_a^c + S_c^b, \int_A^c f + \int_c^B f\right)$$

$$\le \rho\left(S_a^c, \int_A^c f\right) + \rho\left(S_c^b, \int_c^B f\right) \le \varepsilon.$$

So, f is integrable on $[A, B]$, and

$$\int_A^B f = \int_A^c f + \int_c^B f. \quad \text{🍎}$$

Proposition 13.13 *Let X be invariant. If f is integrable and $g = f$ almost everywhere (with respect to the Lebesgue measure), then g is integrable and*

$$\int_A^B f = \int_A^B g.$$

Proof: Let $\int_A^B f = I$. Then to any $\varepsilon > 0$ there exist a map $\delta_0 : [A, B] \to \mathbb{R}^+$ and a positive real number P such that

$$\rho(\sum_\Pi f, I) \le \varepsilon$$

whenever Π is a δ_0-fine partition of any bounded interval $[a, b] \subset \mathbb{R}$ with

$$[A, B] \cap [-P, P] \subset [a, b] \subset [A, B].$$

Put

$$W_j = \{t \in [A, B] : \rho(f(t), g(t)) \in (j-1, j]\}, j \in \mathbb{N}; \quad W = \bigcup_{j=1}^\infty W_j.$$

Let $j \in \mathbb{N}$: since $\lambda(W_j) = 0$, there exists a set $G_j \supset W_j$ which is union of countable many open intervals with total length less than $\dfrac{\varepsilon}{2^j j}$. Define

$$\delta(t) = \begin{cases} \delta_0(t) & \text{if } t \notin W, \\ \delta(t) \text{ such that } (t - \delta(t), t + \delta(t)) \subset G_j & \text{if } t \in W_j. \end{cases}$$

Let $[a, b]$ be any bounded subinterval of $[A, B]$ such that

$$[A, B] \cap [-P, P] \subset [a, b],$$

and $\Pi = \{([x_{i-1}, x_i], t_i) : i = 1, ..., n\}$ be any δ-fine partition of $[a, b]$. Then we get:

$$\rho(\sum_{i=1}^{n}(x_i - x_{i-1})g(t_i), I)$$

$$= \rho(\sum_{t_i \notin W}(x_i - x_{i-1})g(t_i) + \sum_{t_i \in W}(x_i - x_{i-1})g(t_i), I)$$

$$= \rho(\sum_{t_i \notin W}(x_i - x_{i-1})f(t_i) + \sum_{t_i \in W}(x_i - x_{i-1})g(t_i)$$

$$+ \sum_{t_i \in W}(x_i - x_{i-1})f(t_i), I + \sum_{t_i \in W}(x_i - x_{i-1})f(t_i))$$

$$= \rho(\sum_{i=1}^{n}(x_i - x_{i-1})f(t_i) + \sum_{t_i \in W}(x_i - x_{i-1})g(t_i),$$

$$I + \sum_{t_i \in W}(x_i - x_{i-1})f(t_i))$$

$$\leq \rho(\sum_{\Pi} f, I) + \sum_{t_i \in W}(x_i - x_{i-1})\rho(g(t_i), f(t_i))$$

$$\leq \varepsilon + \sum_{j=1}^{\infty}\sum_{t_i \in W_j}(x_i - x_{i-1})\rho(g(t_i), f(t_i))$$

$$\leq \varepsilon + \sum_{j=1}^{\infty} j \sum_{t_i \in W_j}(x_i - x_{i-1})$$

$$\leq \varepsilon + \sum_{j=1}^{\infty} j \frac{\varepsilon}{2^j j} = 2\varepsilon.$$

This concludes the proof.

13.2 Convergence Theorems

Throughout this paragraph, we always suppose that our involved metric semigroup X is invariant. We shall start with a version of the Henstock Lemma (see also [170], pp. 81-83).

Proposition 13.14 *Let $f:[A,B] \to X$ be integrable, $\varepsilon > 0$, and γ be a gauge on $[A,B]$ such that*

$$\rho(\sum_{\Pi} f, \int_{A}^{B} f) \leq \varepsilon$$

whenever Π is any γ-fine partition of $[A,B]$. Let $A_i \subset [A,B]$, $i=1,...,m$, be nonoverlapping intervals (one or two of them may be halflines) and $t_i \in A_i$ be such that

$$A_i \subset \gamma(t_i) \quad (i = 1,...,m).$$

Then

$$\rho(\sum_{i=1,...,m, |A_i|<+\infty} |A_i| f(t_i), \sum_{i=1}^{m}\int_{A_i} f) \leq \varepsilon.$$

Proof: Let int A_i be the interior of A_i, $i = 1,...,m$. Since the A_i's are non-overlapping, the set

$$[A,B]\setminus\bigcup_{i=1}^{m}(\text{int }A_i)$$

is empty or it is the union of non-overlapping intervals (and eventually also halflines) $B_1,\ldots,B_p$. Let $\eta>0$. Since f is integrable on each B_j, for each $j=1,\ldots,p$ there exists a gauge γ_j on B_j such that

$$\gamma_j(x)\subset\gamma(x)\qquad\forall x\in B_j,$$

and

$$\rho(\sum_{\Pi_j}f,\int_{B_j}f)<\frac{\eta}{p+1}$$

for every γ_j-fine partition Π_j of B_j. Let now Π_j be any γ_j-fine partition of B_j. We observe that

$$\Pi:=\{(A_i,t_i):i=1,\ldots,m\}\cup(\cup_{j=1}^{p}\Pi_j)$$

is a γ-fine partition of $[A,B]$. Then we have:

$$\rho(\sum_{i=1,\ldots,m,|A_i|<+\infty}|A_i|f(t_i),\sum_{i=1}^{m}\int_{A_i}f)$$

$$=\rho(\sum_{i=1,\ldots,m,|A_i|<+\infty}|A_i|f(t_i)+\sum_{j=1}^{p}\sum_{\Pi_j}f,\sum_{i=1}^{m}\int_{A_i}f+\sum_{j=1}^{p}\sum_{\Pi_j}f)$$

$$\leq\rho(\sum_{\Pi}f,\int_{A}^{B}f)+\rho(\sum_{i=1}^{m}\int_{A_i}f+\sum_{j=1}^{p}\int_{B_j}f,\sum_{i=1}^{m}\int_{A_i}f+\sum_{j=1}^{p}\sum_{\Pi_j}f)$$

$$\leq\varepsilon+\rho(\sum_{j=1}^{p}\int_{B_j}f,\sum_{j=1}^{p}\sum_{\Pi_j}f)\leq\varepsilon+\sum_{j=1}^{p}\rho(\int_{B_j}f,\sum_{\Pi_j}f)<\varepsilon+\sum_{j=1}^{p}\frac{\eta}{p+1}<\varepsilon+\eta.$$

Since the inequality

$$\rho(\sum_{i=1,\ldots,m,|A_i|<+\infty}|A_i|f(t_i),\sum_{i=1}^{m}\int_{A_i}f)<\varepsilon+\eta$$

holds for any $\eta>0$, then we obtain the result.

Definition 13.15 A sequence of integrable functions $(f_k:[A,B]\to X)_k$ is said to be *equiintegrable* if to any $\varepsilon>0$ there exists a gauge γ on $[A,B]$ such that

$$\rho(\sum_{\Pi}f_k,\int_{A}^{B}f_k)\leq\varepsilon$$

for any γ-fine partition Π and every $k\in\mathbb{N}$.

Theorem 13.16 *Let $(f_k)_k$ be an equiintegrable sequence and let*

$$\lim_{k\to+\infty}\rho(f_k(t),f(t))=0$$

for any $t\in[A,B]$. Then f is integrable, and

$$\lim_{k\to+\infty}\rho(\int_{A}^{B}f_k,\int_{A}^{B}f)=0.$$

Proof: First of all, we observe that, $\forall \varepsilon > 0$, there exist: a non-negative function $E : [A, B] \to \mathbb{R}$, strictly positive on $[A, B] \cap \mathbb{R}$, KH-integrable in $[A, B]$, with

$$\int_A^B E \leq \frac{\varepsilon}{2}$$

(for example,

$$E(t) = \frac{\varepsilon}{2\pi(1+t^2)}, \quad t \in [A, B],$$

with the convention $E(+\infty) = E(-\infty) = 0$); a gauge γ_0 on $[A, B]$, such that

$$\sum_{\substack{i=1,\ldots,n, \\ |I_i| < +\infty}} |I_i| E(t_i) \leq \varepsilon$$

for each γ_0-fine partition Π of $[A, B]$, $\Pi = \{(I_i, t_i) : i = 1, \ldots, n\}$. Let now $\varepsilon > 0$, γ be as in Definition 13.15, $\hat{\gamma} = \gamma \cap \gamma_0$, and Π be any $\hat{\gamma}$-fine partition of $[A, B]$. Then

$$\rho(\sum_\Pi f_k, \sum_\Pi f) = \rho(\sum_{i=1,\ldots,n,|I_i|<+\infty} |I_i| f_k(t_i), \sum_{i=1,\ldots,n,|I_i|<+\infty} |I_i| f(t_i))$$

$$\leq \sum_{i=1,\ldots,n,|I_i|<+\infty} |I_i| \rho(f_k(t_i), f(t_i))$$

$$\leq \sum_{i=1,\ldots,n,|I_i|<+\infty} |I_i| E(t_i) \leq \varepsilon$$

for sufficiently large k (depending on the chosen partition Π). It follows that

$$\lim_{k \to +\infty} \rho(\sum_\Pi f_k, \sum_\Pi f) = 0.$$

Now,

$$\rho(\sum_\Pi f, \int_A^B f_k) \leq \rho(\sum_\Pi f, \sum_\Pi f_k) + \rho(\sum_\Pi f_k, \int_A^B f_k) \leq 2\varepsilon$$

for sufficiently large k (still depending on the fixed partition Π). Thus, in correspondence with $\varepsilon > 0$, we found the gauge $\hat{\gamma}$ on $[A, B]$, which is such that, for every $\hat{\gamma}$-fine partition Π of $[A, B]$, $\exists k^* \in \mathbb{N} : \forall k, i \geq k^*$,

$$\rho(\int_A^B f_k, \int_A^B f_i) \leq \rho(\int_A^B f_k, \sum_\Pi f) + \rho(\sum_\Pi f, \int_A^B f_i) \leq 4\varepsilon. \qquad (13.5)$$

It follows that the sequence $(\int_A^B f_k)_k$ of elements of X is Cauchy. Since (X, ρ) is complete, there exists $I \in X$ such that

$$\lim_{k \to +\infty} \rho(\int_A^B f_k, I) = 0.$$

Fix arbitrarily $\varepsilon > 0$. Then there exists $k_0 \in \mathbb{N}$ such that, $\forall k \geq k_0$,

$$\rho(\int_A^B f_k, I) \le \varepsilon,$$

and there exists a gauge γ^* on $[A, B]$ such that, for each γ^*-fine partition Π of $[A, B]$, there exists $k_{00} \in \mathbb{N}$, $k_{00} \ge k_0$, such that, $\forall k \ge k_{00}$,

$$\rho(\sum_\Pi f_k, \sum_\Pi f) \le \varepsilon.$$

Moreover, we know that

$$\rho(\sum_\Pi f_k, \int_A^B f_k) \le \varepsilon$$

for any γ-fine partition Π and for every $k \in \mathbb{N}$. Therefore, for each $\gamma \cap \gamma^*$-fine partition Π, we get:

$$\rho(\sum_\Pi f, I) \le \rho(\sum_\Pi f, \sum_\Pi f_{k_{00}})$$
$$+ \rho(\sum_\Pi f_{k_{00}}, \int_A^B f_{k_{00}}) + \rho(\int_A^B f_{k_{00}}, I) \le 3\varepsilon.$$

So we have proved that f is integrable and $I = \int_A^B f$. Finally, by proceeding as in (13.5), we have that, $\forall \varepsilon > 0$, $\exists$ gauge $\overline{\gamma}$ on $[A, B]$ such that, for any $\overline{\gamma}$-fine partition Π, $\exists \overline{k} \in \mathbb{N} : \forall k \ge \overline{k}$,

$$\rho(\int_A^B f_k, \int_A^B f) \le \rho(\int_A^B f_k, \sum_\Pi f_k)$$
$$+ \rho(\sum_\Pi f_k, \sum_\Pi f) + \rho(\sum_\Pi f, \int_A^B f) \le 3\varepsilon,$$

what implies that

$$\lim_{k \to +\infty} \rho(\int_A^B f_k, \int_A^B f) = 0.$$

This concludes the proof. 🍎

Theorem 13.17 *Let $[a,b] \subset \mathbb{R}$ be a bounded interval, and $(f_n : [a,b] \to X)_n$ be a sequence of functions, integrable on $[a,b]$ and uniformly convergent to a function f (i.e. for every $\varepsilon > 0$ there exists $n_0 \in \mathbb{N}$ such that $\rho(f_n(t), f(t)) \le \varepsilon$ for each $n \ge n_0$ and any $t \in [a,b]$). Then f is integrable on $[a,b]$ and*

$$\lim_{n \to +\infty} \rho(\int_a^b f_n, \int_a^b f) = 0.$$

Proof: Let $\varepsilon > 0$, and take any two partitions Π_1, Π_2 of $[a,b]$ and $n \ge n_0$, where $n_0 = n_0(\varepsilon)$ is as in the hypotheses. Then

$$\rho(\sum_{\Pi_1} f, \sum_{\Pi_2} f) \le \rho(\sum_{\Pi_1} f, \sum_{\Pi_1} f_n) + \rho(\sum_{\Pi_1} f_n, \sum_{\Pi_2} f_n) + \rho(\sum_{\Pi_2} f_n, \sum_{\Pi_2} f)$$
$$\le 2\varepsilon(b-a) + \rho(\sum_{\Pi_1} f_n, \sum_{\Pi_2} f_n).$$

Since f_n is integrable on $[a,b]$ $\forall n \in \mathbb{N}$, then for every $n \in \mathbb{N}$ there exists a map $\delta_n : [a,b] \to \mathbb{R}^+$ such that

$$\rho(\sum_{\Pi_1} f_n, \sum_{\Pi_2} f_n) \le \varepsilon$$

for any two δ_n-fine partitions Π_1, Π_2 of $[a,b]$. So in particular, taking $n = n_0$, we get that, in correspondence with ε, $\exists$ a map $\delta = \delta_{n_0} : [a,b] \to \mathbb{R}^+$, such that, for any two δ-fine partitions Π_1, Π_2 of $[a,b]$,

$$\rho(\sum_{\Pi_1} f, \sum_{\Pi_2} f) \le 2\varepsilon(b-a) + \varepsilon,$$

hence f is integrable on $[a,b]$, by virtue of the Bolzano-Cauchy condition (Proposition 13.6). Since f is integrable, there exists a map $\delta' : [a,b] \to \mathbb{R}^+$ such that

$$\rho(\sum_{\Pi} f, \int_a^b f) \le \varepsilon$$

for any δ'-fine partition Π of $[a,b]$. Fix $n \ge n_0$ and choose a map $\kappa_n : [a,b] \to \mathbb{R}^+$ such that

$$\rho(\sum_{\Pi} f_n, \int_a^b f_n) \le \varepsilon$$

whenever Π is a κ_n-fine partition of $[a,b]$. Put $\bar{\delta}_n = \min\{\delta', \kappa_n\}$. Then for any $\bar{\delta}_n$-fine partition Π of $[a,b]$ we obtain:

$$\rho(\int_a^b f_n, \int_a^b f) \le \rho(\int_a^b f, \sum_{\Pi} f) + \rho(\sum_{\Pi} f, \sum_{\Pi} f_n) + \rho(\sum_{\Pi} f_n, \int_a^b f_n)$$
$$\le 2\varepsilon + \varepsilon(b-a),$$

hence

$$\lim_{n \to +\infty} \rho(\int_a^b f_n, \int_a^b f) = 0.$$

This concludes the proof.

REFERENCES

[1] S. I. Ahmed and W. F. Pfeffer, "A Riemann integral in a locally compact Hausdorff space", J. Australian Math. Soc., vol. 41, pp. 115-137, 1986.

[2] V. N. Alexiuk and F. D. Beznosikov, "Extension of continuous outer measure on a Boolean algebra" (Russian), Izv. VUZ, vol. 4 (119), pp. 3-9, 1972.

[3] Ch. D. Aliprantis and O. Burkinshaw, Locally solid Riesz spaces. New York: Academic Press, 1978.

[4] Ch. D. Aliprantis and O. Burkinshaw, Positive Operators. New York: Academic Press, 1985.

[5] Ch. D. Aliprantis, D. J. Brown and O. Burkinshaw, Existence and Optimality of Competitive Equilibria. Springer-Verlag, 1990

[6] P. Antosik and C. Swartz, Matrix methods in Analysis. Lecture Notes in Mathematics, vol. 1113, Springer-Verlag, 1985.

[7] P. Antosik and C. Swartz, "The Vitali-Hahn-Saks theorem for algebras", J. Math. Anal. Appl., vol. 106, pp. 116-119, 1985.

[8] P. Antosik and C. Swartz, "The Nikodym convergence theorem for lattice-valued measures", Rev. Roumaine Math. Pures Appl., vol. 37, pp. 299-306, 1992.

[9] K. T. Atanassov, Intuitionistic Fuzzy Sets: Theory and Applications. Studies in Fuzziness and Soft Computing, Physics Verlag, 1999.

[10] A. Avallone and A. Basile, "Integration: uniform structure", J. Math. Anal. Appl., vol. 159. pp. 373-381, 1991.

[11] S. J. Bernau, "Unique representation of Archimedean lattice groups and normal Archimedean lattice rings", Proc. London Math. Soc., vol. 15, pp. 599-631, 1965.

[12] K. P. S. Bhaskara Rao and M. Bhaskara Rao, Theory of Charges. London-New York: Academic Press, 1983.

[13] P. Billingsley, Convergence of Probability Measures, New York: John Wiley and Sons, Inc., 1968.

[14] G. Birkhoff, Lattice theory. Providence, Rhode Island: Am. Math. Soc., 1967.

[15] A. Boccuto, "Riesz Spaces, Integration and Sandwich Theorems", Tatra Mt. Math. Publ., vol. 3, pp. 213-230, 1993.

[16] A. Boccuto, " On Stone-type extensions for l-group-valued measures", Math. Slovaca, vol. 45, pp. 309-315, 1995.

[17] A. Boccuto, "Abstract Integration in Riesz spaces", Tatra Mt. Math. Publ., vol. 5, pp. 107-124, 1995.

[18] A. Boccuto, "Integration in Riesz spaces with respect to (D)-convergence", Tatra Mt. Math. Publ., vol. 10, pp. 33-54, 1997.

[19] A. Boccuto, "Differential and integral calculus in Riesz spaces", Tatra Mt. Math. Publ., vol. 14, pp. 293-323, 1998.

[20] A. Boccuto, "A Perron-type integral of order 2 for Riesz spaces", Math. Slovaca, vol. 51, pp. 185-204, 2001.

[21] A. Boccuto, "Egorov property and weak σ-distributivity in l-groups", Acta Math. (Nitra), vol. 6, pp. 61-66, 2003.

[22] A. Boccuto, D. Candeloro and B. Riečan, "Abstract generalized KH-integrals in Riesz spaces", Real Anal. Exch., 2008/2009, to appear.

[23] A. Boccuto and B. Riečan, "A note on the improper Kurzweil-Henstock integral", Acta Univ. Mat. Belii, vol. 9, pp. 7-12, 2001.

[24] A. Boccuto and B. Riečan, "A note on improper Kurzweil-Henstock integral in Riesz spaces", Acta Math. (Nitra), vol. 5, pp. 15-24, 2002; Addendum to "A note on improper Kurzweil-Henstock integral in Riesz spaces", ibidem, vol. 7, pp. 53-59, 2004.

[25] A. Boccuto and B. Riečan, "A note on a Pettis-Kurzweil-Henstock-type integral in Riesz spaces", Real Anal. Exch., vol. 28, pp. 153-161, 2002/2003.

[26] A. Boccuto and B. Riečan, "On the Henstock-Kurzweil integral for Riesz-space-valued functions defined on unbounded intervals", Czech. Math. J., vol. 54 (129), pp. 591-607, 2004.

[27]   A. Boccuto and B. Riečan, "Convergence theorems for the (SL)-integral in the Riesz space-context", Tatra Mt. Math. Publ., vol. 34, pp. 201-222, 2006.

[28]   A. Boccuto and B. Riečan, "Improper Kurzweil-Henstock integral for metric semigroup-valued functions", Atti Sem. Mat. Fis. Univ. Modena e Reggio Emilia, vol. 54, 75-95, 2006.

[29]   A. Boccuto and B. Riečan, "The Kurzweil-Henstock integral for Riesz Space-Valued maps defined in abstract topological spaces and convergence theorems", PanAmerican Math. J., vol. 16, pp. 63-79, 2006.

[30]   A. Boccuto and B. Riečan, "The symmetric Choquet integral with respect to Riesz space-valued capacities", Czech. Math. J., vol. 58 (133), pp. 289-310, 2008.

[31]   A. Boccuto, B. Riečan and A. R. Sambucini, "Some properties of an improper GHk integral in Riesz spaces", Indian J. Math., vol. 50, pp. 21-51, 2008.

[32]   A. Boccuto and A. R. Sambucini, "On the De Giorgi-Letta integral with respect to means with values in Riesz spaces", Real Anal. Exch., vol. 21, pp. 793-810, 1995/1996.

[33]   A. Boccuto and A. R. Sambucini, "The monotone integral with respect to Riesz space-valued capacities", Rend. Mat. (Roma), vol. 16, pp. 491-524, 1996.

[34]   A. Boccuto and A. R. Sambucini, "The Burkill-Cesari Integral for Riesz Spaces", Rend. Ist. Mat. Univ. Trieste, vol. 28, pp. 33-47, 1996.

[35]   A. Boccuto and A. R. Sambucini, "Comparison between different types of abstract integrals in Riesz spaces", Rend. Circ. Mat. Palermo, Ser. II, vol. 46, pp. 255-278, 1997.

[36]   A. Boccuto and A. R. Sambucini, "The Henstock-Kurzweil integral for functions defined on unbounded intervals and with values in Banach spaces", Acta Math. (Nitra), vol. 7, pp. 3-17, 2004.

[37]   A. Boccuto and V. A. Skvortsov, "Henstock-Kurzweil type integration of Riesz-space-valued functions and applications to Walsh series", Real Anal. Exch., vol. 29, pp. 419-438, 2003/2004.

[38]   A. Boccuto and V. A. Skvortsov, "On Henstock type integrals with respect to abstract derivation bases for Riesz-space-valued functions", J. Appl. Funct. Anal., vol. 1, pp. 251-270, 2006.

[39]   A. Boccuto and V. A. Skvortsov, "Some applications of the Maeda-Ogasawara-Vulikh representation theorem to Differential Calculus in Riesz spaces", Acta Math. (Nitra), vol. 9, pp. 13-24, 2006.

[40]   B. Bongiorno, "The Henstock-Kurzweil integral", in Handbook of Measure Theory, E. Pap ed., Amsterdam: Elsevier, 2002, Chapter 13, pp. 587-615.

[41]   B. Bongiorno, L. Di Piazza and K. Musiał, "An alternate approach to the McShane integral", Real Anal. Exch., vol. 25, pp. 829-848, 1999/2000.

[42]   J. C. Breckenridge, "Burkill-Cesari integrals of quasi additive interval functions", Pacific J. Math., vol. 37, pp. 635-654, 1971.

[43]   J. K. Brooks, "On the Vitali - Hahn - Saks and Nikodym theorems", Proc. Nat. Acad. Sci. U.S.A., vol. 64, pp. 468-471, 1969.

[44]   J. K. Brooks, "Equicontinuous sets of measures and applications to Vitali's integral convergence theorem and control measures", Advances in Math., vol. 10, pp. 165-171, 1973.

[45]   J. K. Brooks, "On a theorem of Dieudonné", Advances in Math., vol. 36, pp. 165-168, 1980.

[46]   J. K. Brooks and R. V. Chacon, "Continuity and compactness of measures", Advances in Math., vol. 37, pp. 16-26, 1980.

[47]   J. K. Brooks and R. S. Jewett, "On finitely additive vector measures", Proc. Nat. Acad. Sci. U.S.A., vol. 67, pp. 1294-1298, 1970.

[48]   J. K. Brooks and A. Martellotti, "On the De Giorgi-Letta integral in infinite dimensions", Atti Sem. Mat. Fis. Univ. Modena, vol. 40, pp. 285-302, 1992.

[49]   J. K. Brooks and J. Mikusiński, "On some theorems in Functional Analysis", Bull. Acad. Pol. Sci., vol. 18, pp. 151-155, 1970.

[50]   G. Buskes, "Extensions of Riesz homomorphisms I.", J. Austral. Math. Soc. Ser. A, vol. 39, pp. 107-120, 1985.

[51]   D. Butnariu and E. P. Klement, "Triangular norm-based measures and their Markov kernel representation", J. Math. Anal. Appl., vol. 162, pp. 111-143, 1991.

[52]   D. Candeloro, "Sui teoremi di Vitali-Hahn-Saks, Dieudonné e Nikodým", Rend. Circ. Mat. Palermo, Ser. II, vol. 8, pp. 439-445, 1985.

[53]   D. Candeloro, "Uniforme esaustività e assoluta continuità", Bollettino U.M.I., vol. 4-B, pp. 709-724, 1985.

[54]   D. Candeloro, "Alcuni teoremi di uniforme limitatezza", Rend. Accad. Naz. Sci. Detta dei XL, vol. 9, pp. 249-260, 1985.

[55]   D. Candeloro, "Riemann-Stieltjes integration in Riesz spaces", Rend. Mat. (Roma), Ser. VII, vol. 16, pp. 563-585, 1996.

[56]   D. Candeloro and G. Letta, "Sui teoremi di Vitali - Hahn - Saks e di Dieudonné", Rend. Accad. Naz. Sci. Detta dei XL, vol. 9, pp. 203-213, 1985.

[57]   S. S. Cao, "The Henstock integral for Banach-valued functions", SEA Bull. Math., vol. 16, pp. 35-40, 1992.

[58]   S. S. Cao, "On the Henstock-Bochner integral", "SEA Bull. Math., Special Issue, pp. 1-3, 1993.

[59]   D. Caponetti and V. Marraffa, "A descriptive definition of a BV integral in the real line", Math. Bohemica, vol. 124, pp. 421-432, 1999.

[60]   L. Cesari, "Quasi-additive set functions and the concept of integral over a variety", Trans. Amer. Math. Soc., vol. 102, pp. 94-113, 1962.

[61]   L. Cesari, "Extension problem for quasi additive set function and Radon-Nikodym derivatives", Trans. Amer. Math. Soc., vol. 102, pp. 114-146, 1962.

[62]   W. Chojnacki, "Sur un théorème de Day, un théorème de Mazur-Orlicz et une généralisation de quelques théorèmes de Silverman", Colloq. Math., vol. 50, pp. 257-262, 1986.

[63]   R. R. Christian, "On order-preserving integration", Trans. Amer. Math. Soc., vol. 86, pp. 463-488, 1957.

[64]   R. Cignoli, I. M. L. D'Ottaviano and D. Mundici, "Algebraic Foundations of Many-Valued Reasoning", Trends in Logic, Studia Logica Library, Vol. 7, Kluwer Acad. Publ., 2000.

[65]   G. Coletti and G. Regoli, "Probabilità qualitative non archimedee e realizzabilità", Rivista di Matematica per le Scienze economiche e sociali, Fasc. I e II, pp. 79-99, 1983.

[66]   M. Congost Iglesias, "Medidas y probabilidades en estructuras ordenadas", Stochastica, vol. 5, pp. 45-68, 1981.

[67]   R. Cristescu, "On integration in ordered vector spaces and on some linear operators", Rend. Circ. Mat. Palermo, Suppl., vol. 33, pp. 289-299, 1993.

[68]   C. Debiève and M. Duchoň, "Integration by parts in vector lattices", Tatra Mt. Math. Publ., vol. 6, pp. 13-18, 1995.

[69]   C. Debiève, M. Duchoň and B. Riečan, "Moment problem in some ordered vector spaces", Tatra Mt. Math. Publ., vol. 12, pp. 247-251, 1997.

[70]   E. De Giorgi and G. Letta, "Une notion générale de convergence faible des fonctions croissantes d'ensemble", Ann. Scuola Sup. Pisa, vol. 33, pp. 61-99, 1977.

[71]   P. De Lucia and P. Morales, "Equivalence of Brooks-Jewett, Vitali-Hahn-Saks and Nikodym convergence theorems for uniform semigroup-valued additive functions on a Boolean ring", Ricerche Mat., vol. 35, pp. 75-87, 1986.

[72]   P. De Lucia and P. Morales, "Some consequences of the Brooks-Jewett theorem for additive uniform semigroup-valued functions", Conf. Semin. Mat. Univ. Bari, vol. 227, pp. 23 p., 1988.

[73]   D. Denneberg, Non-additive measure and integral. Kluwer, 1994.

[74]   J. Diestel, "Uniform integrability: an introduction", Rend. Ist. Mat. Univ. Trieste, vol. 23, pp. 41-80, 1991.

[75]   J. Diestel and J. J. Uhl, Vector measures. Amer. Math. Soc. Surveys, vol. 15, Providence, 1977.

[76]   L. Di Piazza, "The Pettis and the McShane integrals for vector valued functions", Real Anal. Exch., Summer Symposium 2003, pp. 87-91, 2003.

[77]   L. Di Piazza, "Kurzweil-Henstock type integration on Banach spaces", Real Anal. Exch., vol. 29, pp. 543-555, 2003/2004.

[78]   L. Di Piazza and V. Marraffa, "The McShane, PU and Henstock integrals of Banach valued functions", Czech. Math. J., vol. 52 (127), pp. 609-633, 2002.

[79]   L. Di Piazza and V. Marraffa, "An equivalent definition of the vector-valued McShane integral by means of partitions of unity", Studia Math., vol. 151, pp. 175-185, 2002.

[80]   L. Di Piazza and D. Preiss, "When do McShane and Pettis integral coincide?", Ill. J. Math., vol. 47, pp. 1177-1187, 2003.

[81]   J. Dixmier, "Sur certains espaces considérés par M. H. Stone", Summa Brasil. Math., vol. 2, pp. 151-182, 1951.

[82]   I. Dobrakov, "On integration in Banach spaces,I", Czech. Math. J., vol. 20 (95), pp. 511-536, 1970.

[83]   I. Dobrakov, "On integration in Banach spaces,II", Czech. Math. J., vol. 20 (95), pp. 680-695, 1970.

[84]   I. Dobrakov, "On submeasures I", Dissert. Math., vol. 112, pp. 1-39, 1972.

[85]   I. Dobrakov, "On integration in Banach spaces,III", Czech. Math. J., vol. 29 (104), pp. 478-499, 1979.

[86]   I. Dobrakov, "On integration in Banach spaces,IV", Czech. Math. J., vol. 30 (105), pp. 259-279, 1980.

[87]   I. Dobrakov, "On integration in Banach spaces,V", Czech. Math. J., vol. 30 (105), pp. 610-628, 1980.

[88]   I. Dobrakov, "On integration in Banach spaces,VI", Czech. Math. J.,vol. 35 (110), pp. 173-187, 1985.

[89]   I. Dobrakov, "On integration in Banach spaces,VII," Czech. Math. J., vol. 38 (113), pp. 434-449, 1988.

[90]   I. Dobrakov, "On integration in Banach spaces,VIII," Czech. Math. J., vol. 37 (112), pp. 487-506, 1987.

[91]   I. Dobrakov, "On integration in Banach spaces,IX", Czech. Math. J., vol. 38 (113), pp. 589-601, 1988.

[92]   I. Dobrakov, "On integration in Banach spaces,X", Czech. Math. J., vol. 38 (113), pp. 713-725, 1988.

[93]   I. Dobrakov, "On integration in Banach spaces,XI", Czech. Math. J., vol. 39 (115), pp. 8-24, 1990.

[94]   I. Dobrakov, "On integration in Banach spaces,XII", Czech. Math. J., vol. 40 (115), pp. 424-440, 1990.

[95]   I. Dobrakov, "On integration in Banach spaces,XIII", Czech. Math. J., vol. 40 (115), pp. 566-582, 1990.

[96]   L. Drewnowski, "Topological rings of sets, continuous set functions, integration I, II", Bull. Acad. Pol. Sci., vol. 20, pp. 269-286, 1972.

[97]   L. Drewnowski, "Equivalence of Brooks-Jewett, Vitali-Hahn-Saks and Nikodým theorems", Bull. Acad. Pol. Sci., vol. 20, pp. 725-731, 1972.

[98]   M. Duchoň, J. Haluška and B. Riečan, "On the Choquet integral for Riesz space valued measures", Tatra Mt. Math. Publ., vol. 19, pp. 75-89, 2000.

[99]   M. Duchoň and B. Riečan, "On the Kurzweil-Stieltjes integral in ordered spaces", Tatra Mt. Math. Publ., vol. 8, pp. 133-142, 1996.

[100]   M. Duchoň and B. Riečan, "Fuzzy moment problem", Tatra Mt. Math. Publ., vol. 14, pp. 193-197, 1998.

[101]   N. Dunford and J. T. Schwartz, Linear Operators I; General Theory. New York: Interscience, 1958.

[102]   A. Dvurečenskij, "Loomis-Sikorski theorem for σ-complete MV-algebras and -groups", J. Australian Math. Soc. Series A, vol. 67, pp. 1-17, 1999.

[103]   A. Dvurečenskij and S. Pulmanová, New Trends in Quantum Structures. Kluwer, 2000.

[104]   A. Dvurečenskij and S. Pulmanová, "Conditional probability on σ-MV-algebras", Fuzzy Sets and Systems, vol. 155, pp. 102-118, 2005.

[105]   R. Dyckerhoff and K. Mosler, "Stochastic dominance with nonadditive probabilities", ZOR Methods and Models of Operations Research, vol. 37, pp. 231-256, 1993.

[106]   B. Faires, "On Vitali-Hahn-Saks-type theorems", Ann. Inst. Fourier (Grenoble), vol. 26, pp. 99-114, 1976.

[107]   M. Federson, "The monotone convergence theorem for multidimensional abstract Kurzweil vector integrals", Czech. Math. J., vol. 52 (127), pp. 429-439, 2002.

[108]   M. Federson, "Substitution formulas for the Kurzweil and Henstock vector integrals", Math. Bohemica, vol. 127, pp. 15-26, 2002.

[109]   M. Federson, "Some peculiarities of the Henstock and Kurzweil integrals of Banach space-valued functions", Real Anal. Exch., vol. 29, pp. 439-460, 2003/2004.

[110]   M. Federson and R. Bianconi, "Linear integral equations of Volterra concerning the integral of Henstock", Real Anal. Exch., vol. 25, pp. 389-417, 1999/2000.

[111]   W. Filter, "A note on measurable functions", Arch. Math., vol. 47, pp. 535-536, 1986.

[112]   W. Filter, "A note on Archimedean Riesz spaces and their extended order duals", Libertas Math., vol. 6, pp. 101-106, 1986.

[113]   W. Filter, "Dual spaces of $C_\infty(X)$ ", Rend. Circ. Mat. Palermo, vol. 35, pp. 135-158, 1986.

[114]   W. Filter, "On measure representations of Riesz spaces", Atti Sem. Mat. Fis. Univ. Modena, vol. 35, pp. 77-94, 1987.

[115]   W. Filter, "Normal Measures on Stonian Spaces with σ-Compact and Paracompact Support", Math. Nachr., vol. 136, pp. 91-98, 1988.

[116]   W. Filter, "Hypercomplete Riesz spaces", Atti Sem. Mat. Fis. Univ. Modena, vol. 38, pp. 227-240, 1990.

[117]   W. Filter, "Hypercompletions of Riesz spaces", Proc. Amer. Math. Soc., vol. 109, pp. 775-780, 1990.

[118]   W. Filter, "Riesz spaces of step functions", Ricerche Mat., vol. 39, pp. 247-257, 1990.

[119]   W. Filter, "Representations of Riesz spaces as spaces of measures I", Czech. Math. J., vol. 42, pp. 415-432, 1992.

[120]   W. Filter, "Representations of Riesz spaces as spaces of measures II", Czech. Math. J., vol. 42, pp. 635-648, 1992.

[121]   W. Filter, "Representations of Archimedean Riesz spaces - a survey", Rocky Mt. J. Math., vol. 24, pp. 771-851, 1994.

[122]   W. Filter, "Some remarks on localness in Riesz spaces of continuous functions", Ricerche Mat., vol. 43, pp. 347-355, 1994.

[123]   W. Filter, "Equivalence Relations on Stonian Spaces", Advances in Math., vol. 123, pp. 120-143, 1996.

[124]   P. C. Fishburn, "The axioms of Subjective Probability", Statistical Science, vol. 1, pp. 335-345, 1986.

[125]   P. C. Fishburn, "The axioms and algebra of ambiguity", Theory and Decision, vol. 34, pp. 119-137, 1993.

[126]   E. E. Floyd, "Boolean algebras with pathological order properties", Pacific J. Math., vol. 5, pp. 687-689, 1955.

[127]   C. K. Fong, "A continuous version of Orlicz-Pettis theorem via vector-valued Henstock-Kurzweil integrals", Czech. Math. J., vol. 52 (127), pp. 531-536, 2002.

[128]   G. Fox and P. Morales, "Théorèmes de Nikodym et de Vitali - Hahn - Saks pour les mesures à valeurs dans un sémigroupe uniforme", in Measure theory and its applications, Proc. Conf., Sherbrooke/Can. 1982, Lecture Notes in Mathematics, vol. 1033, 1983, pp. 199-208.

[129]   D. H. Fremlin, Topological Riesz Spaces and Measure Theory. London: Cambridge Univ. Press, 1974.

[130]   D. H. Fremlin, "A direct proof of the Matthes-Wright integral extension theorem", J. London Math. Soc., vol. 11, pp. 276-284, 1975.

[131]   D. H. Fremlin, "The Henstock and McShane integrals of vector-valued functions", Ill. J. Math., vol. 38, pp. 471-479, 1994.

[132]   D. H. Fremlin, "The generalized McShane integral", Ill. J. Math., vol. 39, pp. 39-67, 1995.

[133]   D. H. Fremlin, "Integration of Vector-Valued Functions", Atti Sem. Mat. Fis. Univ. Modena, vol. 42, pp. 205-211, 1994.

[134]   D. H. Fremlin and J. Mendoza, "On the integration of vector-valued functions", Ill. J. Math., vol. 38, pp. 127-147, 1994.

[135]   F. J. Freniche, "The Vitali-Hahn-Saks theorem for Boolean algebras with the subsequential interpolation property", Proc. Amer. Math. Soc., vol. 92, pp. 262-366, 1984.

[136]   B. Fuchssteiner and W. Lusky, Convex cones. North-Holland Publ. Co., 1981.

[137] T. Gesternkorn and J. Manko, "Probabilities of intuitionistic fuzzy events", in Issues in Intelligent Systems: Paradigms, O. Hryniewicz et al. eds, 2005, pp. 53-58.

[138] I. Gilboa and D. Schmeidler, "Additive representation of non-additive measures and the Choquet integral", Ann. Oper. Research, vol. 52, pp. 43-65, 1994.

[139] Z. Gong and C. Wu, "The McShane integral of fuzzy-valued functions", Southeast Asian Bull. Math., vol. 24, pp. 365-373, 2000.

[140] Z. Gong and C. Wu, "On the problem of characterizing derivatives for the fuzzy-valued functions", Fuzzy Sets and Systems, vol. 127, pp. 315-322, 2002.

[141] Z. Gong and C. Wu, "Bounded variation, absolute continuity and absolute integrability for fuzzy-number-valued functions", Fuzzy Sets and Systems, vol. 129, pp. 83-94, 2002.

[142] R. A. Gordon, "The McShane integral of Banach-valued functions", Ill. J. Math., vol. 34, pp. 557-567, 1990.

[143] R. A. Gordon, The integrals of Lebesgue, Denjoy, Perron and Henstock. Providence: Amer. Math. Soc., 1994.

[144] M. Grabisch, T. Murofushi and M. Sugeno (Eds.), Fuzzy Measures and Integrals: Theory and Applications. Studies in Fuzziness and Soft Computing, vol. 40, Physics Verlag, 2000.

[145] G. H. Greco, "Integrale monotono", Rend. Sem. Mat. Univ. Padova, vol. 57, pp. 149-166, 1977.

[146] P. Grzegorzewski and E. Mrówka, "Probability of intuitionistic fuzzy events", in Soft Methods in Probability, Statistics and Data Analysis, P. Grzegorzewski et al. eds, 2002, pp. 105-115.

[147] E. D. Habil, "Brooks-Jewitt and Nikodym convergence theorems for orthoalgebras that have the weak subsequential interpolation property", Int. J. Theor. Phys., 34, pp. 465-491, 1995.

[148] J. Haluška, "On the generalized continuity of the semivariation in locally convex spaces", Acta Univ. Carolin. Math. Phys., vol. 32, pp. 23-28, 1991.

[149] J. Haluška, "On integration in complete vector lattices", Tatra Mt. Math. Publ., vol. 3, pp. 201-212, 1993.

[150] J. Haluška, "On convergences of functions in complete bornological locally convex spaces", Revue Roumaine Math. Pures Appl., vol. 38, pp. 327-337, 1993.

[151] J. Haluška, "On lattices of set functions in complete bornological locally convex spaces", Simon Stevin, vol. 67, pp. 27-48, 1993.

[152] J. Haluška, "On the continuity of the semivariation in locally convex spaces", Math. Slovaca, vol. 43, pp. 185-192, 1993.

[153] R. Henstock, Theory of Integration. London: Butterworth, 1963.

[154] R. Henstock, The General Theory of Integration. Oxford: Clarendon Press, 1991.

[155] R. Henstock, "The construction of path integrals", Math. Japonica, vol. 39, pp. 15-18, 1994.

[156] M. C. Isidori, A. Martellotti and A. R. Sambucini, "Integration with respect to orthogonally scattered measures", Math. Slovaca, vol. 48, pp. 253-269, 1998.

[157] M. C. Isidori, A. Martellotti and A. R. Sambucini, "The Bochner and the monotone integrals with respect to a nuclear-valued finitely additive measure", Math. Slovaca, vol. 48, pp. 377-390, 1998.

[158] M. C. Isidori, A. Martellotti and A. R. Sambucini, "The monotone integral", Atti Sem. Mat. Fis. Univ. Modena, Supplemento al Vol., vol. 46, pp. 803-825, 1998.

[159] S. Izumi, "An Abstract Integral VII", Proc. Imp. Acad. Tokyo, vol. 18, pp. 53-56, 1942.

[160] J. Jakubík, "Weak σ-distributivity of lattice ordered groups", Math. Bohemica, vol. 126, pp. 151-159, 2001.

[161] D. Kannan, An Introduction to Stochastic Processes. Elsevier North Holland, Inc., 1979.

[162] L. V. Kantorovič and G. P. Akilov, Functional Analysis. Nauka (Russian), 1977; English transl. Oxford: Pergamon Press, 1982.

[163] J. Kurzweil, "Generalized ordinary differential equations and continuous dependence on a parameter", Czech. Math. J., vol. 7 (82), pp. 418-446, 1957.

[164]   J. Kurzweil, Nicht absolut-konvergente Integrale. Leipzig: Teubner, 1980.

[165]   J. Kurzweil, Henstock-Kurzweil integration: its relation to topological vector spaces. Singapore: World Scientific, 2000.

[166]   J. Kurzweil, Integration between the Lebesgue integral and the Henstock-Kurzweil integral: its relation to locally convex vector spaces. Singapore: World Scientific, 2002.

[167]   J. Kurzweil and Š. Schwabik, "On McShane integrability of Banach space-valued functions", Real Anal. Exch.. vol. 29, pp. 763-780, 2003/2004.

[168]   P. Y. Lee, Lanzhou Lectures on Henstock integration. World Scientific Publishing Co., 1989.

[169]   P. Y. Lee and R. Výborný, "Kurzweil-Henstock Integration and the Strong Lusin Condition", Boll. Un. Mat. Ital., vol. 7-B, pp. 761-773, 1993.

[170]   P. Y. Lee and R. Výborný, The integral: An easy approach after Kurzweil and Henstock. Cambridge Univ. Press., 2000.

[171]   K. Lendelová, "IF-probability on MV-algebras", Notes on Intuitionistic Fuzzy Sets, vol. 11, pp. 66-72, 2005.

[172]   Z. Lipecki, "Riesz-type representation theorems for positive operators", Math. Nachr., vol. 131, pp. 351-356, 1987.

[173]   X. Liu and G. Zhang, "Lattice-valued fuzzy measure and lattice-valued fuzzy integral", Fuzzy Sets and Systems, vol. 62, pp. 319-332, 1994.

[174]   W. A. J. Luxemburg and J. J. Masterson, "An extension of the concept of the order dual of a Riesz space", Canad. J. Math., vol. 19, pp. 488-498, 1967.

[175]   W. A. J. Luxemburg and A. C. Zaanen, Riesz Spaces, I. North-Holland Publishing Co., 1971.

[176]   F. Maeda and T. Ogasawara, "Representation of vector lattices", J. Sci. Hiroshima Univ. Ser. A, vol. 12, pp. 17-35, 1942. (Japanese)

[177]   P. Maličký, "The monotone limit convergence theorem for elementary functions with values in a vector lattice", Comment. Math. Univ. Carolin., vol. 27, pp. 53-67, 1986.

[178]   P. Maličký, "A vector lattice variant of the ergodic theorem", in Proc. 14th Winter School on Abstract Analysis (Srní 1986), Suppl. Rend. Circ. Mat. Palermo, Ser. 2, pp. 391-397, 1987.

[179]   P. Maličký, "Random variables with values in a vector lattice" Acta Math. Univ. Comen., vol. 52-53, pp. 249-263, 1987.

[180]   P. Maličký, "On the conditional expectation of vector lattice valued variables", Acta Math. Univ. Comen., vol. 56-57, pp. 115-127, 1989.

[181]   W. Marik, "Integration in topologischen Gruppen", Ph.D. thesis, 1979.

[182]   W. Marik, "Monoidwertige Integrale", Czech. Math. J., vol. 37 (112), pp. 351-375, 1987.

[183]   V. Marraffa, "The McShane and Kurzweil-Henstock integrals of functions taking values in a locally convex space", Real Anal. Exch., Summer Symposium 2003, pp. 77-80, 2003.

[184]   A. Martellotti, "On integration with respect to lctvs-valued finitely additive measures", Rend. Circ. Mat. Palermo, Serie II, vol. 43, pp. 181-214, 1994.

[185]   M. Matloka, "On an integral of fuzzy mappings", Busefal, vol. 31, pp. 45-55, 1987.

[186]   M. Matloka, "On a fuzzy integral", in Proc. Polish Symp. Interval and Fuzzy Mathematics, Poznań 1987, pp. 163-170.

[187]   P. McGill, "Integration in vector lattices", J. Lond. Math. Soc., vol. 11, pp. 347-360, 1975.

[188]   R. M. McLeod, The generalized Riemann integral. Carus Math. Monograph 20, Math. Assoc. Amer., 1980.

[189]   E. J. McShane, A Riemann-type integral that includes Lebesgue- Stieltjes, Bochner and stochastic integrals. Mem. Amer. Math. Soc. 88, 1969.

[190]   E. J. McShane, "A unified theory of integration", Amer. Math. Monthly, vol. 80, pp. 349-358, 1973.

[191]   E. J. McShane, Unified Integration. Academic Press, 1983.

[192]   P. Muldowney, A general theory of integration in function spaces, including Wiener and Feynman integration. New York: John Wiley and Sons, Inc., 1987.

[193]   D. Mundici, "Interpretation of AFC*-algebras in Łukasiewicz sentential calculus", J. Funct. Anal., vol. 65, pp. 889-894, 1986.

[194]   D. Mundici, "Tensor products and the Loomis-Sikorski theorem for MV-algebras", Adv. Appl. Math., vol. 22, pp. 227-248, 1999.

[195]   T. Murofushi and M. Sugeno, "Choquet integral models and independence concepts in multiattribute utility theory", Int. J. Uncertainty, Fuzziness and Knowledge-Based Systems, vol. 8, pp. 385-415, 2000.

[196]   K. Musiał, "Topics in the theory of Pettis integration", Rend. Ist. Mat. Univ. Trieste, vol. 23, pp. 177-262, 1991.

[197]   K. Musiał, "Pettis integral", in Handbook of Measure Theory, E. Pap ed., Amsterdam: Elsevier, 2002, Chapter 12, pp. 531-586.

[198]   S. Nanda, "On integration of fuzzy mapping", Fuzzy Sets and Systems, vol. 32, pp. 95-101, 1989.

[199]   M. Orihara and G. Sunouchi, "An Abstract Integral VIII", Proc. Imp. Acad. Tokyo, vol. 18, pp. 535-538, 1942.

[200]   W. Orlicz and R. Urbański, "A generalization of the Brooks-Jewett theorem", Bull. Acad. Pol. Sci, Sér. Sci. Math. vol. 28, pp. 55-59, 1980.

[201]   E. Pap, "A generalization of a theorem of Dieudonné for k-triangular set functions", Acta Sci. Math. (Szeged), vol. 50, pp. 159-167, 1986.

[202]   E. Pap, "The Vitali-Hahn-Saks theorems for k-triangular set functions", Atti Sem. Mat. Fis. Univ. Modena, vol. 35, pp. 21-32, 1987.

[203]   E. Pap, "The Brooks-Jewett theorem for non-additive set functions", Univ. u Novom Sadu Zb. Rad. Prirod. - Mat. Fak. Ser. Mat., vol. 21, pp. 75-81, 1991.

[204]   E. Pap, Null-additive set functions. Kluwer Acad. Publ. and Ister Science, 1995.

[205]   A. L. Peressini, Ordered topological vector spaces. Harper and Row, 1967.

[206]   B. J. Pettis, "On integration in vector spaces", Trans Amer. Math. Soc., vol. 44, pp. 277-304, 1938.

[207]   W. F. Pfeffer, "Lectures on Geometric Integration and the Divergence Theorem", Rend. Ist. Mat. Univ. Trieste, vol. 23, pp. 263-314, 1991.

[208]   W. F. Pfeffer, The Riemann approach to integration. Cambridge Univ. Press, 1993.

[209]   W. F. Pfeffer, "The Lebesgue and Denjoy-Perron integrals from a deterministic point of view", Ricerche Mat., vol. 48, no. 2, pp. 211-223, 1999.

[210]   R. Potocký, "On random variables having values in a vector lattice", Math. Slovaca, vol. 27, pp. 267-276, 1977.

[211]   R. Potocký, "On the expected value of vector lattice-valued random variables", Math. Slovaca, vol. 36, pp. 401-405, 1986.

[212]   P. Pucci, "Integrali di Riemann e di Burkill-Cesari", Rend. Mat. (Roma), Ser. VII, vol. 3, pp. 253-275, 1983.

[213]   M. Renčová and B. Riečan, "Probability on IF-sets: an elementary approach", in First Int. Workshop on IFS, Generalized Nets and Knowledge Engineering, pp. 8-17, 2006.

[214]   G. Riccobono, "A PU-integral on an abstract metric space", Math. Bohemica, vol. 122, pp. 83-95, 1997.

[215]   G. Riccobono, "Convergence theorems for the PU-integral", Math. Bohemica, vol. 125, pp. 77-86, 2000.

[216]   B. Riečan, "An extension of the Daniell integration scheme", Mat. Čas., vol. 25, no. 3, pp. 211-219, 1975.

[217]   B. Riečan, "On the Kurzweil integral in compact topological spaces", Radovi Mat., vol. 2, pp. 151-163, 1986.

[218]   B. Riečan, "Remark on an integral of M. Matloka", Math. Slovaca, vol. 38, pp. 341-344, 1988.

[219]   B. Riečan, "On the Kurzweil integral for functions with values in ordered spaces, I", Acta Math. Univ. Comenian., vol. 56 -57, pp. 411-424, 1989.

[220]   B. Riečan, "On operators valued measures in lattice ordered groups", Atti Sem. Mat. Fis. Univ. Modena, vol. 40, pp. 151-154, 1992.

[221]   B. Riečan, "A descriptive definition of the probability on intuitionistic fuzzy sets", in EUSFLAT 2003, M. Wagenecht, R. Hampet eds., 2003, pp. 263-266.

[222]   B. Riečan, "Kurzweil-Henstock integral with values in metric semigroups", in Reports on Real Analysis, Conference - Rowy 2003 (J. Jedrzejewski ed.), Wyzsza szkoła Informatyki w Łodzi, Łódź, 2004, pp. 173-174.

[223]   B. Riečan, "On the probability on IF-sets and MV-algebras", Notes on IFS, vol. 11, pp. 21-25, 2005.

[224]   B. Riečan, "On a problem of Radko Mesir: general form of IF-probabilities", Fuzzy Sets and Systems, vol. 152, pp. 1485-1490, 2006.

[225]   B. Riečan, "Probability theory on intuitionistic fuzzy sets", Lecture Notes in Computer Science, A volume in Honour of Daniele Mundici's 60th birthday, 2007.

[226]   B. Riečan and D. Mundici, "Probability on MV-algebras", in Handbook of Measure Theory, E. Pap ed., Amsterdam: Elsevier, 2002, Chapter 21, 869-909.

[227]   B. Riečan and T. Neubrunn, Integral, Measure and Ordering. Kluwer Acad. Publ. and Ister Science, 1997.

[228]   B. Riečan and P. Volauf, "On a technical lemma in lattice ordered groups", Acta Math. Univ. Comen., vol. 44-45, pp. 31-35, 1984.

[229]   B. Riečan and M. Vrábelová, "On the Kurzweil integral for functions with values in ordered spaces, II", Math. Slovaca, vol. 43, pp. 471-475, 1993.

[230]   B. Riečan and M. Vrábelová, "On integration with respect to operator valued measures in Riesz spaces", Tatra Mt. Math. Publ., vol. 2, pp. 149-165, 1993.

[231]   B. Riečan and M. Vrábelová, "On the Kurzweil integral for functions with values in ordered spaces, III", Tatra Mt. Math. Publ., vol. 8, pp. 93-100, 1996.

[232]   B. Riečan and M. Vrábelová, "The Kurzweil construction of an integral in ordered spaces", Czech. Math. J., vol. 48 (123), pp. 565-574, 1998.

[233]   S. M. Ross, Introduction to Probability Models. Academic Press, Inc., 1985.

[234]   A. R. Sambucini, "Integrazione per seminorme in spazi localmente convessi", Rivista di Matematica Univ. Parma, vol. 3, pp. 371-381, 1994.

[235]   A. R. Sambucini, "Un teorema di Radon-Nikodym in spazi localmente convessi rispetto all'integrazione per seminorme", Rivista di Matematica Univ. Parma, vol. 4, pp. 49-60, 1995.

[236]   M. Scarsini, "Dominance conditions in non-additive expected utility theory", J. of Math. Econ., vol. 21, pp. 173-184, 1992.

[237]   W. Schachermayer, "On some classical measure-theoretic theorems for non-sigma-complete Boolean algebras", Dissert. Math., vol. 214, pp. 1-33, 1982.

[238]   D. Schmeidler, "Integral representation without additivity", Proc. Am. Math. Soc., vol. 97, pp. 255-261, 1986.

[239]   Š. Schwabik, Generalized ordinary differential equations. Singapore: World Scientific, 1992.

[240]   Š. Schwabik, "On non-absolutely convergent integrals", Math. Bohemica, vol. 121, pp. 369-383, 1996.

[241]   Š. Schwabik, "Abstract Perron-Stieltjes integral", Math. Bohemica, vol. 121, pp. 425-447, 1996.

[242]   Š. Schwabik, "Linear Stieltjes integral equations in Banach spaces", Math. Bohemica, vol. 124, pp. 433-457, 1999.

[243]   Š. Schwabik, "Operator-valued functions of bounded semivariation and convolutions", Math. Bohemica, vol. 126, pp. 745-777, 2001.

[244]   Š. Schwabik, "A note on integration by parts for abstract Perron-Stieltjes integrals", Math. Bohemica, vol. 126, pp. 613-629, 2001.

[245]   Š. Schwabik, "Integration in Banach spaces", in Summer Symposium 2003, Real Anal. Exch., pp. 77-80, 2003.

[246]   Š. Schwabik and G. Ye, "On the strong McShane integral of functions with values in a Banach space", Czech. Math. J., vol. 51 (126), pp. 819-828, 2001.

[247]   Š. Schwabik and G. Ye, "Topics in Banach space integration", in Series in Real Analysis, vol. 10, World Scientific Publishing Co., 2005.

[248]   Š. Schwabik and I. Vrkoč, "On Kurzweil-Henstock equiintegrable sequences", Math. Bohemica, vol. 121, pp. 189-207, 1996.

[249]   J. Šipoš, "Integral with respect to a pre-measure", Math. Slovaca, vol. 29, pp. 141-155, 1979.

[250]   J. Šipoš, "Non linear integrals", Math. Slovaca, vol. 29, pp. 257-270, 1979.

[251]   J. Šipoš, "Extension of partially ordered group-valued measure-like set functions", Čas. pro. pěst. mat., vol. 108, pp. 113-121, 1983.

[252]   J. Šipoš, "On extension of group valued measures", Math. Slovaca, vol. 40, pp. 279-286, 1990.

[253]   W. V. Smith, "Convergence in measure of integrands", Rev. Roum. Math. Pures et Appl., vol. 26, pp. 899-903, 1981.

[254]   C. Swartz, Introduction to gauge integrals. World Scientific, 2001.

[255]   Gy. Szabó and Á. Száz, "The Net Integral and a Convergence Theorem", Math. Nachr., vol. 135, pp. 53-65, 1988.

[256]   Gy. Szabó and Á. Száz, "Defining nets for integration", Publ. Math. Debrecen, vol. 36, pp. 237-252, 1989.

[257]   Á. Száz, "The fundamental Theorem of Calculus in an abstract setting", Tatra Mt. Math. Publ., vol. 2, pp. 167-174, 1993.

[258]   M. Talagrand, Pettis integral and measure theory. Mem. Amer. Math. Soc. 367, 1984.

[259]   M. Urbaníková, "Limit theorems for B-lattice random variables", Math. Slovaca, vol. 52, pp. 99-108, 2002.

[260]   P. Volauf, "Extension and regularity of l-group valued measures", Math. Slovaca, vol. 27, pp. 47-53, 1977.

[261]   P. Volauf, "On the lattice group valued submeasures", Math. Slovaca, vol. 40, pp. 407-411, 1990.

[262]   M. Vonkomerová, "On the extension of positive operators", Math. Slovaca, vol. 31, pp. 251-262, 1981.

[263]   P. Vrábel, "Lower integral on lattice ordered groups", Zborník Pedagogickej fakulty v Nitre, Matematika 2, Bratislava: SPN, pp. 65-80, 1982.

[264]   P. Vrábel, "Lower integral on MV-algebras", Acta Math. (Nitra), vol. 2, pp. 51-58, 1995.

[265]   P. Vrábel, "Integral on MV-algebras", Tatra Mt. Math. Publ., vol. 12, pp. 21-25, 1997.

[266]   M. Vrábelová, "The fundamental theorem of calculus in ordered spaces", Acta Math. (Nitra), vol. 3, pp. 31-38, 1998.

[267]   M. Vrábelová, "Fubini's theorem in ordered spaces", Acta Math. (Nitra), vol. 5, pp. 37-42, 2002.

[268]   M. Vrábelová, "Double Kurzweil-Henstock integral in Riesz spaces", Acta Math. (Nitra), vol. 7, pp. 45-52, 2004.

[269]   M. Vrábelová, "Fubini's theorem in metric semigroups", Acta Math. (Nitra), vol. 8, pp. 31-37, 2005.

[270]   M. Vrábelová, "On the extension of subadditive measures in lattice ordered groups", Czech. Math. J., vol. 57, pp. 95-103, 2007.

[271]   B. Z. Vulikh, "Une définition du produit dans les espaces semi-ordonnés linéaires", Dokl. Akad. Nauk SSSR, vol. 26, pp. 850-854, 1940.

[272]   B. Z. Vulikh, "Sur les propriétés du produit dans les espaces semi-ordonnés linéaires", Dokl. Akad. Nauk SSSR, vol. 26, pp. 855-859, 1940.

[273]   B. Z. Vulikh, Introduction to the theory of partially ordered spaces. Groningen: Wolters-Noordhoff Sci. Publ., 1967.

[274]   R. Výborný, "Some applications of Kurzweil-Henstock integration", Math. Bohemica, vol. 118, pp. 425-441, 1993.

[275]   G. Warner, "The Burkill-Cesari integral", Duke Math. J., vol. 35, pp. 61-78, 1968.

[276]   H. Weber, "Compactness in spaces of group-valued contents, the Vitali-Hahn-Saks theorem and Nikodym's boundedness theorem", Rocky Mt. J. Math., vol. 16, pp. 253-275, 1986.

[277]   J. D. M. Wright, "Stone-algebra-valued measures and integrals", Proc. London Math. Soc., vol. 19, pp. 107-122, 1969.

[278]   J. D. M. Wright, "The measure extension problem for vector lattices", Ann. Inst. Fourier (Grenoble), vol. 21, pp. 65-85, 1971.

[279]   J. D. M. Wright, "Measures with values in a partially ordered vector space", Proc. London Math. Soc., vol. 25, pp. 675-688, 1972.

[280]   J. D. M. Wright, "An extension theorem", J. London Math. Soc., vol. 7, pp. 531-539, 1973.

[281]   C. Wu and Z. Gong, "On Henstock integrals of interval-valued functions and fuzzy-valued functions", Fuzzy Sets and Systems, vol. 115, pp. 377-391, 2000.

[282]   C. Wu and Z. Gong, "On Henstock integral of fuzzy-number-valued functions (I)", Fuzzy Sets and Systems, vol. 120, pp. 523-532, 2001.

[283]   C. Wu and Z. Gong, "On Henstock integral of fuzzy-number-valued functions (II): the descriptive characteristic of (FH) integral and its convergence theorems", 2003, manuscript.

[284]   C. Wu and M. Ma, "On embedding problem of fuzzy number space: Part I", Fuzzy Sets and Systems, vol. 44, pp. 33-38, 1991.

[285]   C. Wu and M. Ma, "On embedding problem of fuzzy number space: Part II", Fuzzy Sets and Systems, vol. 45, pp. 189-202, 1992.

[286]   C. Wu, X. Yao and S. S. Cao, "The vector-valued integrals of Henstock and Denjoy", Sains Malaysiana, vol. 24, pp. 13-22, 1995.

[287]   J. Xu and P. Y. Lee, "Stochastic Integrals of Itô and Henstock", Real Anal. Exch., vol. 18, pp. 352-366, 1992/1993.

[288]   G. Ye and Š. Schwabik, "The McShane and the Pettis integral of Banach space-valued functions defined on $\mathbb{R}^m$", Ill. J. Math., vol. 46, pp. 1125-1144, 2002.

[289]   A. C. Zaanen, Riesz spaces, II. North - Holland Publishing. Co., 1983.

INDEX

www.ingramcontent.com/pod-product-compliance
Lightning Source LLC
Chambersburg PA
CBHW050836220326
41598CB00006B/376